Geographies of Mobility

This book seeks to bring together different philosophical, theoretical, and methodological approaches to the study of human mobility within the discipline of geography. With five thematic sections – conceptualizing and analyzing mobility, inequalities of mobility, politics of mobility, decentering mobility, and qualifying abstraction – and 27 substantive chapters by leading researchers in the field, it provides a comprehensive overview of the latest thinking about human mobility and related issues. The contributors discuss mobility issues as diverse as everyday mobilities of young people, migrants and refugees, and sex workers; the relationships between citizenship and mobility; and the potential and pitfalls of big data for understanding mobility. This, coupled with a broad international focus, means that *Geographies of Mobility* will not only encourage and enrich dialogue on a theme that is of major importance to varied geographic research communities, but will also be of great interest to students and researchers across the wider social sciences. This book was originally published as a special issue of *Annals of the American Association of Geographers*.

Mei-Po Kwan is a Professor in the Department of Geography and Geographic Information Science at the University of Illinois at Urbana-Champaign, USA. Her research interests include human mobility, environmental health, sustainable transport and cities, and geographic information science. She has received many prestigious honors and awards for her groundbreaking contributions to these areas.

Tim Schwanen is an Associate Professor of Transport Studies and the Director of the Transport Studies Unit in the School of Geography and the Environment at the University of Oxford, UK. His research focuses on questions around mobility, cities, climate change, well-being, and inequality.

Geographies of Mobility

Recent Advances in Theory and Method

Edited by
Mei-Po Kwan and Tim Schwanen

Routledge
Taylor & Francis Group

LONDON AND NEW YORK

First published 2017 by Routledge

2 Park Square, Milton Park, Abingdon, Oxfordshire OX14 4RN
52 Vanderbilt Avenue, New York, NY 10017

Routledge is an imprint of the Taylor & Francis Group, an informa business

First issued in paperback 2018

Chapters 1-20, 22-27© 2017 American Association of Geographers
Chapter 21 © 2017 Gina Porter

British Library Cataloguing in Publication Data
A catalogue record for this book is available from the British Library

ISBN 13: 978-1-138-29026-6 (hbk)
ISBN 13: 978-0-367-13352-8 (pbk)

Typeset in Goudy
by diacriTech, Chennai

Publisher's Note
The publisher accepts responsibility for any inconsistencies that may have arisen during the conversion of this book from journal articles to book chapters, namely the possible inclusion of journal terminology.

Disclaimer
Every effort has been made to contact copyright holders for their permission to reprint material in this book. The publishers would be grateful to hear from any copyright holder who is not here acknowledged and will undertake to rectify any errors or omissions in future editions of this book.

Contents

CONTENTS

CONTENTS

Citation Information

The chapters in this book were originally published in the *Annals of the American Association of Geographers*, volume 106, issue 2 (March 2016). When citing this material, please use the original page numbering for each article, as follows:

CITATION INFORMATION

CITATION INFORMATION

Chapter 27

Another Tale of Two Cities: Understanding Human Activity Space Using Actively Tracked Cellphone Location Data

Yang Xu, Shih-Lung Shaw, Ziliang Zhao, Ling Yin, Feng Lu, Jie Chen, Zhixiang Fang, and Qingquan Li

Annals of the American Association of Geographers, volume 106, issue 2 (March 2016) pp. 489–502

For any permission-related enquiries please visit:
http://www.tandfonline.com/page/help/permissions

Notes on Contributors

Stuart C. Aitken is a Professor of Geography and June Burnett Chair in the Department of Geography at San Diego State University, USA.

Tom Baker is a Lecturer in the School of Environment at the University of Auckland, Auckland, New Zealand.

Asha Best is a PhD candidate in the Department of American Studies at Rutgers University–Newark, USA.

Ling Bian is a Professor in the Department of Geography, University at Buffalo, the State University of New York, USA.

Amit Birenboim is a PhD candidate in the Department of Geography at The Hebrew University of Jerusalem, Israel.

David Bissell is a Senior Lecturer in the Research School of Social Sciences at the Australian National University, Australia.

Derek Breese is an Analyst at the U.S. Census Bureau, USA.

Irene Casas is an Assistant Professor of Geographic Information Sciences at the College of Liberal Arts of Louisiana Tech University, USA.

Jie Chen is an Assistant Researcher in the State Key Laboratory of Resources and Environmental Information System, Institute of Geographic Sciences and Natural Resources, China.

Julie Cidell is an Associate Professor in the Department of Geography and GIS at the University of Illinois, Urbana–Champaign, USA.

Ian R. Cook is a Senior Lecturer in the Department of Social Sciences and Languages at the Northumbria University, Newcastle, UK.

Meghan Cope is a Professor in the Department of Geography at the University of Vermont, USA.

Noelani Eidse is a PhD candidate in the Department of Geography at McGill University, Canada.

Zhixiang Fang is a Professor at the State Key Laboratory of Information Engineering in Surveying, Mapping and Remote Sensing, Wuhan University, China.

Sarah Hendricks is a Visiting Professor in the Department of Behavioral Sciences at the University of Tennessee, USA.

Yujie Hu is a PhD candidate in the Department of Geography and Anthropology, Louisiana State University, USA.

Mei-Po Kwan is a Professor in the Department of Geography and Geographic Information Science at the University of Illinois at Urbana–Champaign, USA.

Devon Lechtenberg is an independent scholar in Chicago, USA.

Brian H. Y. Lee is an Assistant Professor in Transportation Systems at the University of Vermont, USA.

Qingquan Li is a Professor at Shenzhen University, Guangdong, China.

Adela C. Licona is an Associate Professor in the Department of English at the University of Arizona, USA.

Feng Lu is a researcher in the State Key Laboratory of Resources and Environmental Information System, Institute of Geographic Sciences and Natural Resources Research, China.

Marta Maria Maldonado is an Associate Professor of Ethnic Studies in the School of Language, Culture and Society at Oregon State University, USA.

David J. Marshall is a Postdoctoral Research Associate in the Department of Geography at Durham University, UK.

Naomi Maynard is a postgraduate student in the Department of Geography at Durham University, UK.

Eugene McCann is a Professor in the Department of Geography at Simon Fraser University, Canada.

Sara McLafferty is the Department Head and a Professor in the Department of Geography and Geographic Information Science at the University of Illinois at Urbana–Champaign, USA.

Byron Miller is an Associate Professor of Geography and the Coordinator of the Urban Studies Program at the University of Calgary, USA.

Deborah Naybor is an Assistant Professor in the Department of Environmental Studies at Paul Smith's College, USA.

Natalie Oswin is an Associate Professor of Geography at McGill University, Canada.

Virginia Parks is the Madeline McKinnie Endowed Professor in the Department of Urban and Environmental Policy at Occidental College, Los Angeles, USA.

Jason Ponto is a doctoral candidate in the Department of Sociology at the University of Calgary, USA.

Jessie P. H. Poon is a Professor in the Department of Geography at University at Buffalo, USA.

Gina Porter is a Professor in the Department of Anthropology, Durham University, UK.

Valerie Preston is a Professor in the Department of Geography at York University, Toronto, Canada.

Marie Price is a Professor of Geography and International Affairs in the Department of Geography at the George Washington University, USA.

Amy E. Ritterbusch is an Associate Professor in the School of Government Alberto Lleras Camargo at University of Andes, Columbia.

Ian Rowen is a PhD candidate in the Department of Geography at the University of Colorado, USA.

Tim Schwanen is an Associate Professor of Transport Studies and the Director of the Transport Studies Unit in the School of Geography and the Environment at the University of Oxford, UK.

Shih-Lung Shaw is the Alvin and Sally Beaman Professor and Arts and Sciences Excellence Professor in the Department of Geography at the University of Tennessee, USA.

Mimi Sheller is a Professor of Sociology in the Department of Sociology and the Director of the Center for Mobilities Research and Policy at Drexel University, USA.

Noam Shoval is a Full Professor in the Department of Geography at The Hebrew University of Jerusalem, Israel.

Justin Spinney is a Lecturer in Human Geography in the School of Planning and Geography at Cardiff University, UK.

Lynn A. Staeheli is a Professor of Human Geography in the Department of Geography at Durham University, UK.

Elaine Stratford is an Associate Professor in the Discipline of Geography and Spatial Sciences, School of Land and Food, University of Tasmania, Australia.

Cristina Temenos is a Postdoctoral Fellow in the Humanities Center at Northeastern University, Boston, USA.

Sarah Turner is a Professor in the Department of Geography at McGill University, Canada.

Lorraine van Blerk is a Professor in Human Geography in the School of Social Sciences Geography at the University of Dundee, UK

Fahui Wang is the James J. Parsons Professor and Chair of the Department of Geography and Anthropology, Louisiana State University, USA.

Kevin Ward is a Professor in Geography, School of Environment, Education and Development, University of Manchester, UK.

Yang Xu is a PhD candidate in the Department of Geography at the University of Tennessee, Knoxville, USA.

Ling Yin is an Associate Professor at Shenzhen Institute of Advanced Technology, Chinese Academy of Sciences, China.

Ziliang Zhao is a PhD candidate in the Department of Geography at the University of Tennessee, USA.

Shiran Zhong is a PhD candidate in the Department of Geography, University at Buffalo, the State University of New York, USA.

Introduction
Geographies of Mobility

Mei-Po Kwan* and Tim Schwanen[†]

*Department of Geography and Geographic Information Science, University of Illinois at Urbana–Champaign
[†]School of Geography and the Environment, University of Oxford

This introductory piece sets the context for the special issue and explains its rationale. It offers a series of reflections on the rise of the mobilities turn and its relations with preexisting research traditions, most notably transportation geography. Rather than placing different approaches in opposition and favoring one over others, we contend that all need to be seen as situated, partial, and also generative modes of abstraction. Each of these approaches makes mobility exist in specific and ultimately simplified and selective ways. In addition, we argue that geography as a pluralistic discipline will benefit from further conversations between modes of conceptualizing, theorizing, and examining mobility. We outline five lines along which such conversations can be structured: conceptualizations and analysis, inequality, politics, decentering and decolonization, and qualifying abstraction. The article concludes with discussion on three fruitful directions for future research on mobility.

此一引介文章, 为此特刊提供脉络, 并解释其逻辑依据。本文提供一系列对于能动性转向兴起的反思, 及其与既有的研究传统之间的关系, 其中多半是运输地理学。有异于将不同的方法相互对立并偏好其中一种方法, 我们主张, 所有的方法皆需被视为情境化、不完全、且同时具有生产力的抽象化模式。每一种方法, 皆使能动性存在于特定且最终是简化且选择性的方式。此外, 我们主张, 地理学作为多元的领域, 将会进一步从概念化、理论化、以及检视能动性的各种模式之间的进一步对话中获益。我们概述此般对话可进行建构的五大方向⬚概念化与分析, 不均等, 政治, 去中心化与去殖民, 以及限定抽象化。本文于结论中, 探讨未来能动性研究的三种成果丰硕之方向。

Esta parte introductoria pone el contexto para el número especial y explica su razón de ser. Ofrece una serie de reflexiones sobre el ascenso del giro de las movilidades y sus relaciones con las tradiciones de investigación preexistentes, más notablemente con la geografía del transporte. Más que formular diferentes enfoques en oposición y favoreciendo a uno sobre los demás, planteamos que todos los enfoques deben verse como situados, parciales y también como modos generativos de abstracción. Cada uno de estos enfoques hacen que la movilidad exista de maneras específicas y, en últimas, simplificadas y selectivas. Argüimos, además, que como disciplina pluralista la geografía se beneficiará de conversaciones avanzadas entre los modos de conceptualizar, teorizar y examinar la movilidad. Presentamos un esquema de cinco líneas a lo largo de las cuales puedan estructurarse tales conversaciones: conceptualizaciones y análisis, desigualdad, política, disgregación y colonización, y abstracción calificada. El artículo concluye con la discusión de tres direcciones productivas de investigación futura sobre movilidad.

It is now ten years since Sheller and Urry's (2006) seminal paper announced a new mobilities paradigm in the social sciences. Complementing and at times competing with established traditions of studying transport, daily travel, tourism, migration, and other forms of (im)mobility, research influenced by the ideas summarized by Sheller and Urry (2006) has taken flight in geography (for useful overviews, see Lorimer 2007; Cresswell 2011, 2012, 2014; Cresswell and Merriman 2011; Adey et al. 2014; Merriman 2015, forthcoming). The mobilities turn has had numerous beneficial effects on the discipline, including widespread acceptance of its key tenet that mobility is endemic to life, society, and space rather than exceptional and the attention it has drawn to a greater range of mobilities than previously considered worthy of academic geographers' attention. Perhaps its greatest achievement has been to elevate mobility to a class of core geographic concepts to which space, place, network, scale, and territory also belong. It is not surprising, therefore, that two of the contributions to this special issue (Cidell and Lechtenberg this issue; Miller and Ponto this issue) explicitly address the question of how mobility relates to those other core concepts.

1

Conceptualizing Mobility

Perhaps inevitably, the elevation of mobility to iconic status in academic geography's panoply of core concepts has only diversified understandings and definitions of what has always been a fuzzy term. Were anybody to claim that mobility used to be a straightforward term prior to the mobilities turn, they could easily be proven wrong. A survey[1] of articles published in the *Annals of the Association of American Geographers* (1911–2010) suggests that in the last four decades of the twentieth century, the term *mobility* was used predominantly to denote residential movements by human individuals and households. Even in that period, however, the term had multiple uses: It was used in relation to individuals' daily and weekly trip-making (Wheeler 1972), the upward social mobility of individuals (Breese 1963), the ongoing movements of cattle and herders (Kollmorgen 1969), and even in Foucauldian fashion as a synonym for energy and force (Sack 1976). Nonetheless, it is before 1960 and from 2000 onward that meanings and referents—that is, the mobility of whom and what—are more diverse. Here we restrict attention to the pre-1960 period, as usage of the term at the start of the twenty-first century might well have been influenced by the early pulses of the mobilities turn.

In the first half-century of scholarship published in the *Annals*, the term *mobility* was used in relation to many different referents: from faunal life (Joerg 1914) and plants (Gleason 1922) to technology (Ginsburg 1957), armies (Frey 1941; Whittlesey 1945), cotton and other commodities (Platt 1927; Marschner 1944; Murphey 1954), oil and nuclear energy (Hoffman 1957), centers of dominance in economic areas (Sauer 1941), and indeed human individuals in subject positions as diverse as customers of shopping centers (Platt 1928) and the "North American Indian" (Dryer 1915, 122). The whole world is indeed on the move and has always been so; what has changed with globalization is the intensity of movement and the geographic scale over which many of those movements occur.

Moreover, in early *Annals* articles, mobility does not merely denote actual movement. The term has been used to denote potential movement or a capacity to become mobile, as in Smith's (1943) discussion of nomad mobilities that folds together everyday mobility and migration or in Hall's (1955) discussion of MacKinder's conceptualization of the transition from horse and camel to railroad mobility in what is now known as Russia and central Asia. Meanwhile, Sauer (1941) seemed to equate mobility with a certain level of energy and dynamism, which is common in more contemporary interpretations, and Whittlesey offered what today would be recognized as a relational understanding: mobility as a capacity to move afforded by the interactions between vessel and ocean (Whittlesey 1945) and between horse or motorized vehicle and state of the road (Whittlesey 1956). Our point is, of course, not to argue that nothing has changed in recent decades in either the realities we study or the worldviews, conceptualizations, and methodologies with which we try to make sense of those realities. It is rather that there are resemblances and connectivities between recent and older thinking about mobility in geography that can easily go unrecognized. Indeed, the suggestion of a linear progression from simple to more sophisticated understandings of mobility in geography caused by the mobilities turn should be avoided.

At the same time, it is also clear that conceptualizations of mobility and immobility have become richer and more diverse over the past decade (e.g., Adey 2006; Cresswell 2006, 2010; Merriman 2007; Hanson 2010; Bissell and Fuller 2011; McCann 2011; Ziegler and Schwanen 2011; Söderström et al. 2013; Adey et al. 2014). Arguably the most influential has been Cresswell's (2006, 2010) understanding of mobility as the fragile entanglement of physical movement, the socially shared meanings ascribed to such movement, and the experienced and embodied practice of movement. This conceptualization is also utilized by Eide, Turner, and Oswin and by Ritterbush in this special issue. It highlights effectively that mobility is more than a functional task imposed by the separation of objects—people, locations, services, and so forth—in space and time and that attempts to reduce mobility to merely the level of functionality amount to its depoliticization.

Cresswell's conceptualization has nonetheless been criticized. Frello (2008) and Enders, Manderscheid, and Mincke (2016) rejected its tripartite nature with reference to Foucault's (1972) archaeological method. They argued that the rules of discourse formation dictate first what can appear and be classified as movement and, second, who is in the position to legitimately claim understanding of movement. There is, then, no extra-discursive, empirical separation of movement from its other—rest, stillness, sojourn, mooring, stasis, and so on—and any such differentiation is a doing that enacts what it purports to describe (Law 2004). Mobility, on this reading, is an ever-changing object of knowledge that is coconstituted by practices involving geographers and other social scientists, alongside all sorts of other agents.

Mobilities and Transportation

Cresswell's conceptualization can also be seen to reinforce a particular representation of transportation geography and transportation studies more widely. This representation separates rather than brings together research on mobilities and transportation within geography (for further discussion of this relation, see Shaw and Hesse 2010; Bissell, Adey, and Laurier 2011; Shaw and Sidaway 2011; Cidell and Prytherch 2015; Schwanen 2016). According to Cresswell (2010), transportation research has by and large failed to illuminate two of the three pillars under his conceptualization. In examining "how often [movement] happens, at what speeds, and where [as well as] who moves and how identity might make a difference," transportation researchers "have not been so good at telling us about the representations and meanings of mobility either at the individual level or the societal level [or about] how mobility is actually embodied and practised" (Cresswell 2010, 19). Implied here is an opposition rather than a contrast (Stengers 2011): transportation versus mobilities research. It would also seem that transportation geography is not merely partial and situated—as any practice of academic knowledge production inevitably is (Haraway 1991)—but severely limited. Defending transportation geography is not our aim here,[2] but Cresswell's account is problematic on two accounts. Not only is transportation geography internally heterogeneous and are parts of it closely connected to and coevolving with the mobilities turn (e.g., Kwan 2007; Schwanen and Kwan 2008; Goetz, Vowles, and Tierney 2009; Bissell, Adey, and Laurier 2011; Shaw and Docherty 2014; Cidell and Prytherch 2015; Wilsmeier and Monios 2015), but parts of the subdiscipline can also be seen to generate new understandings of mobility and not merely movement.

It is not simply the case that in conceptualizations like Cresswell's movement is separated from discourse. At a more fundamental level, mobility is bifurcated between an objective, primary realm of brute fact—movement—and a further reality of secondary qualities and human "additions"—meaning, sensation, perception, feeling, and so forth (cf. Whitehead 1920; Stengers 2011). Where in the current era of big data, physicists pride themselves on cutting through the "biases" resulting from human additions and finally uncovering the "laws" dictating movement (e.g., González, Hidalgo, and Barabasi 2008; Simini et al. 2012), Cresswell and various other mobility scholars criticize transportation researchers for, to paraphrase Latour (2005), substituting the cold fact of movement for the rich meanings or embodiment of mobility. Their critique is a version of what, after McCormack (2012), can be called the default understanding of abstraction as "a malign process of generalization and simplification through which the complexity of the world is reduced at the expense of the experience of those who live in the concrete reality of this world [and that] reproduc[es] disembodied habits of knowing, techniques of alienation, and fail[s] to recognize corporeal difference" (717). But what if the lived and the abstract cannot be placed in dualistic opposition? What if the object–subject bifurcation of mobility into objective and subjective elements is suspended and resisted? What if movement, meaning, and practice are understood as truly entangled and mutually implicated in ways that language struggles to make graspable?

This alternative imagining allows us to think differently about various ways and traditions of researching mobility and to turn oppositions into contrasts. It suggests that those transportation geographers who appear to reduce mobility to movement and those mobilities scholars who seemingly privilege meaning or practice are in fact creating different abstractions—here more affirmatively understood as selections and simplifications—through their particular methodological practices. In so doing they allow mobility as an ontologically uncertain, complex, and emergent process to be articulated and exist in new and differentiated ways. Whereas research on the embodied experience and politics of skateboarding (Stratford this issue) or Latin@ (im)mobilities (Maldonado, Licona, and Hendricks this issue) brings out unique aspects of mobility, studies using Global Positioning System (GPS) tracking technology, regression modeling, and Monte Carlo simulation (Hu and Wang this issue; Naybor, Poon, and Casas this issue) articulate mobility in wholly different ways that are likely to elude other methodological practices. In principle, then, the specific practices of all communities of geographers studying mobility are generative rather reductive (Latour 2005). This most emphatically does not mean that "anything goes" (Feyerabend 1975), as all articulations should be plausible to peer groups in academia and increasingly beyond; they must be sufficiently robust, logically coherent, and inscribed into one or more traditions of research that they simultaneously prolong and change (Stengers 2000, 2005). Emphasizing the generative qualities of research makes clear that mobility is always more than, and in excess of, what a single study or a particular tradition of research can make understandable.

In many ways, different approaches to understanding and examining mobility are the consequence of differences in modes of abstraction. For Whitehead (1926), practices of abstraction were necessary and inevitable. He cast abstraction in a much more positive light than geographers tend to do nowadays because, as a mathematician turned philosopher, he understood that thought, research, and plausible articulations of mobility become impossible without selection and simplification; what matters is how abstraction is practiced (Stengers 2011; Schwanen 2015). Whether practices of abstraction are good or appropriate is difficult to tell because there is no external yardstick—logical positivism's absolute truth—against which abstractions can be evaluated. Any evaluation is necessarily relational and dependent on the purpose of analysis, the researchers' peer group(s), and wider dynamics in how academics and others understand the world.[3] Hence, as feminist theorists have long since reminded us (Haraway 1991; Mouffe 1999; Longino 2002), any such evaluation is also shaped in profound ways by asymmetric and unevenly changing power relations.

Yet, the complexities of evaluating and comparing modes of understanding and examining mobility should not result in what Barnes and Sheppard (2010), after Bernstein (1988), called fragmenting pluralism—a situation in which researchers are only able to communicate within narrow, homogenizing communities whose members share similar habits of thought, dispositions, and practices of abstraction. Barnes and Sheppard set a high standard and sought to avoid a range of (rather common) ways in which researchers situated in a particular approach or tradition engage with other modes of abstraction (see also Kwan 2004). They believed that paying lip service, superficial appropriation, and polemics are best eschewed as well. Theirs is a call for engaged pluralism—a conversation across dividing lines and uneven positions that is as open as possible, that is not rationalized by elimination of the passions, and that marginalizes or excludes no mode of abstraction.

We realize that Barnes and Sheppard's ideal of engaged pluralism for geographical research on mobility is far from straightforward. Yet, this special issue seeks to make a modest contribution to the creation of a pathway toward the habituation and institutionalization of such engaged pluralism. It does so in three ways: by offering a forum in one of the discipline's flagship journals that brings together the many different ways in which geographers currently study mobility, by identifying sometimes fragile lines of connection across the variegated body of research on mobility, and by outlining some avenues for future research where further conversations and debates would be fruitful.

As the eighth of a series of annual special issues of the *Annals* that highlight geographic research around a significant global theme, this special issue is certainly not the first forum for a plurality of geographic approaches to the study of mobility (see, e.g., Uteng and Cresswell 2008; Schwanen and Páez 2010; Ernste, Martens, and Schapendonk 2012; Cidell and Prytherch 2015). We believe, however, that this attempt is unique in scale, openness, and heterogeneity of contributors. It started with a broad call for papers issued in September 2013, asking for abstracts to be submitted to the Association of American Geographers (AAG) journal office. Contributions were sought from a broad spectrum of scholars who address social, cultural, political, environmental, economic, theoretical, and methodological issues related to human mobility. These include geographic research in areas such as: (im)mobility and social differentiation and inequality; (im)mobility of the oppressed, subjugated, and persecuted; (im)mobility and social exclusion; experience of (im)mobility; politics of (im)mobility; commuting; leisure travel; tourism; mobility by different transport modes; sustainable mobility; mobility and resilience; disasters, natural hazards, and mobility; mobility, wellbeing, and health; mobility, energy consumption, and greenhouse gas emissions; space–time modeling and geographic information system (GIS)-based analysis of mobility; mobility research methods; and other relevant areas.

The response to the call for papers was overwhelming: We received 230 abstracts in total. The selection process was difficult because we sought to achieve several goals, including diversity in theme, perspective, approach, method, and regional focus; contribution to geography through innovative theoretical, methodological, or empirical work; and adherence to the theme of mobility. Exclusions remain inevitable, however. Partly because of the requirement for abstracts and papers to be submitted in English and the communication channels used to disseminate the call for papers, researchs from non-Anglophone speakers and countries (and not only the Global South) remains underrepresented. Also, as a result of a focus on human mobility that relates to relatively short timescales, papers that only consider the mobility of artefacts or migration in isolation from people's everyday trips, business travel, or tourism have not been included in this special issue.

Lines of Connection

The strength of special issues is that they enable new insights to emerge from bringing together the individual contributions. In this way, emerging themes, lines of connection, and differentiations within a research community become visible and new opportunities for conversation across shifting position arise. Although it is not a representative sample of all geographic engagements with mobility due to various considerations and reasons, this special issue suggests at least five lines of connection across the heterogeneous ways in which geographers study mobility. As it becomes clear later, these five lines are interwoven in multiple ways. They are also not the only ways in which the article are linked; other linkages could have been drawn out as well. For instance, quite a few articles deal with commuting as a more regular and repetitive form of mobility (Bissell this issue; Hu and Wang this issue; Naybor, Pool, and Casas this issue; Parks this issue; Preston and McLafferty this issue; Zhong and Bian this issue) and with questions of health and well-being in relation to mobility (Baker et al. this issue; Naybor Poon, and Casas this issue; Ritterbush this issue; van Blerk this issue; Zhong and Bian this issue). The fivefold division that follows offers a useful way of organizing the articles in this special issue.

Conceptualizing and Analyzing Mobility

Addressing general and broad theoretical, conceptual, analytical, and methodological issues is an important concern for many geographers interested in mobility studies. As mentioned earlier, mobility has now become a significant core geographic concept alongside space, place, network, scale, and territory. Two of the contributions to this special issue address how geographic work might connect theorizations of space and spatialities in geography with the rich conceptualizations of mobility that have emerged as a result of the mobilities turn. Cidell and Lechtenberg (this issue), for instance, draw on the work of a Czech geographer active in the twentieth century—Kamil

Skrbek—to develop a theoretic framework for connecting the spatialities of transportation geography and mobility studies. They explore four kinds of spaces—spaces of movement, spaces of transportation, structural transportation space, and areas of transportation—and suggest that these notions could offer new analytic tools and the possibility for bringing together the two fields through one conceptual framework. Arguing that sociospatial theory is still largely rooted in a sedentarist perspective, and exploring ways for coherently integrating various dimensions of sociospatiality, Miller and Ponto (this issue) examine the connection between mobility and the four distinct sociospatialities identified by Jessop, Brenner, and Jones (2008): territory, place, scale, and networks. Based on an examination of the practice of automobility, Miller and Ponto argue that mobility is "a social, cultural, and political achievement, inherently power-laden and recursively bound up in the production of territory, place, scale, and networks."

Addressing analytical and methodological issues in human mobility studies that use big data, Kwan (this issue) highlights important changes in the geographic knowledge production process associated with the shift from using traditional "small data" to using big data and explores how computerized algorithms might considerably influence research results. She extends and goes beyond earlier arguments (Kitchin and Lauriault 2015) that big data is socially produced, power-laden, and oligoptic by showing that its use can introduce more rather than less uncertainty in geographical studies of mobility. Big data certainly does not speak for itself and its utilization makes mobility exist in selective, partial, and often problematic ways. She calls into question the notion of data-driven geography, which ignores the potentially significant influence of algorithms on research results. Instead she suggests that it is more appropriate to refer to this new kind of geographic inquiry as algorithm-driven geographies (or algorithmic geographies), as the production of geographic knowledge is now far more dependent on computerized algorithms than before. Birenboim and Shoval (this issue) discuss the opportunities and limitations of smartphone data for geographic scholarship on mobility. Many advantages of such data are summarized, but the authors also point to various risks, including selectivity in sampling, geoprivacy and data confidentiality, and data collection techniques that enact the mobility they purport to describe because participants adjust their mobility practices.

Together these two articles highlight significant methodological issues in human mobility research that uses big data.

Inequalities in Mobility

Inequality and exclusion are classic concerns in transportation geography and research (Hanson and Kwan 2008; Lucas 2012; Schwanen et al. 2015; Weber and Kwan 2015) and are equally prominent in the mobilities literature (Uteng and Cresswell 2008; Ohnmacht, Maksim, and Bergman 2009; Söderström et al. 2013; Adey et al. 2014). Despite many differences in exact focus and conceptualization, it has long been recognized that mobility or mobilities are both generating and an outcome of inequalities and exclusion. One of the most insightful strands of literature in this regard is the work by feminist scholars on home–work relations and strategies for overcoming the space–time constraints imposed by competing claims on one's time and for navigating the social norms and emotions associated with care and employment (e.g., Hanson and Pratt 1995; England 1996; Kwan 1999, 2000; Jarvis 2005). Another such strand is the work on race or ethnicity and mobility, much of which has been influenced by Kain's (1968) spatial mismatch hypothesis (e.g., McLafferty and Preston 1992; Ihlanfeldt 1994) but has since moved beyond this idea to address other concerns, including social exclusion and sociospatial segregation (e.g., Uteng 2009; Farber et al. 2015). Both strands of work are represented in this special issue through articles on commuting as a racial mobility project (Parks this issue), on the ongoing evolution of gender and racial differences in commuting in New York (Preston and McLafferty this issue), and on the activity and travel patterns of widowed women in rural Uganda (Naybor, Poon, and Casas this issue).

The emphasis on gender, race, and their intersections with other processes of social differentiation is complemented by an explicit orientation on other social identities that have more recently attracted attention in the literature on mobility—youth, migration and refugee status and sexuality. In keeping with the wider children's geographies literature, there is now a vibrant body of work on the mobility of children and young people across different modes of abstraction (e.g., Kullman 2010; Buliung, Selima, and Faulkner 2012) to which this special issue adds in various ways (Aitken this issue; Cope and Lee this issue; Van Blerk this issue). Cope and Lee, for instance, qualify now popular arguments in the transportation literature that young people are the driving force behind "peak car" (Goodwin and Van Dender 2013)—the idea that across the Global North car ownership and use are no longer growing and are possibly declining. Using a mixed-method approach, Cope and Lee show the continued importance of the car alongside smartphones and other digital devices in fulfilling young people's mobility needs, particularly in areas with low population densities. Attention for the everyday mobility of migrants and refugees and of LGBT+ individuals is much more recent and nascent (Bose 2014; Nash and Gorman-Murray 2014), but the contributions by Maldonado, Licona and Hendricks, and Ritterbush demonstrate how migrant or refugee status and sexuality are coproduced and coevolve with inequalities in and exclusions from mobility, with forms of involuntary immobility as limit cases.

Mobilities scholars have shown convincingly that inequalities in mobility are not only linked to social identity; differences in network capital (Urry 2007) and motility (Kaufmann 2002) create social stratifications that are only weakly correlated with gender, class, age, and so forth. Sheller (this issue) contributes to work in this tradition through a study of how communities in Haiti seek to resist the uneven distribution of network capital in postearthquake Haiti.

Politics of Mobility

The studies mentioned under the previous heading also fit under this one, particularly if politics of mobility is defined as "the ways in mobilities are both productive of social relations [involving the production and distribution of power] and produced by them" (Cresswell 2010, 21). The contributions in this special issue extend understanding of such politics in various ways. Eidse, Turner, and Oswin (this issue) profitably draw on Cresswell's six elements—force, tempo, rhythm, route, experience, and friction—of a politics of mobility, combining this with Kerkvliet's (2009) notion of everyday politics in their study of street vendors in Hanoi. Struggles over who belongs in streetscapes, where, when, and how are also at the heart of Stratford's article, which combines Cresswell's six elements with Lefebvre's right to the city and thinking on play and generosity through a focus on street skating.

Other articles extend the burgeoning literature on mobilities and citizenship (Cresswell 2006, 2013;

Spinney, Aldred, and Brown 2015). In different ways, Aitken (this issue), Price and Breese (this issue), and Staeheli, Marshall, and Maynard (this issue), as well as Maldonado, Licona, and Hendricks (this issue), show how citizenship as an assemblage of roots and routes (Cresswell 2013) "is produced at many different 'sites' and increasingly [from] the relations between these" (Spinney, Aldred, and Brown 2015, 326). Where Aitken (this issue) and Price and Breese (this issue) consider the relation between individual and nation-state through an analysis of minority populations in Slovenia and Latino migrants in the United States, respectively, Staeheli, Marshall, and Maynard (this issue) focus on citizenship beyond the state and as emerging from transactions and circulations through the example of international conferences as space-times where young citizens are formed. The theme of geopolitics is also picked up by Rowen (this issue), who analyzes the relationships between tourism and state-level geopolitics through a study of tourism mobilities that help to reconfigure the relationships of the People's Republic of China with Hong Kong and Taiwan.

Bissell's (this issue) contribution takes the politics of mobility theme in yet other directions. His concern is that the conventional focus on subject-centered analysis concerned with particular figures—the employed mother, the migrant, the sex worker, the citizen, the tourist, and so on—risks drawing attention away from the micropolitics associated with the ongoing churn of events and encounters during particular movements. His Deleuzian approach—illustrated through autoethnographic research on a commute between Sydney and Wollongong—offers a useful complement to other macropolitical work in the special issue and elsewhere on how gender, race, and migration status shape and distribute (im)mobility.

Decentering Mobility

Of the twenty-six main articles in this special issue, four concentrate on East Asia (including China); three on Africa; one each on Latin America, the Caribbean, and Eastern Europe; and another one—by Cidell and Lechtenberg—draws explicitly on notions postulated by Czech geographer Kamil Skrbek. If Staeheli, Marshall, and Maynard's and Best's contributions are added,[4] it can be argued that half of the contributions have a clear link with settings outside North America, Western Europe, and Australia and New Zealand. This is clear evidence that, as a consequence

of and enabled by the globalized and globalizing mobilities of people (not least academic geographers!), information, and ideas, geographic scholarship is undergoing a long overdue shift away from its conventional orientation toward the Global North.

This decentering of orientation is beneficial for multiple reasons. It opens up new questions and concerns across traditions of studying mobility within geography. For instance, both the analysis by Naybor, Poon, and Casas (this issue), which is firmly set in the transportation geography tradition, and the mobilities articles by van Blerk and Ritterbush highlight the complex relationships between mobility and livelihoods. Naybor, Poon, and Casas show how the lifting of constraints on mobility makes it easier for widowed women in rural Uganda to earn a living, thereby empowering a group that is otherwise at risk of social marginalization. In contrast, van Blerk (this issue) and Ritterbush (this issue) each study sex workers, with the former working with young women in Ethiopia and the latter with transgender women in Bogota. Both show how livelihood and identity can trap sex workers in particular places but are also made possible by—indeed necessitate—moves away from familiar places, relatives, and friends. Mobility and immobility become imbued with multiple and profoundly ambiguous meanings and affectivities in the process.

Second, a focus on mobility outside Global North settings can easily demonstrate the spatial and historical contingency of understandings of mobility. This is convincingly shown in Porter's (this issue) contribution on the interconnections of physical mobility and mobile phone usage in rural areas in Tanzania and Malawi. Her research suggests that the widely reported conclusion in Western studies that increased mobile phone use has not generated a major reduction in travel activity does not hold in parts of rural Africa. There, the friction of distance is (still) so much larger because walking remains by far the most dominant way of getting around, mobility is costly in financial terms, and traffic accidents take many lives.

Finally, although the connection is certainly not inevitable, a decentering away from Global North settings might facilitate the diffusion of postcolonial and decolonial thinking[5] throughout geographic scholarship on mobility. Past research has certainly engaged with postcolonial theory (e.g., Sheller 2003; Roy 2012), but it is fair to say that such thinking has been taken up less in both transportation geography and mobilities scholarship than in other parts of geography. That postcolonial theory can strengthen mobility research in geography is in this collection most clearly demonstrated by Best's (this issue) study of dollar cabs operated by Caribbean immigrants in Brooklyn, New York. More conventionally seen as an informal, semiclandestine form of transport in urgent need of regulation or as a last resort for poor people without access to "regular" public transport or private car use, dollar cabs become ambiguous elements in contemporary New York that open new understandings of speed, time, and everyday life in transnational migrant communities as well as the racialized nature of automobility if examined through a postcolonial lens. Best's paper therefore offers an interesting complement and contrast to other papers on mobility, race, and migrant status in this special issue (e.g., Maldonado, Licona, and Hendricks; Parks; Preston and McLafferty).

Qualifying Abstractions of Mobility

From an affirmative perspective on abstraction (Whitehead 1926; McCormack 2012; Schwanen 2015), all articles in this special issue selectively engage and simplify mobility as an ontologically uncertain, complex, and emergent process. They might do so in radically different ways but always innovatively. It can, in fact, be said that the articles qualify abstractions: Not only do these articles make mobility to exist in particular ways through their conceptualizations, theorizations, and methodological practices, but they also identify and "add" particular qualities of and to mobility that in previous and other contemporaneous research remain more or less unarticulated. The set of articles under this heading of the special issue is fairly arbitrary but gathers contributions that innovatively bring out aspects of mobility by linking various understandings of mobility to specific strands of literature and theory, methodological tools, and new data-producing technologies from other parts of the discipline and beyond.

Some papers articulate specific facets of mobility by drawing on specific bodies of literature in geography and beyond. Thus, Spinney (this issue) seeks to understand governmental interventions to encourage cycling through the lens of biopolitics and Harvey's (2001) work on spatial fixes. Baker et al. (this issue), in contrast, focus on the mobility of ideas as "emerg[ing] from people and their relations with others" and more specifically on policies, thereby combining understandings from the new mobilities paradigm with similar developments in urban planning and anthropology. Hu and Wang (this issue) evaluate excess commuting and

examine the temporal trends of commuting patterns in both time and distance, using a Monte Carlo simulation-based approach that takes into account the effects of land use patterns. Zhong and Bian's (this issue) article offers an interesting contrast with that by Baker et al., although the former engages less with a particular body of social theory than the now widely deployed analytical frameworks (in transportation geography) of network science and graph theory. Like Baker et al., Zhong and Bian examine how a particular object—in their case influenza—diffuses spatially through the movements of people. Although not explicitly interested in meaning and power, these processes are still shaping and implicitly considered in Zhong and Bian's analysis of individuals' daily travel between homes and workplaces. The final contribution speaks to the question of how big data—in particular those generated with and through mobile phones—enable new practices of abstraction to geographers interested in mobility and other issues (see also Graham and Skelton 2013; Kitchin and Lauriault 2015; Rae and Singleton 2015). Xu et al. (this issue) show how big data collected with mobile phones can be used to extend and improve time-geographic analyses of human activity spaces.

Avenues for Further Research

Bringing together many different ways of conceptualizing, theorizing, and empirically examining mobility and thereby creating new connections across modes of abstraction, the special issue also points to various themes and developments that could stimulate further conversations across traditions and modes of abstraction within geographic scholarship on mobility. Here we limit ourselves to identifying three such themes and developments.

Health and Well-Being

As suggested earlier, health and well-being is a theme that runs across a number of article in the special issue. In many ways it is central to geographic scholarship on mobilities because exposure to factors that influence health and well-being, access to or use of health care facilities, and spread of disease are inextricably connected to human movement at various spatial and temporal scales (Gatrell 2011; Kwan 2012, 2013; Schwanen and Wang 2014; Chen and Kwan 2015). As people move around in their daily life and over their life course, they are under the influence of many different places (or geographic contexts) and come into contact with different persons or social groups, in particular during time they spend outside of their residential neighborhood and during travel between different locations. Thus, people's exposure to environmental and social influences that affect their health and well-being changes over space and time in a highly complex manner. Moreover, particular forms of mobility might be more or less healthy because of the level of physical activity involved—witness the large literature in transportation and health literature on cycling and walking as "active travel" (Gatrell 2013; Schwanen 2016)—and because they can induce and stimulate experiences of belongingness, freedom and autonomy, and self-esteem (Hanson 2010; Nordbakke and Schwanen 2014).

Several areas seem especially fruitful for future research on the relationships between mobility and health and well-being. First, moving beyond the traditional notion of static, area-based geographic context (e.g., the residential neighborhood) to take into account the effects of people's mobility on their health and well-being will be an important research area. Adopting dynamic conceptualizations of geographic context and developing methods for collecting and analyzing dynamic data of human movement and environmental influences will be essential tasks (Kwan 2012, 2013). Second, future work on the relationships between mobility and health and well-being "should consider both the objective and the subjective" dimensions of well-being; researchers should also pay attention "to the multiple ways in which well-being and its linkages to mobility are context-dependent and shaped by the particularities of time and place" (Nordbakke and Schwanen 2014, 104). Third, because researchers from a wide range of disciplines have contributed important insights to understanding the complex relationships of mobility with health and well-being, future research can benefit considerably from adopting interdisciplinary perspectives that integrate diverse elements of various conceptualizations and methods. Fourth, the experiences of mobility and well-being seem drastically different for different social groups (e.g., older people). It is important for future research to be attentive to the effects of social difference that are relatively understudied (e.g., sexuality, religion, migrant or refugee status), while also heeding what Bissell (this issue) calls the micropolitics of mobility to which people of any social group might be exposed during everyday movements. Finally, the discursive constitution of certain forms of mobility as

healthy or unhealthy and the effects that such constitution has on mobility practices and experiences in different places deserve further scrutiny. For instance, can positioning urban cycling as healthy by policymakers and public health officials trump well-known barriers to cycling such as traffic safety risks, poor air quality, and inadequate physical infrastructure? To what extent do discourses that link automobility to obesity make car use a guilty pleasure or even—as with smoking (Tan 2013)—a form of resistance for particular groups in specific geographical contexts?

Further Decentering and Decolonization

This special issue suggests a broader trend of decentering of geographic scholarship on mobility away from the Global North. For various reasons, though, this process needs to be taken much further. First, from a policy and governance point of view, mobility poses one of the biggest challenges in regions outside the Western world. It is in emerging economies and developing countries that both overall mobility levels and inequalities in mobilities are growing most rapidly and seen to cause difficult ethical questions about priorities. For instance, should governments in these countries actively encourage (motorized) mobility to enhance (economic) development and individuals' life chances at the cost of slower emission reduction and decarbonization? Should governments condone growing sociospatial polarization in the short term in the hope that trickle-down effects will improve overall welfare in the long run, or should they reduce inequalities in mobility and guarantee "mobility rights" from the start? Geographers should not only address such questions but also critically interrogate their framing and unpack the often taken-for-granted assumptions on which they are based.

Second, research on mobilities beyond the Global North is for the most part conducted by scholars born in or at least trained in the center—academic institutions in the Western world or heavily influenced by Western thought. Conversations on the geographies of mobility would be greatly enriched if they became more "worlded" in the way urban theory is now starting to be (McCann, Roy, and Ward 2013; Sheppard, Leitner, and Maringanti 2013; Sheppard et al. 2015). The result will be the coming into being of geographies of mobility that durably reconfigure familiar distributions of core and periphery, theory and empirics. It would also enable the generation of mobility theories that are no longer formulated predominantly in the West or on the basis of European-American ideas and practices regarding methods, data, and analysis. Also, Western theories would not simply be exported as if they were universal tools for making sense of other parts of the world that are taken to be little more than fields where materials can be harvested to test and refine theories formulated from a Western standpoint. This form of geographic scholarship on mobility would not be a hegemonic project seeking to provide somehow superior alternative knowledges but options (Mignolo 2011)—that is, modes of abstraction that neither take as given the epistemological, political, economic, and social domination of Euro-American ideas, institutions, and habits nor seek to marginalize and displace other conceptual and methodological practices. It would engage in dialogues and, through engaging other modes of abstraction in "publicly recognized forums for the criticism of evidence, of methods, and of assumptions and reasoning" (Longino 2002, 129), seek to induce change in those other modes.

Combining Big and Small Data

Geographers and transportation researchers have studied human mobility for decades. Many past studies used detailed data collected through activity-travel diary surveys, which provide rich and detailed information about many attributes of respondents' activities and trips. Geographers have also incorporated GPS data as an important element in this research (e.g., Shoval and Isaacson 2007; Shen, Kwan, and Chai 2013; Shoval et al. 2014). As collecting this kind of detailed data is costly and time-consuming, the rapid increase in the volume, diversity, and intensity of inexpensive data from various big data sources in recent years has stimulated new developments in human mobility research (e.g., González, Hidalgo, and Barabasi 2008). Although this research has yielded interesting findings (e.g., people make more short trips than long ones, and they return to certain locations regularly), what can be observed from big data about actual human movement is rather limited or can be highly misleading, as Kwan (this issue) argues. Studies using big data sets also tend to overestimate people's mobility and underestimate their daily travel distance.

An important area for future research is thus how traditional "small data," including qualitative data, can be used together with big data for overcoming the limitations of the latter (see also Kitchin and Lauriault 2015). For instance, activity-travel diaries record the details of respondents' activities and trips according to

the temporal sequence in which they are undertaken on a particular survey day (Hanson and Hanson 1981; Kwan 1999). Qualitative or mixed mobile methods, such as ride-along interviews and GPS and video-based geonarratives, have also been used to capture people's experiences while moving around (e.g., Kwan and Ding 2008; Bell et al. 2015; Curtis et al. 2015). Because none of the rich and nuanced data collected through traditional or qualitative methods are available in popular big data sets, these data can be used to complement or enrich the analysis performed with big data in human mobility research. Future research can also explore the intersection between geographic studies of mobility and the broad concerns in digital humanities and the development of mobile methods in mobilities research. In this way, further connections can be forged between different approaches to the study of mobility in geography and engaged pluralism can become the norm that is enacted in research practices rather than an abstract vision for the future.

Acknowledgments

We thank all individuals who submitted abstracts and manuscripts for consideration. We deeply appreciate all of the authors for their excellent contributions and their efforts in adhering to the tight timetable and length restriction we are required to impose. We are also grateful to the editorial board members and reviewers who provided constructive and timely comments. Publication of this special issue would be impossible without the superb assistance of Jennifer Cassidento, Miranda Lecea, and Robin Maier of the AAG Office.

Funding

Mei-Po Kwan was supported by the following grants when writing this article: NSF IIS-1354329, NSF BCS-1244691, and NSFC 41529101. Tim Schwanen was supported by grant ES/N011538/1 by the ESRC.

Notes

1. We conducted a search of the journal's back catalogue by using *mobility* as the search term for the first 100 volumes of the *Annals* on the JSTOR Web site. More than 400 articles and commentaries were returned, going as far back as 1914.
2. Although the field is more vibrant, engaging, and concerned with more topical issues than many geographers believe (Schwanen 2016), it remains slow in engaging with the wider philosophical and theoretical debates

elsewhere in the discipline. It continues to struggle with the legacy of the quantitative revolution of the 1950s and 1960s and finds it difficult to reconcile the concerns of cultural and critical geography with the pressures exerted by cross-disciplinary dialogues with engineering, economics, and business studies and the unequal power relations characterizing those dialogues (see also Hanson 2000; Ng et al. 2014).

3. Whitehead himself was particularly concerned about inertia and obsolescence in abstractions. His philosophical project in the 1920s and 1930s can be seen as a fallible attempt to revise abstractions in thought dating back to the seventeenth and eighteenth centuries but no longer appropriate in light of the works of Darwin, Einstein, Bohr, Heisenberg, and others.

4. Staeheli, Marshall, and Maynard (this issue) discuss ethnographic fieldwork conducted at an international youth conference that took place in Sri Lanka, and Best (this issue) studies dollar cabs operated by Caribbean immigrants in Brooklyn, New York.

5. See Mignolo (2011) for discussion of the differences between postcolonial and decolonial thought. Although both seek to confront the legacies of colonialism, these bodies of thought have different origins and genealogies. Postcolonial scholarship emerged from the experience of British colonization in the Middle East and South Asia and has been influenced heavily by postmodernity and poststructuralism. Originating from the Caribbean and Latin America, decoloniality seeks to make visible, critique, and move beyond historic and contemporary forms of epistemic, social, political, and economic domination that place Eurocentric concepts and practices at the apex of civilization.

References

Adey, P. 2006. If mobility is everything then it is nothing: Towards a relational politics of (im)mobilities. *Mobilities* 1 (1): 75–94.

Adey, P., D. Bissell, K. Hannam, P. Merriman, and M. Sheller, eds. 2014. *The Routledge handbook of mobilities.* London and New York: Routledge.

Barnes, T. J., and E. Sheppard. 2010. "Nothing includes everything": Towards engaged pluralism in Anglophone economic geography. *Progress in Human Geography* 34 (2): 193–214.

Bell, S. L., C. Phoenix, R. Lovell, and B. W. Wheeler. 2015. Using GPS and geo-narratives: A methodological approach for understanding and situating everyday green space encounters. *Area* 47 (1): 88–96.

Bernstein, R. J. 1988. Pragmatism, pluralism and the healing of wounds. *Proceedings and Addresses of the American Philosophical Association* 63:5–18.

Bissell, D., P. Adey, and E. Laurier. 2011. Introduction to the special issue on geographies of the passenger. *Journal of Transport Geography* 19 (5): 1007–09.

Bissell, D., and G. Fuller. 2011. *Stillness in a mobile world.* London and New York: Routledge.

Bose, P. S. 2014. Refugees in Vermont: Mobility and acculturation in a new immigrant destination. *Journal of Transport Geography* 36:151–59.

Breese, G. 1963. Development problems in India. *Annals of the Association of American Geographers* 53 (3): 253–65.

Buliung, R., S. Selima, and G. Faulkner. 2012. Guest editorial: Special section on child and youth mobility—Current research and nascent themes. *Journal of Transport Geography* 20 (1): 31–33.

Chen, X., and M.-P. Kwan. 2015. Contextual uncertainties, human mobility, and perceived food environment: The uncertain geographic context problem in food access research. *American Journal of Public Health* 105 (9): 1734–37.

Cidell, J., and D. Prytherch, eds. 2015. *Transport, mobility, and the production of urban space.* London and New York: Routledge.

Cresswell, T. 2006. *On the move: Mobility in the modern Western world.* London and New York: Routledge.

———. 2010. Towards a politics of mobility. *Environment and Planning D: Space and Society* 28 (1): 17–31.

———. 2011. Mobilities I: Catching up. *Progress in Human Geography* 35 (4): 550–58.

———. 2012. Mobilities II: Still. *Progress in Human Geography* 36 (5): 645–53.

———. 2013. Citizenship in worlds of mobility. In *Critical mobilities,* ed. O. Söderström, S. Randeria, D. Ruedin, G. D'Amato, and F. Panese, 105–24. London and New York: Routledge.

———. 2014. Mobilities III: Moving on. *Progress in Human Geography* 38 (5): 712–21.

Cresswell, T., and P. Merriman. 2011. *Geographies of mobilities: Practices, spaces, subjects.* Farnham, UK: Ashgate.

Curtis, A., J. W. Curtis, E. Shook, S. Smith, E. Jefferis, L. Porter, C. Felix, and P. R. Kerndt. 2015. Spatial video geonarratives and health: Case studies in post-disaster recovery, crime, mosquito control and tuberculosis in the homeless. *International Journal of Health Geographics* 14 (1): 22.

Dryer, C. R. 1915. Natural economic regions. *Annals of the Association of American Geographers* 5:121–25.

Enders, M., K. Manderscheid, and C. Mincke, eds. 2016. *The mobilities paradigm: Discourses and ideologies.* Farnham, UK: Ashgate.

England, K. V. L. 1996. *Who will mind the baby? Geographies of child care and working mothers.* London and New York: Routledge.

Ernste, H., K. Martens, and J. Schapendonk. 2012. The design, experience and justice of mobility. *Tijdschrift voor Economische en Sociale Geografie* 103 (5): 509–15.

Farber, S., M. O'Kelly, H. Miller, and T. Neutens. 2015. Measuring segregation using patterns of daily travel behavior: A social interaction based model of exposure. *Journal of Transport Geography* 49:26–38.

Feyerabend, P. 1975. *Against method: Outline of an anarchist theory of knowledge.* London: Verso.

Foucault, M. 1972. *The archaeology of knowledge,* trans. A. M. Sheridan Smith. London: Tavistock.

Frello, B. 2008. Towards a discursive analytics of movement: On the making and unmaking of movement as an object of knowledge. *Mobilities* 3 (1): 25–50.

Frey, J. W. 1941. Petroleum utilization in peacetime and in wartime. *Annals of the Association of American Geographers* 31 (2): 113–18.

Gatrell, A. C. 2011. *Mobilities and health.* Aldershot, UK: Ashgate.

———. 2013. Therapeutic mobilities: Walking and "steps" to wellbeing and health. *Health and Place* 22:98–106.

Ginsburg, N. 1957. Natural resources and economic development. *Annals of the Association of American Geographers* 47 (3): 197–212.

Gleason, H. A. 1922. The vegetational history of the middle west. *Annals of the Association of American Geographers* 12:39–85.

Goetz, A. R., T. Vowles, and S. Tierney. 2009. Bridging the qualitative–quantitative divide in transport geography. *The Professional Geographer* 61 (3): 323–35.

González, M. C., C. A. Hidalgo, and A. L. Barabasi. 2008. Understanding individual human mobility patterns. *Nature* 453 (7196): 779–82.

Goodwin, P., and K. Van Dender. 2013. "Peak car"—Themes and issues. *Transport Reviews* 33 (3): 243–54.

Graham, M., and T. Skelton. 2013. Geography and the future of big data, big data and the future of geography. *Dialogues in Human Geography* 3 (3): 255–61.

Hall, A. R. 1955. Mackinder and the course of events. *Annals of the Association of American Geographers* 45 (2): 109–26.

Hanson, S. 2000. Transportation: Hooked on speed, eyeing sustainability. In *Companion to economic geography,* ed. E. Sheppard and T. J. Barnes, 468–83. Oxford, UK: Blackwell.

———. 2010. Gender and mobility: New approaches for informing sustainability. *Gender, Place and Culture: A Journal of Feminist Geography* 17 (1): 5–23.

Hanson, S., and P. Hanson. 1981. The travel-activity patterns of urban residents: Dimensions and relationships to sociodemographic characteristics. *Economic Geography* 57:332–47.

Hanson, S., and M.-P. Kwan, eds. 2008. *Transport: Critical essays in human geography.* Aldershot, UK: Ashgate.

Hanson, S., and G. Pratt. 1995. *Gender, work, and space.* London and New York: Routledge.

Haraway, D. 1991. *Simians, cyborgs, and women: The reinvention of nature.* London and New York: Routledge.

Harvey, D. 2001. *Space of capital: Towards a critical geography.* London and New York: Routledge.

Hoffman, G. F. 1957. The role of nuclear power in Europe's future energy balance. *Annals of the Association of American Geographers* 47 (1): 15–40.

Ihlanfeldt, K. R. 1994. The spatial mismatch between jobs and residential locations within urban areas. *Journal of Policy Development and Research* 1 (1): 219–44.

Jarvis, H. 2005. Moving to London time: Household coordination and the infrastructure of everyday life. *Time and Society* 14 (1): 133–54.

Jessop, B., N. Brenner, and M. Jones. 2008. Theorizing sociospatial relations. *Environment and Planning D* 26 (3): 389–401.

Joerg, W. L. G. 1914. The subdivision of North America into natural regions: A preliminary inquiry. *Annals of the Association of American Geographers* 4:55–83.

Kain, J. F. 1968. Housing segregations, negro employment, and metropolitan decentralization. *Quarterly Journal of Economics* 82:175–97.

Kaufmann, V. 2002. *Re-thinking mobility.* Farnham, UK: Ashgate.

Kerkvliet, B. J. T. 2009. Everyday politics in peasant societies (and ours). *Journal of Peasant Studies* 36 (1): 227–43.

Kitchin, R., and T. P. Lauriault. 2015. Small data in the era of big data. *GeoJournal* 80 (4): 463–75.

Kollmorgen, W. M. 1969. The woodsman's assaults on the domain of the cattleman. *Annals of the Association of American Geographers* 59 (2): 215–39.

Kullman, K. 2010. Transitional geographies: Making mobile children. *Social and Cultural Geography* 11 (8): 827–44.

Kwan, M.-P. 1999. Gender, the home–work link, and space–time patterns of non-employment activities. *Economic Geography* 75 (4): 370–94.

———. 2000. Gender differences in space–time constraints. *Area* 32 (2): 145–56.

———. 2004. Beyond difference: From canonical geography to hybrid geographies. *Annals of the Association of American Geographers* 94 (4): 756–63.

———. 2007. Mobile communications, social networks, and urban travel: Hypertext as a new metaphor for conceptualizing spatial interaction. *The Professional Geographer* 59 (4): 434–46.

———. 2012. The uncertain geographic context problem. *Annals of the Association of American Geographers* 102 (5): 958–68.

———. 2013. Beyond space (as we knew it): Toward temporally integrated geographies of segregation, health, and accessibility. *Annals of the Association of American Geographers* 103 (5): 1078–86.

Kwan, M.-P., and G. Ding. 2008. Geo-narrative: Extending geographic information systems for narrative analysis in qualitative and mixed-method research. *The Professional Geographer* 60 (4): 443–65.

Latour, B. 2005. What is given in experience? *Boundary* 32 (1): 223–37.

Law, J. 2004. *After method: Mess in social science research.* London and New York: Routledge.

Longino, H. 2002. *The fate of knowledge.* Princeton, NJ: Princeton University Press.

Lorimer, H. 2007. Cultural geography: Worldly shapes, differently arranged. *Progress in Human Geography* 31 (1): 89–100.

Lucas, K. 2012. Transport and social exclusion: Where are we now? *Transport Policy* 20:105–13.

Marschner, F. K. 1944. Structural properties of medium- and small-scale maps. *Annals of the Association of American Geographers* 34 (1): 1–46.

McCann, E. 2011. Urban policy mobilities and global circuits of knowledge: Toward a research agenda. *Annals of the Association of American Geographers* 101 (1): 107–30.

McCann, E., A., Roy, and K. Ward. 2013. Urban pulse—Assembling/worlding cities. *Urban Geography* 34 (5): 581–89.

McCormack, D. P. 2012. Geography and abstraction: Towards an affirmative critique. *Progress in Human Geography* 36 (6): 715–34.

McLafferty, S., and V. Preston. 1992. Spatial mismatch and labor market segmentation for African-American and Latina women. *Economic Geography* 68 (4): 406–31.

Merriman, P. 2007. *Driving spaces: A cultural-historical geography of England's M1 motorway*. Malden, MA: Blackwell.

———. 2015. Mobilities I: Departures. *Progress in Human Geography* 39 (1): 87–95.

———. Forthcoming. Mobilities II: Cruising. *Progress in Human Geography*.

Mignolo, W. D. 2011. *The darker side of western modernity: Global futures, decolonial options*. Durham, NC: Duke University Press.

Mouffe, C. 1999. Deliberative democracy or agnostic pluralism? *Social Research* 66 (3): 745–58.

Murphey, R. 1954. The city as a center of change: Western Europe and China. *Annals of the Association of American Geographers* 44 (4): 349–62.

Nash, C. J., and Gorman-Murray, A. 2014. LGBT neighbourhoods and "new mobilities": Towards understanding transformations in sexual and gendered urban landscapes. *International Journal of Urban and Regional Research* 38 (3): 756–72.

Ng, A. K. Y., C. Ducruet, W. Jacobs, J. Monios, T. Notteboom, J.-P. Rodrigue, B. Slack, K.-C. Tam, and G. Wilmsmeier. 2014. Port geography at the crossroads with human geography: Between flows and spaces. *Journal of Transport Geography* 41:84–96.

Nordbakke, S., and T. Schwanen. 2014. Well-being and mobility: A theoretical framework and literature review focusing on older people. *Mobilities* 9 (1): 104–29.

Ohnmacht, T., H. Maksim, and M. M. Bergman, eds. 2009. *Mobilities and inequality*. Farnham, UK: Ashgate.

Platt, R. S. 1927. A classification of manufactures, exemplified by Porto Rican industries. *Annals of the Association of American Geographers* 17 (2): 79–91.

———. 1928. A detail of regional geography: Ellison Bay community as an industrial organism. *Annals of the Association of American Geographers* 18 (2): 81–126.

Rae, A., and A. Singleton. 2015. Putting big data in its place: A *Regional Studies and Regional Science* perspective. *Regional Studies and Regional Science* 2 (1): 1–15.

Roy, A. 2012. Ethnographic circulations: Space–time relations in the worlds of poverty management. *Environment and Planning A* 44 (1): 31–41.

Sack, R. D. 1976. Magic and space. *Annals of the Association of American Geographers* 66 (2): 309–22.

Sauer, C. O. 1941. Foreword to historical geography. *Annals of the Association of American Geographers* 31 (1): 1–24.

Schwanen, T. 2015. Understanding process: Can transport research come to terms with temporality? In *Handbook on transport and development*, ed. R. Hickman, D. Bonilla, M. Givoni, and D. Banister, 660–74. Cheltenham, UK: Edward Elgar.

———. 2016. Geographies of transport: Reinventing a field? *Progress in Human Geography* 40 (1): 126–137.

Schwanen, T., and M.-P. Kwan. 2008. The Internet, mobile phone and space–time constraints. *Geoforum* 39 (3): 1362–77.

Schwanen, T., K. Lucas, N. Akyelken, D. C. Solsona, J.-A. Carrasco, and T. Neutens. 2015. Rethinking the links between social exclusion and transport disadvantage through the lens of social capital. *Transportation Research: Policy and Practice* 74:123–35.

Schwanen, T., and A. Páez. 2010. The mobility of older people—An introduction. *Journal of Transport Geography* 18 (5): 591–95.

Schwanen, T., and D. Wang. 2014. Well-being, context, and everyday activities in space and time. *Annals of the Association of American Geographers* 104 (4): 833–51.

Shaw, J., and I. Docherty. 2014. *The transport debate*. Bristol, UK: Policy Press.

Shaw, J., and M. Hesse. 2010. Transport, geography and the "new" mobilities. *Transactions of the Institute of British Geographers* 35 (3): 305–12.

Shaw, J., and J. D. Sidaway. 2011. Making links: On (re) engaging with transport and transport geography. *Progress in Human Geography* 35 (4): 502–20.

Sheller, M. 2003. *Consuming the Caribbean: From arawaks to zombies*. London and New York: Routledge.

Sheller, M., and J. Urry. 2006. The new mobilities paradigm. *Environment and Planning A* 38 (2): 207–26.

Shen, Y., M.-P. Kwan, and Y. Chai. 2013. Investigating commuting flexibility with GPS data and 3D geovisualization: A case study of Beijing, China. *Journal of Transport Geography* 32 (1): 1–11.

Sheppard, E., V. Gidwani, M. Goldman, H. Leitner, A. Roy, and A. Maringanti. 2015. Introduction: Urban revolutions in the age of global urbanism. *Urban Studies* 52 (11): 1947–61.

Sheppard, E., H. Leitner, and A Maringanti. 2013. Urban pulse—Provincializing global urbanism: A manifesto. *Urban Geography* 34 (7): 893–900.

Shoval, N., and M. Isaacson. 2007. Sequence alignment as a method for human activity analysis in space and time. *Annals of the Association of American Geographers* 97 (2): 282–97.

Shoval, N., M.-P. Kwan, K. H. Reinau, and H. Harder. 2014. The shoemaker's son always goes barefoot: Implementations of GPS and other tracking technologies for geographic research. *Geoforum* 51:1–5.

Simini, F., M. C. González, A. Maritan, and A.-L. Barabási. 2012. A universal model for mobility and migration patterns. *Nature* 484:96–100.

Smith, J. R. 1943. Grassland and farmland as factors in the cyclical development of Eurasian history. *Annals of the Association of American Geographers* 33 (3): 135–61.

Söderström, O., S. Randeria, D. Ruedin, G. D'Amato, and F. Panese, eds. 2013. *Critical mobilities*. London and New York: Routledge.

Spinney, J., R. Aldred, and K. Brown. 2015. Geographies of citizenship and everyday (im)mobility. *Geoforum* 64:325–32.

Stengers, I. 2000. *The invention of modern science*, trans D. W. Smith. Minneapolis: University of Minnesota Press.

———. 2005. Events and histories of knowledge. *Review* 28 (2): 143–59.

———. 2011. *Thinking with Whitehead: A free and wild creation of concepts*, trans. M. Chase. Cambridge, MA: Harvard University Press.

Tan, Q. H. 2013. Smoking spaces as enabling spaces of well-being. *Health and Place* 24:173–82.

Urry, J. 2007. *Mobilities*. Cambridge, UK: Polity.

Uteng, T. P. 2009. Gender, ethnicity, and constrained mobility: Insights into the resultant social exclusion. *Environment and Planning A* 41 (5): 1055–71.

Uteng, T. P., and T. Cresswell, eds. 2008. *Gendered mobilities*. Aldershot, UK: Ashgate.

Weber, J., and M.-P. Kwan. 2015. Mobility and travel activity patterns. In *International encyclopedia of the social & behavioral sciences*. 2nd ed., ed. J. D. Wright, 636–39. Oxford, UK: Elsevier.

Wheeler, J. O. 1972. Trip purposes and urban activity linkages. *Annals of the Association of American Geographers* 62 (4): 641–54.

Whitehead, A. N. 1920. *The concept of nature*. Cambridge, UK: Cambridge University Press.

———. 1926. *Science and the modern world*. Cambridge, UK: Cambridge University Press.

Whittlesey, D. 1945. The horizon of geography. *Annals of the Association of American Geographers* 35 (1): 1–36.

———. 1956. Southern Rhodesia—An African compage. *Annals of the Association of American Geographers* 46 (1): 1–97.

Wilsmeier, G., and J. Monios. 2015. The production of capitalist "smooth" space in global port operations. *Journal of Transport Geography* 47:59–69.

Ziegler, F., and T. Schwanen. 2011. "I like to go out to be energised by different people": An exploratory analysis of mobility and wellbeing in later life. *Ageing and Society* 31 (5): 758–81.

Developing a Framework for the Spaces and Spatialities of Transportation and Mobilities

Julie Cidell* and Devon Lechtenberg[†]

*Department of Geography and GIS, University of Illinois
[†]Independent Scholar, Chicago, IL

Recent debates over space and spatiality have touched on a variety of different fields within geography but have rarely considered transportation or mobilities. Mobile objects and actors can be seen as occupying space only temporarily; as constructing space through their travels (including waiting to travel); as being present and absent before, after, and during their travels; or a host of other possibilities that complicate the geographer's basic question: What is where? A theoretical consideration of the spatialities of transportation and mobilities could offer new analytic tools not only for scholars in these areas but for those interested in the production of space more broadly. It also offers the possibility to bring together transportation geography and mobilities under one framework. This article draws on the work of Kamil Skrbek, a Czech geographer active in the twentieth century, to explore four kinds of spaces: spaces of movement, spaces of transportation, structural transportation space, and areas of transportation. These four types include different extents infrastructure, vehicles, and passengers; places through which these objects pass or bypass; social, cultural, and economic discourses and meanings; and lived experiences of those within and outside of various transportation networks. Using the example of the shipment of crude oil by rail within North America, we explicate how these different spaces are relevant to the issue and what they offer to an analysis of this important topic that existing approaches cannot.

晚近有关空间及空间性的辩论, 已触及地理学中的各种不同领域, 但却仍鲜少考量运输或能动性。移动的物体与行动者, 可被视为仅只是暂时性地占据空间; 透过他们的移动建构空间 (包含等待移动); 作为其移动之前、之后和期间的在场与缺席; 或作为复杂化地理学者的基本问题 "何物于何处?" 的其他可能性之宿主。对于运输和能动性的空间性之理论考量, 不仅能够提供在这些领域中的学者崭新的分析工具, 同时能够更广泛地提供给对空间生产有兴趣的学者使用。此般考量, 亦提供了将运输地理与能动性同时置放在单一架构下的可能性。本文运用一位活跃于二十世纪的捷克地理学家斯卡贝克 (Kmil Skrbek) 的研究, 探讨四种空间: 活动的空间, 运输的空间, 结构的运输空间, 以及运输范围。这四种类型, 包含不同的范围基础建设、运输工具及乘客; 这些物体行经或绕过的地方; 社会、文化和经济论述及意义; 以及在各种运输网络之内与之外的生活经验。我们运用北美原油铁道运输的案例, 阐述这些不同的空间如何与该议题有关, 以及它们对于分析此般重要议题提供了什麼既有方法未能提供的分析。

Los recientes debates sobre espacio y espacialidad han tocado una variedad de campos diferentes dentro de la geografía, pero raramente han tomado en cuenta el transporte y las movilidades. Los objetos y actores móviles pueden visualizarse solamente como entes que ocupan espacio temporalmente; como actores que construyen espacio por medio de sus viajes (incluyendo la espera para viajar); como sujetos que están presentes y ausentes antes, después y durante sus viajes; o como un conjunto de otras posibilidades que complican el básico interrogante del geógrafo: ¿Qué es dónde? Una consideración teórica de las espacialidades del transporte y las movilidades podría ofrecer nuevas herramientas analíticas no solo para eruditos en estas áreas sino, con mayor amplitud, para los interesados en la producción de espacio. Ofrece también la posibilidad de integrar la geografía del transporte y las movilidades dentro de un solo marco. Este artículo se apoya en el trabajo de Kamil Skrbek, un geógrafo checo activo en el siglo XX, para explorar cuatro tipos de espacio: espacios de movimiento, espacios de transporte, espacio estructural del transporte, y áreas de transporte. Estos cuatro tipos incluyen diferentes extensiones de infraestructura, vehículos y pasajeros; lugares a través de los cuales cruzan o se desvían estos objetos; discursos y significados sociales, culturales y económicos; y experiencias vitales de quienes estuvieron dentro o fuera de varias redes de transporte. Usando el ejemplo del envío de petróleo crudo por tren en Norteamérica, explicamos cómo estos diferentes espacios importan para este tema y qué ofrecen para un análisis de tan importante tópico que no puedan ofrecer los enfoques existentes.

The mobilities turn in the social sciences has introduced a focus on the lived, qualitative experiences of travel and transportation, drawing on cultural studies, sociology, and geography to explicate what "life on the move" entails for people and places (Sheller and Urry 2006). At the same time, transportation geography has not only witnessed continued strength in its tradition of quantitative, policy-relevant work but has begun to incorporate more critical perspectives (Goetz, Vowles, and Tierney 2009; see Hanson [2003, 2006]; Shaw and Hesse [2010]; Cidell and Prytherch [2015] for more details). Although there continue to be calls to bring together these two approaches to the study of movement and transport (Shaw and Hesse 2010; Shaw and Sidaway 2010), we find that both have been neglected by the broader geographic community in theorizing space and spatiality. In this article, we offer one approach for thinking through the spatial relationships of transportation and mobility, based on our elaboration of the work of Kamil Skrbek.

For many geographers outside the subfield, the spatial concepts relevant to transportation are nodes and corridors—roads, airports, rails, and other visible sites of hard infrastructure—rather than the wider social and political spaces that infrastructure and its users help to construct (but see Schwanen 2015). At the same time, although flows of people and goods are the hallmark of globalization, little attention was paid in the initial literature on globalization as to how those flows actually happen and what their consequences are for the places they pass through. Ironically, this was happening at the same time as a rich literature was developing on multiple spatialities, which also largely neglected the unique spatialities of transportation. For example, mobile objects and actors can be seen as occupying space only temporarily, as constructing space through their travels (including waiting to travel); as being present and absent before, after, and during their travels; or a host of other possibilities that complicate the geographer's basic question: What is where? More recent work has begun to explicate how the spaces of transportation that are typically taken for granted are themselves shaped by and meaningful to human activity, from motorways that appear "placeless" but are actually places in their own right (Merriman 2004), to the seemingly open but in fact tightly defined and regulated phenomenon known as airspace (Williams 2011), to the shifting character of trainspace that depends on who or what is traveling through it (Cidell 2012b).

Rather than dividing up the spatialities of transportation and mobility by mode, however, it would be more productive to develop a framework that could be applied across modes and experiences to more explicitly understand how spaces and places are constructed by and help construct flows of people and goods (e.g., Cidell and Prytherch 2015). In seeking out a basis for a theoretical framework for the spatiality of transportation and mobility, we turn to a perspective from outside the Anglo-American academy, answering the calls from geographers to incorporate geographic work from other traditions and in languages other than English to "enhance cross-fertilization, thereby enriching our discipline" (Garcia-Ramon 2004, 367; see also Kitchin 2003; Timár 2007; Bajerski 2010; Fall and Minca 2012).We found this basis in the spatial terms for transportation described by the Czech geographer Kamil Skrbek (1977). Skrbek was born in 1913 in Příkrý in the Bohemian possessions of the Habsburg monarchy. He possessed an RNDr. degree, which, although not equivalent to a full doctorate, is higher than the *magister* or master's degree. He published several books on transportation geography from the 1960s to the 1980s, mostly in what is today Slovakia. Aside from these few facts, we know very little of him, although we can surmise that in all likelihood he is no longer an active scholar.

In this article, we first briefly review recent debates over spatiality in the geographic literature, followed by discussions of how the transportation and mobilities literature have considered spatiality. We then develop our framework through four of Skrbek's spaces: spaces of movement, spaces of transportation, areas of transportation, and structural transportation spaces. After explicating each of these, we illustrate their use through the example of crude oil rail shipments in North America. The recent and ongoing exploitation of oil sands and oil shale in the central United States and Canada has led to new spatial patterns in the shipment of crude oil to oil refineries, including pipelines and rail lines (see Barry [2013] for more details). Considering the spatialities of transportation based on Skrbek helps us to understand conflicts over these shipments and the potential hazards they pose to the areas they travel through, hazards that are different because of their mobile and temporary nature (Cidell 2012b). The framework we develop here can be used to theorize spatiality and transport or mobility in a variety of settings, and we consider some of those as well as the implications for geographers in transportation geography and related fields.

Space, Spatiality, and Transportation

Since the 1980s, debates over the nature of space and spatialities have expanded beyond geography to include multiple cognate disciplines in three key ways. First, the scale debates considered whether spatial scales are pregiven or constructed, whether scales are a matter of analysis or practice, or whether scales exist at all (e.g., Howitt 1998; Marston 2000; Swyngedouw 2004; Marston, Jones, and Woodward 2005; Jonas 2006). At the same time, many social science and humanities disciplines have taken a spatial turn, contributing their own theorizations of space and spatiality (e.g., Finnegan 2008; Peters and Kessl 2009; Arias 2010; Richardson et al. 2013). Finally, relational spaces and topologies have also had a significant impact, considering the extent to which traditional spatialities such as scale and territory need to still be considered (e.g., Dicken and Malmberg 2001; Amin 2002; Massey 2005; Jones 2009).

There have been two broad categories of response to these debates: to consider what we might call the "big four"—territory, place, scale, and networks—in combination with each other (Jessop, Brenner, and Jones 2008) or to seek out new spatialities altogether. With regard to the former, the idea is to not privilege one spatiality over another but to let them emerge from the data (Kortelainen 2010). Combining multiple spatialities can also lead to a more accurate understanding, such as bringing together networks with territory to understand how the nation-state is constructed and maintained (Häkli 2008). Others find these combinations inadequate and have developed new kinds of spatialities: fire space (Mol and Law 1994; Law and Mol 2001), fluid space (Bear and Eden 2008), positionality (Sheppard 2002), and vertebration (Prytherch 2010), for example.

Transportation and mobility present unique challenges to spatial theorization because of the sometimes-present, sometimes-absent nature of their subjects, which might be why they have been rarely considered in the preceding debates. Transport geographers, however, have long incorporated space into their analyses of flows of people and goods. Although transportation collapses space and time by altering the positionality of places relative to each other, it does so unevenly, shaping those places in the process (Banister and Berechman 2001; Knowles 2006). In particular, the changing spatial form of transportation networks, including the tendency to concentrate activity in key hubs, has been a major factor in the establishment,

growth, and decline of cities around the world (Fleming and Hayuth 1994; Smith and Timberlake 1995; O'Kelly 1998). The rise of geographic information systems for transportation (GIS-T) has enabled ever more sophisticated spatial analyses of these networks and practices across time and space (Miller 1999; Thill 2000; Neutens et al. 2012), with the potential for incorporation of new technologies to further understand movement through space (Shoval et al. 2014). Traditionally, such analyses have considered transportation as a derived demand, with the distance between various economic and social activities driving the need for travel. The distinction between accessibility and mobility is key here, with the ability to get to necessary goods, services, and social connections (accessibility) depending on the ability of people and goods to travel between different locations (mobility). The understanding of transportation as derived from other activities, however, has changed with the growth of logistics and supply chain management (Hesse and Rodrigue 2004; Rodrigue 2006) and the need to consider more sustainable forms of transport to decouple accessibility from mobility (Kenyon, Lyons, and Rafferty 2002; Banister 2008; Hull 2008).

The mobilities turn out of sociology has taken a different approach to the understanding of movement through space. This literature was largely inspired by the social sciences' treatment of sedentarism as the default situation, instead theorizing the ways in which mobility is part of our social condition (Urry 2007). The mobilities lens has been turned on migration, transportation, virtual travel, and other forms of bodily movement of humans and nonhumans alike. Although some have argued that mobilities themselves represent a new spatiality (Leitner, Sheppard, and Sziarto 2008), we consider that only a starting point (see later). Mobile subjects create and shape multiple kinds of spaces, based on Urry's categorization of five different kinds of mobilities: corporeal, objects, imaginative, virtual, and communicative (Urry 2007). For example, the embodied nature of many mobilities means that how space is occupied and produced needs to be considered, including the temporal and material aspects of walking, cycling, driving, traveling by ferry, and so on (Spinney 2006; Middleton 2009; Vannini 2011). The space of a train car or a highway is ever changing, in terms of both the interior and exterior, and existing concepts like networks or places cannot capture that fluidity (Robertson 2007; Watts 2008). Borders are set in place with increasingly complex topologies, reinforced in response to actual or desired mobility, and their intended and actual porosities shape flows as well as places (Budd, Bell, and Brown 2009; Frétigny 2013). Objects in motion pose different challenges and opportunities from objects that are still, and these flows and pauses structure space accordingly (Bissell 2007; Cidell 2012a; Hui 2012). Brand new kinds of spaces are produced by mobile subjects, including airspace (Millward 2008; Williams 2011) and trainspace (Cidell 2011), distinguished from standard spatialities by their flickering, flexible presence and their construction through a combination of sedentary and mobile processes. Finally, the production of identities through mobility is also tied to spatiality, another reason why sedentarism should not be privileged over mobility either theoretically or in practice (Kesselring 2006; Prout 2009; Ho 2011).

Despite the healthy growth in both the transportation and mobilities literatures (and in the overlap and conversations between the two), the larger geographic literature has not significantly incorporated either into theorizations about spatiality, mostly relegating transportation to networks or corridors and not considering in detail the broader spaces that mobilities help to construct. In this article, we draw on Skrbek's work to consider four different kinds of spaces relevant to both transport and mobilities as a way to encourage geographers and scholars in related fields to consider the difference that mobility and motion make to how we understand space and place.

Skrbek's Spaces

Skrbek (1977) described a number of terms in his monograph *Dopravná Geografia Československa a Svetadielov I* (*Transportation Geography of Czechoslovakia and the Continents: Volume 1*), which, if further developed, could serve as new analytical tools for transportation geographers and mobilities scholars (see Table 1). Skrbek's first and most basic concept, a space of movement (*prostor pobyhu*), denotes any traversable space that witnesses a minimum of movement. This movement is not necessarily possible on a regular or predictable basis, nor is it supported by infrastructure or facilities that would enable the regular, predictable, and safe usage of the said space for movement. It is this raw and untamed character that distinguishes spaces of movement from others. When the previously mentioned supporting facilities and institutions are emplaced, a space of transportation or transportation space (*dopravní prostor*) is created. A space of transportation is closely related to yet distinct from an area of

Table 1. Summary of Skrbek's spaces

Type of space	Defining characteristics	Examples
Space of movement	Devoid of development Movement is very difficult	Unexplored or remote and sparsely populated areas Great Plains prior to the introduction of rail
Space of transportation (transportation space)	Developed infrastructure Regular and facilitated movement Indeterminate boundary	Regularly traveled airspaces, roads, sidewalks, waterways, etc. Large networks
Area of transportation	Developed infrastructure Regular and facilitated movement Defined boundary Comprehensive governance Typically larger scale	Mainly theoretical Large multimodal networks Centrally managed transportation networks
Structural transportation space	Uniform traffic of one or more modes Infrastructure and facilities	Built-up transportation corridors Kennedy and Dan Ryan Expressway corridors in Chicago
Transportation network	A collection of transportation routes	Layout of city streets U.S. interstate highway system German Autobahn network
Transportation zone	A centrally managed space of transportation	Distance-based pricing regimes on metro systems
Transportation gateway	A choke point of transportation	Mountain passes Major rail and highway junctions
Transportation landscape	A localized landscape unmistakably shaped by transportation infrastructure	Airports Large rail yards Port facilities Localized highway facilities in urban areas Large interchanges

transportation (*dopravní oblast*). Skrbek, seeming to employ the philosophical distinction between propositions, writes that spaces of transportation should be used in a "*synthetic* and general sense," whereas areas of transportation should be used in "an *analytical* and special sense" (italics added). The authors' reading of Skrbek is that the vital distinction between the two terms is found in the undefined extent of a space of transportation as compared to the defined extent of the area of transportation. If an area of transportation is "special" because it is a transportation space of determined extent (geographic or otherwise), then a space of transportation with an undetermined extent, even if identical to an area of transportation in every other way, is still general in nature (e.g., compare the City of Chicago in the jurisdictional sense vs. Chicago in the general sense).

Skrbek went on to describe more concrete spatial terms for transportation, including structural transportation space, transportation corridors (or zones), transportation networks, transportation gateways, and transportation landscapes (Table 1). With the exception of transportation zones, we propose that these additional terms are in fact subordinate concepts to transportation space. Structural transportation space (*strukturální dopravní prostor*) consists of a uniformity of traffic created by one or more specific modes of transportation within individual transportation spaces. A transportation network (*dopravní sítě*) consists of a number of mostly linear transportation structures arranged in a planned web. In addition to defining features such as network density, frequency, and volume of traffic, Skrbek indicated that time can also serve as a feature that can shape a transportation zone. A transportation gateway (*dopravní brana*) is created by either the physical or built environment forming conditions through which multiple transportation corridors must pass and converge. Transportation landscapes (*dopravní krajina*) are the built environments of transportation infrastructure that come to dominate the visible landscape. A transportation zone (*dopravní zóna*) is a space of transportation that is highly centralized in terms of planning and management. It often assumes a circular shape around the focal point of its operation, like different ticket prices depending on distance traveled. Given these characteristics of transportation zones, they might be best classified as subordinate to the concept of areas of transportation.

The concepts embodied by the previously mentioned terms have been addressed to varying degrees by either transportation geography or mobilities studies, and sometimes both. Transportation geography has traditionally favored the discussion of concrete concepts as described by structural transportation spaces, transportation networks, transportation zones, and transportation gateways. To a lesser extent, the concept reflected in the term *space of transportation* has also been addressed. The wealth of research within transportation geography is solidly grounded in empirical data and thus lends itself to not only describing known transportation spaces in the real world but also determining whether or not they might eventually be classified as areas of transportation. Determining the defining characteristics of an area of transportation proves to be challenging, though. Which processes or entities determine the geographic extent of an area of transportation, the most important characteristic in distinguishing between areas and spaces? Are these processes economic, political, or social? Furthermore, are the contents of the area just as important as its boundary for the purposes of defining it? Mobilities studies have much to offer in seeking an answer to these questions. Mobilities studies have often focused on the experience of transportation by various means at the scale of the individual. The focus of mobilities scholars on modern forms of transportation with its rich qualitative description would be complemented by the use of Skrbek's terms, which largely reflect the context of modern transportation. The case study in the next section illustrates one way to do so.

Transporting Oil

Our case study complements recent work on the materiality of commodity flows that emphasizes the unique challenges facing communities and places that coexist with major flows of freight (e.g., Barry 2013; Cowen 2014). The development of both the oil sands in Alberta and oil shale in the Western United States have led to rapid increases in the volume and frequency of shipments of crude oil from its extraction point to existing refineries. While approval for the Keystone XL pipeline remains under review, oil producers have turned to rail. Given the 2013 Lac-Mégantic, Quebec, explosion that killed forty-seven people, however, in addition to derailments in Lynchburg, Virginia, Philadelphia, and Cassleton, North Dakota, in that same year resulting in oil spills, fires, or both, North Americans are slowly becoming more aware of the hazards that rail lines pose to major cities and residential neighborhoods.

The Great Plains could be considered a space of movement for the Keystone XL pipeline, as the low population and development densities of the Plains enable the pipeline's proposed path to be altered to reduce the exposure of the Ogallala Aquifer and the Nebraska Sandhills to potential leaks (Table 2). The pipeline would then integrate the transportation spaces of oil pipeline infrastructure into a wider regional transportation space encompassing all modes of transportation. One of the major objections to the pipeline, however, is that the aquifer has its own space of movement and were a spill to occur, oil could easily spread through the only source of water for the agricultural center of the country.

Without the pipeline, rail has become the preferred method for oil companies to transport their product to refineries. This change in the use of the rail network needs no environmental review because North American railroads are privately owned, using existing infrastructure. The existing space of transportation of the North American rail network has shaped where oil shipments go (Table 2). Those oil refineries not adjacent to oil fields are predominantly located in coastal areas to handle imported oil. Connecting the space of extraction to the space of production means using the space of transportation in the form of the North American rail network. In many cases, this means coming in the "back door" to major metropolitan areas such as the San Francisco Bay Area or Chicago. Although hazardous materials in the form of refined oil products have been exported from these locations, the greater hazard posed by unrefined oil entering these cities has become a site of conflict (see, e.g., Oakley [2014], on the Bay Area).

It is in addressing these potential hazards that Skrbek's area of transportation becomes relevant.

Table 2. Case study examples of Skrbek's spaces

Type of space	Case study examples
Space of movement	Great Plains prior to modernization
	Possible pipeline locations
Space of transportation	North American "trainspace"
Area of transportation	DOT-regulated space
	CN-owned rail network
Structural transportation space	Rail lines and adjacent neighborhoods
	Railroad crossings
Transportation network	North American rail network
Transportation gateway	Chicago
	San Francisco Bay Area
Transportation landscape	Port facilities in San Francisco Bay Area
	Rail yards in Chicago

An area of transportation is defined not only by the outer boundary but also by the transportation activities and affected sites contained within it. In practice, this means that systems of governance must exist that both define its outer boundaries and sufficiently address its internal activities. For example, Lac-Mégantic is located along the rail line that connects the North Dakota oil shale with refineries on Canada's eastern coast. As in many small towns in North America that grew because of the access that rail transportation offered, the rail line runs directly through the center of the town, which was why the derailment and subsequent explosion were so devastating. Journalists and activists have noted the extent of exposure to similar hazards in cities across the continent, including Albany, New York (Mouawad 2014), Seattle, Washington (Wilhelm 2013), Richmond, California (Oakley 2014), and Winnipeg, Manitoba (Carter 2014). (See Forest-Ethics [2014] for a map of the areas within the United States that are within a mile of railroads carrying oil.) Responses include trying to block such shipments or reroute them, provide additional information to local emergency responders to increase preparedness, or strengthen the train cars to reduce the possibility of a spill or explosion should a derailment occur. Of these, the first one only moves the problem around, and the second has become a matter of debate due to potential security issues. Following Canada's lead in requiring thicker train cars for carrying crude oil would reduce the hazard these rail cars pose, no matter where they are passing through. Thus, the realization of an area of transportation consists not only of maintaining boundaries and the order of transportation network but of activities as well.

Mobilities scholars have written on the nature of the structural transportation space of airports, train stations, and city streets, considering how place-making occurs even inside these supposedly uniform and flow-focused kinds of spaces. Skrbek's concept of a uniform transportation space is further belied in the case of oil by rail by the nature of the cargo being carried. On the one hand, rail lines and rail yards can be considered as uniform spaces of transportation, given their devotion to that mode and exclusion of other activities. On the other hand, though, as conflicts over crude oil shipments and other hazardous materials show (Cidell 2012b), the nature of the material being carried through that space matters. A commuter rail line is a different form of structural transportation space than a freight line, even if they share the same track. Adjacent neighborhoods and communities are drawn into those spaces differently depending on the nature of the train traffic (Cidell 2011). Spaces and flows are thus tied together, one constructing the other, requiring close attention to how structural transportation spaces are produced and what effects they have on their surroundings and the places they connect.

Conclusion

The connection between spaces and flows has been increasingly important in the geographical literature (Massey 2005; Hall and Hesse 2012), but analytically, it has been difficult to conceptualize that relationship. This difficulty stems from the differing emphases of transportation geography and mobilities studies, with transportation geography's strong predisposition toward examining large-scale processes and the focus of the mobilities literature on phenomena at scales as small as the body. If placed together within a coherent framework, the strengths of each approach would collectively produce a more holistic approach to the study of spaces and flows. This can be done by using Skrbek's spatial terms as a starting point.

Skrbek's spatial terms offer us a means of understanding conflicts over how to get crude oil from its production sites in the North American interior to refineries on the coasts in two key ways. First, the distinction between spaces of transportation and areas of transportation enables us to make a connection between the generic, abstract spaces of the global freight network and the territorially defined spaces of cities and towns that intersect with that network. Second, despite the fixed nature of transportation infrastructure found in a structural transportation space, the varying character of that space depends on what is traveling through it.

Our elaboration of Skrbek's spatial terms offers a strong framework for achieving further progress in transportation and mobilities studies by combining their individual strengths and mitigating their individual weaknesses. Our case study highlights the difference between, for example, a commuter rail line and a freight line carrying oil as different hazard spaces with different regulatory implications. The hierarchy of Skrbek's terms helps us achieve this by providing a means of coordinating various scales (macro, meso, and micro) within the analysis. Further research should focus on who is defining these different spaces

(railroad executives, federal regulators, community activists, etc.) and how they are doing so. At the same time, our case study shows that not all of Skrbek's spaces are applicable to every situation, which indicates the flexbility of this framework for a broad range of issues in transportation and mobility. The study of these spaces of transportation produces empirical evidence that can be used to refine a holistic definition of an area of transportation, which would be applicable regardless of mode, mobility, or governance structure. Thus, the use of these terms as analytical tools has great potential for transportation geography and mobilities. Finally, we have provided a way for geographers more broadly speaking to conceptualize mobility in its relationship to place, something that has been largely missing from disciplinary debates on space and spatiality.

Acknowledgments

The authors would like to thank Andrew Goetz and the three anonymous referees for their comments and suggestions on earlier drafts of this article. Any remaining errors are the responsibility of the authors. This research was not externally funded.

References

Amin, A. 2002. Spatialities of globalisation. *Environment and Planning A* 34 (3): 385–400.

Arias, S. 2010. Rethinking space: An outsider's view of the spatial turn. *GeoJournal* 75 (1): 29–41.

Bajerski, A. 2010. Anglo-amerykańska dominacja w geografii: Główne wątki dyskusji prowadzonej w ramach geografii krytycznej [Anglo-American domination in geography: The main threads of a discussion conducted within the framework of critical geography]. *Przeglad Geograficzny* 82 (2): 143–58.

Banister, D. 2008. The sustainable mobility paradigm. *Transport Policy* 15 (2): 73–80.

Banister, D., and Y. Berechman. 2001. Transport investment and the promotion of economic growth. *Journal of Transport Geography* 9 (3): 209–18.

Barry, A. 2013. *Material politics: Disputes along the pipeline.* Malden, MA: Wiley Blackwell.

Bear, C., and S. Eden. 2008. Making space for fish: The regional, network and fluid spaces of fisheries certification. *Social & Cultural Geography* 9 (5): 487–504.

Bissell, D. 2007. Animating suspension: Waiting for mobilities. *Mobilities* 2 (2): 277–98.

Budd, L., M. Bell, and T. Brown. 2009. Of plagues, planes and politics: Controlling the global spread of infectious diseases by air. *Political Geography* 28 (7): 426–35.

Carter, A. 2014. Winnipeg derailment renews safety concerns about crude oil shipments. *Global News* 20 June. http://globalnews.ca/news/1407694/winnipeg-derailment-renews-safety-concerns-about-crude-oil-shipments/ (last accessed 11 November 2015).

Cidell, J. 2011. Fear of a foreign railroad: Transnationalism, trainspace, and (im)mobility in the Chicago suburbs. *Transactions of the Institute of British Geographers* 37 (4): 593–608.

———. 2012a. Flows and pauses in the urban logistics landscape: The municipal regulation of shipping container mobilities. *Mobilities* 7 (2): 233–45.

———. 2012b. Just passing through: The risky mobilities of hazardous materials transport. *Social Geography* 7 (1): 13–22.

Cidell, J., and D. Prytherch. 2015. *Transportation and mobility in the production of urban space.* London and New York: Routledge.

Cowen, D. 2014. *The deadly life of logistics: Mapping violence in global trade.* Minneapolis: University of Minnesota Press.

Dicken, P., and A. Malmberg. 2001. Firms in territories: A relational perspective. *Economic Geography* 77 (4): 345–63.

Fall, J., and C. Minca. 2012. Not a geography of what doesn't exist, but a counter-geography of what does: Rereading Giuseppe Dematteis' *Le Metafore della Terra. Progress in Human Geography* 37 (4): 542–63.

Finnegan, D. A. 2008. The spatial turn: Geographical approaches in the history of science. *Journal of the History of Biology* 41 (2): 369–88.

Fleming, D. K., and Y. Hayuth. 1994. Spatial characteristics of transportation hubs: Centrality and intermediacy. *Journal of Transport Geography* 2 (1): 3–18.

ForestEthics. 2014. Oil train blast zone. http://explosive-crude-by-rail.org/ (last accessed 15 October 2015).

Frétigny, J. B. 2013. La frontière à l'épreuve des mobilités aériennes: Étude de l'aéroport de Paris Charles-de-Gaulle [The border put to the test of air mobilities: The example of Paris Charles de Gaulle airport]. *Annales de géographie* 2:151–74.

Garcia-Ramon, M.-D. 2004. On diversity and difference in geography: A southern European perspective. *European Urban and Regional Studies* 11 (4): 367–70.

Goetz, A. R., T. M. Vowles, and S. Tierney. 2009. Bridging the qualitative–quantitative divide in transport geography. *The Professional Geographer* 61 (3): 323–35.

Häkli, J. 2008. Regions, networks and fluidity in the Finnish nation-state. *National Identities* 10 (1): 5–20.

Hall, P. V., and M. Hesse, eds. 2012. *Cities, regions and flows.* London and New York: Routledge.

Hanson, S. 2003. Transportation: Hooked on speed, eyeing sustainability. In *A companion to economic geography,* ed. E. Sheppard and T. Barnes, 469–83. Malden, MA: Blackwell.

———. 2006. Viewpoint: Imagine. *Journal of Transport Geography* 14:232–33.

Hesse, M., and J. P. Rodrigue. 2004. The transport geography of logistics and freight distribution. *Journal of Transport Geography* 12 (3): 171–84.

Ho, E. L. E. 2011. "Claiming" the diaspora: Elite mobility, sending state strategies and the spatialities of citizenship. *Progress in Human Geography* 35 (6): 757–72.

Howitt, R. 1998. Scale as relation: Musical metaphors of geographical scale. *Area* 30 (1): 49–58.

Hui, A. 2012. Things in motion, things in practices: How mobile practice networks facilitate the travel and use of leisure objects. *Journal of Consumer Culture* 12 (2): 195–215.

Hull, A. 2008. Policy integration: What will it take to achieve more sustainable transport solutions in cities? *Transport Policy* 15 (2): 94–103.

Jessop, B., N. Brenner, and M. Jones. 2008. Theorizing sociospatial relations. *Environment and Planning D: Society and Space* 26 (3): 389.

Jonas, A. E. 2006. Pro scale: Further reflections on the "scale debate" in human geography. *Transactions of the Institute of British Geographers* 31 (3): 399–406.

Jones, M. 2009. Phase space: Geography, relational thinking, and beyond. *Progress in Human Geography* 33 (4): 487–506.

Kenyon, S., G. Lyons, and J. Rafferty. 2002. Transport and social exclusion: Investigating the possibility of promoting inclusion through virtual mobility. *Journal of Transport Geography* 10 (3): 207–19.

Kesselring, S. 2006. Pioneering mobilities: New patterns of movement and motility in a mobile world. *Environment and Planning A* 38 (2): 269–79.

Kitchin, R. 2003. Cuestionando y desestabilizando la hegemonía angloamericana y del inglés en geografía [Questioning and destabilizing the hegemony of Anglo-America and of English in geography]. *Documents d'Anàlisi Geogràfica* 42:17–36.

Knowles, R. D. 2006. Transport shaping space: Differential collapse in time–space. *Journal of Transport Geography* 14 (6): 407–25.

Kortelainen, J. 2010. Old-growth forests as objects in complex spatialities. *Area* 42 (4): 494–501.

Law, J., and A. Mol. 2001. Situating technoscience: An inquiry into spatialities. *Environment and Planning D* 19:609–21.

Leitner, H., E. Sheppard, and K. M. Sziarto. 2008. The spatialities of contentious politics. *Transactions of the Institute of British Geographers* 33 (2): 157–72.

Marston, S. A. 2000. The social construction of scale. *Progress in Human Geography* 24 (2): 219–42.

Marston, S. A., J. P. Jones, and K. Woodward. 2005. Human geography without scale. *Transactions of the Institute of British Geographers* 30 (4): 416–32.

Massey, D. 2005. *For space.* Thousand Oaks, CA: Sage.

Merriman, P. 2004. Driving places: Marc Augé, non-places, and the geographies of England's M1 motorway. *Theory, Culture & Society* 21 (4–5): 145–67.

Middleton, J. 2009. "Stepping in time": Walking, time, and space in the city. *Environment and Planning A* 41 (8): 1943–61.

Miller, H. J. 1999. Measuring space–time accessibility benefits within transportation networks: Basic theory and computational procedures. *Geographical Analysis* 31 (1): 1–26.

Millward, L. 2008. *Women in imperial airspace, 1922–1937.* Montreal and Kingston, ON, Canada: McGill-Queen's University Press.

Mol, A., and J. Law. 1994. Regions, networks and fluids: Anaemia and social topology. *Social Studies of Science* 24 (4): 641–71.

Mouawad, J. 2014. Bakken crude, rolling through Albany. *New York Times* 27 February 2104. http://www.nytimes.com/2014/02/28/business/energy-environment/bakkancrude-rolling-through-albany.html (last accessed 11 November 2015).

Neutens, T., M. Delafontaine, D. M. Scott, and P. De Maeyer. 2012. An analysis of day-to-day variations in individual space–time accessibility. *Journal of Transport Geography* 23:81–91.

Oakley, D. 2014. Demonstration over "bomb trains" hauling crude oil in Richmond. *San Jose Mercury News* 31 May 2014. http://www.mercurynews.com/my-town/ci_25873144/demonstration-over-bomb-trains-hauling-crude-oil-richmond (last accessed 11 November 2015).

O'Kelly, M. E. 1998. A geographer's analysis of hub-and-spoke networks. *Journal of Transport Geography* 6 (3): 171–86.

Peters, M. A., and F. Kessl. 2009. Space, time, history: The reassertion of space in social theory. *Policy Futures in Education* 7 (1): 20–30.

Prout, S. 2009. Security and belonging: Reconceptualising Aboriginal spatial mobilities in Yamatji country, Western Australia. *Mobilities* 4 (2): 177–202.

Prytherch, D. 2010. 'Vertebrating' the region as networked space of flows: Learning from the spatial grammar of Catalanist territoriality. *Environment and Planning A* 42:1537–54.

Richardson, D. B., N. D. Volkow, M. P. Kwan, R. M. Kaplan, M. F. Goodchild, and R. T. Croyle. 2013. Spatial turn in health research. *Science* 339 (6126): 1390–92.

Robertson, S. 2007. Visions of urban mobility: The Westway, London, England. *Cultural Geographies* 14 (1): 74–91.

Rodrigue, J. P. 2006. Transportation and the geographical and functional integration of global production networks. *Growth and Change* 37 (4): 510–25.

Schwanen, T. 2015. Geographies of transport I: Reinventing a field? *Progress in Human Geography*. Advance online publication. doi: 10.1177/0309132514565725

Shaw, J., and M. Hesse. 2010. Transport, geography and the "new" mobilities. *Transactions of the Institute of British Geographers* 35 (3): 305–12.

Shaw, J., and J. D. Sidaway. 2010. Making links: On (re)engaging with transport and transport geography. *Progress in Human Geography* 35 (4): 502–20.

Sheller, M., and J. Urry. 2006. The new mobilities paradigm. *Environment and Planning A* 38:207–26.

Sheppard, E. 2002. The spaces and times of globalization: Place, scale, networks, and positionality. *Economic Geography* 78 (3): 307–30.

Shoval, N., M. P. Kwan, K. H. Reinau, and H. Harder. 2014. The shoemaker's son always goes barefoot: Implementations of GPS and other tracking technologies for geographic research. *Geoforum* 51:1–5.

Skrbek, K. 1977. *Dopravná geografia Československa a svetadielov* [Transportation geography of Czechoslovakia and the continents: Volume 1]. Bratislava, Slovakia: Alfa.

Smith, D. A., and M. Timberlake. 1995. Conceptualising and mapping the structure of the world system's city system. *Urban Studies* 32 (2): 287–302.

Spinney, J. 2006. A place of sense: A kinaesthetic ethnography of cyclists on Mont Ventoux. *Environment and Planning D* 24 (5): 709–32.

Swyngedouw, E. 2004. Scaled geographies: Nature, place and the politics of scale. In *Scale and geographic inquiry*, ed. E. Sheppard and R. McMaster, 129–53. Oxford, UK: Blackwell.

Thill, J. C. 2000. Geographic information systems for transportation in perspective. *Transportation Research Part C: Emerging Technologies* 8 (1): 3–12.

Timár, J. 2007. Differences and inequalities: The "double marginality" of east central European feminist geography. *Documents d'Anàlisi Geogràfica* 49:73–98.

Urry, J. 2007. *Mobilities*. London: Polity.

Vannini, P. 2011. Constellations of ferry (im)mobility: Islandness as the performance and politics of insulation and isolation. *Cultural Geographies* 18 (2): 249–71.

Watts, L. 2008. The art and craft of train travel. *Social & Cultural Geography* 9 (6): 711–26.

Wilhelm, S. 2013. Opening the spigot: Why Washington could become a major global fuel supplier. *Puget Sound Business Journal* 6 September 2013. http://www.bizjournals.com/seattle/news/2013/09/05/opening-the-spigot-why-washingtons.html (last accessed 11 November 2015).

Williams, A. J. 2011. Reconceptualising spaces of the air: Performing the multiple spatialities of UK military airspaces. *Transactions of the Institute of British Geographers* 36 (2): 253–67.

Mobility Among the Spatialities

Byron Miller* and Jason Ponto[†]

*Department of Geography, University of Calgary
[†]Department of Sociology, University of Calgary

Despite the explosive growth of mobilities research, much sociospatial theory continues to be rooted in a sedentarist perspective, failing to incorporate the insights of this burgeoning field. Mobilities research, in contrast, often considers a variety of sociospatial relations, yet stops short of coherent integration with other dimensions of sociospatiality. In this article, we examine the mobilities turn in light of Jessop, Brenner, and Jones's (2008) TPSN framework, which recognizes the polymorphic nature of sociospatial relations. We discuss the interrelationships between mobility and the four distinct sociospatialities identified by Jessop, Brenner, and Jones: territory (T), place (P), scale (S), and networks (N). Each of these sociospatialities is coimplicated with mobility: Territory concerns the malleable areal and bordered structure of the state and the uneven freedoms granted, and constraints imposed on, objects and bodies as they attempt to move through and across political jurisdictions; place emphasizes the embedded and performative nature of mobility and considers place-appropriate and place-transgressive activity; scale concerns movement associated with the tangled and politicized processes of scale production and examines how mobility is affected by the uneven scaling of power, resources, opportunity, and identity; networks address flows of bodies, objects, and knowledge across space, through specific channels. To illustrate the coimplicated relationships among mobility and territory, place, scale, and networks, we examine the practice of automobility, stressing the ontological contingency of mobility: Neither mobility nor fixity can be assumed. Mobility is, rather, a social, cultural, and political achievement, inherently power-laden and recursively bound up in the production of territory, place, scale, and networks.

尽管能动性研究有着爆炸性的增长, 但社会空间理论多半仍持续根植于固着的视角, 无法纳入此一成长中的领域之洞见。相对而言, 能动性研究经常考量各种社会空间关系, 但却未能达到与社会空间性的其他面向的一致性整合。我们于本文中, 根据杰索普 (Jessop)、布瑞纳 (Brenner) 与琼斯 (Jones) (2008) 的 TPSN 架构来检视能动性转向, 该架构承认社会空间关系的多重形态本质。我们探讨能动性与杰索普、布瑞纳和琼斯所指认的四大特殊社会空间性之间的关联性: 领域 (T), 地方 (P), 尺度 (S), 以及网络 (N)。这些社会空间性中的每一项皆与能动性共同有关: 领域考量国家具延展性的区域和划界的结构, 以及当物体与身体企图穿越政治管辖范围时, 赋予他们的不均自由和加诸其上的限制; 地方强调能动性的镶嵌及展演本质, 并考量适合地方与逾越地方的活动; 尺度考量与尺度生产的纠结及政治化过程相关的行动, 并检视能动性如何受到权力、资源、机会和认同的不均尺度化所影响; 网络应对身体、物件和知识横跨空间、透过特定管道的流动。为了描绘能动性和领域、地方、尺度及网络之间的共同连结关系, 我们检视汽车能动性的实践, 强调能动性在本体论上的偶然性: 能动性或固着性皆无法被预设。更确切而言, 能动性是社会、文化与政治的达成, 本质上是充满权力, 且在递归上与领域、地方、尺度和网络的生产密切相关。

A pesar del crecimiento protuberante de la investigación sobre movilidades, gran parte de la teoría socioespacial se mantiene arraigada dentro de una perspectiva sedentaria, fallando en la incorporación de las perspicacias logradas por este campo en expansión. La investigación de movilidades, en contraste, considera a menudo una variedad de relaciones socioespaciales, aunque no avanza en mayor grado en la integración coherente con las otras dimensiones de la socioespacialidad. En este artículo examinamos el giro de las movilidades a la luz del esquema TLER [TPSN, en inglés] de Jessop, Brenner y Jones (2008), que reconoce la naturaleza polimórfica de las relaciones socioespaciales. Discutimos las interrelaciones entre la movilidad y las cuatro socioespacialidades identificadas por Jessop, Brenner y Jones: territorio (T), lugar (L), escala (E) y redes (R). Cada una de estas socioespacialidades está coimplicada con la movilidad: Al territorio concierne la estructura areal maleable y delimitada del estado, y las desiguales libertades concedidas, lo mismo que las limitaciones impuestas sobre objetos y cuerpos en cuanto estos intentan moverse en y a través de jurisdicciones políticas; el lugar enfatiza la naturaleza incrustada y representacional de la movilidad y considera la actividad apropiada y transgresora en términos de lugar; a la escala concierne el movimiento asociado con los procesos enredados y politizados de la producción de escala y examina cómo se afecta la movilidad por la desigual escala del poder, los recursos, la oportunidad y la identidad; las redes o cadenas se encargan de los flujos de cuerpos, objetos y conocimiento a través del espacio, por medio de canales específicos. Para ilustrar las relaciones coimplicadas entre movilidad y territorio, lugar, escala y redes, examinamos la práctica de automovilidad, destacando la contingencia ontológica de la movilidad: Ni la movilidad ni la fijeza pueden asumirse. Más que nada, la movilidad es un logro social, cultural y político, inherentemente cargado de poder y recursivamente atado a la producción de territorio, lugar, escala y redes.

Mobility is movement through space. Since the "mobilities turn" (Urry 2000, 2007; Adey 2006, 2010; Cresswell 2006, 2010; Sheller and Urry 2006; McCann 2010; McCann and Ward 2011a), scholarship across the social sciences and humanities has examined movement in important new ways, extending far beyond traditional concerns with and approaches to transportation, migration, and diffusion. The "new mobilities paradigm" (Sheller and Urry 2006) has responded to the so-called sedentarist practice of privileging the resting state of bodies and objects by emphasizing the inherent but contested dynamism of social relations. As a result, recent scholarship has produced new and insightful analyses of the experiences, practices, and politics of mobile subjects and objects. For example, Cresswell (2006, 2010) developed an experiential, interpretive, and practice-oriented approach to "constellations of mobility" that contrasts with notions of sedentary order. Kaufmann, Bergman, and Joye (2004) considered mobility as an issue of social justice. McCann and Ward (2011a) investigated the relationships of actors, institutions, and practices involved in the production and mutation of urban policy as it travels through policy networks. Adey (2006) analyzed airports as relational sociotechnical nodes, simultaneously bound up with processes of territorialization. Jensen (2011) analyzed ways in which techniques and rationalities of government make mobility knowable and visible, and Böhm et al. (2006) highlighted truth, power, and subjectivity at play in "regimes" of automobility. Mobilities are not just about movement of physical objects but about a much broader palette of research questions and concerns. Mobilities research raises questions about a variety of spatialities, their implications for mobile practices, and how mobile practices are, in turn, implicated in other spatialities.

Mobility, the TPSN Framework of Sociospatial Relations, and Power Relations

Mobility has long been recognized as a central characteristic of capitalist societies. Harvey's ([1982] 2007) political–economic notion of the *spatial fix* highlights the necessity of fixing immobile objects in space (e.g., transportation infrastructure, energy grids, communication networks, production facilities, etc.), to enable the movement of goods, services, labor, and capital. In dynamic and contested processes the territories, places, scales, and networks through which capitalism is organized are reorganized, on an ongoing basis, to open up new spaces of capital circulation and accumulation. In Harvey's account the production of mobility proceeds hand in hand with the reorganization of other spatialities. Indeed, mobility is increasingly recognized as a key "spatiality" among those frequently addressed in contemporary geographical research (Leitner, Sheppard, and Sziarto 2008). That multiple spatialities are implicated in mobility is increasingly clear. As Hannam, Sheller, and Urry (2006) asserted, "Mobilities cannot be described without attention to the necessary spatial, infrastructural and institutional moorings that configure and enable mobilities" (3).

Although mobility is a material process, it cannot be reduced to embodied physical movement. Approaching mobility from a poststructuralist perspective, Cresswell (2006) examined mobility as experience and representation (or nonrepresentation) of becoming, "linked to a world of practice, of anti-essentialism, anti-foundationalism, and resistance to established forms of ordering and discipline" (47). Considering not only physical movement, Cresswell's (2010) historically and geographically variable "*constellations of mobility*—particular patterns of movement, representations of movement, and ways of practicing movement that make sense together" (18) points toward a more integrative approach to mobility.

In the world of sociospatial relations, mobility is never a given. Rather, as Massey (1994) explained, "Mobility, and control over mobility, both reflects and reinforces power" (150). But by what means is control over mobility exercised? As several authors have noted (Leitner and Miller 2007; Jessop, Brenner, and Jones 2008; Leitner, Sheppard, and Sziarto 2008; Miller 2013), sociospatial processes almost always involve multiple spatialities. Jessop, Brenner, and Jones (2008) suggested that territory (T), place (P), scale (S), and networks (N) are centrally important and "must be viewed as mutually constitutive and relationally intertwined dimensions of sociospatial relations" (389).

25

They asserted, moreover, that failure to "explore the interconnections among the various dimensions of sociospatial relations . . . [can lead to] a variety of theoretical deficits, methodological hazards, and empirical blind spots" (398). Their work, fortunately, can be extended to consider how the production of mobilities and immobilities is related to the production of territory, place, scale, and networks.

A variety of scholars, such as de Certeau (1984) and Deleuze and Guattari (1987), identify territorial strategies as means by which a relatively static social order can be established; territory functions as a "spatialization of domination" (Cresswell 2006, 47). Indeed, Elden (2010) argued that "territory can be understood as a political technology: it comprises techniques for measuring land and controlling terrain" (13). It is a spatial technology of power that can be employed for various purposes, including the control of mobility. Others point to scale as a critical spatial technology. Smith (1992), for instance, pointed to the ways in which rescaling alters social processes when he observed that "the scale of struggle and the struggle over scale are two sides of the same coin" (74). Rescaling can have wide-reaching implications for the availability of resources, the capacity to act, and agents' legitimacy—all of which can affect mobility. The notion of networks is not necessarily one of unconstrained mobility but rather of "channelized" flows and exchange, a point Deleuze and Guattari (1987) made in their discussion of "striated space" and others made in discussions of network structure (Castells 2000; Taylor 2003). Even place can be a basis for exclusion, with obvious implications for mobility (Sennett 1970; Sibley 1995). Every spatiality can be considered a spatial technology of power. "Places, scales, territories, regions and networks are produced, altered, and in some cases dismantled . . . to advance the interests of particular actors. These spatialities, moreover, exist materially . . . as well as in the form of narratives and imaginaries that may frame and motivate particular courses of action or inaction" (Miller 2013, 332). Action entails, by necessity, movement, and any change in the spatialities that enable and constrain action will, by necessity, affect mobility.

The production of mobilities and immobilities must be understood in relation to the production of territory, place, scale, and networks. At the risk of oversimplification, we offer the following definitions of each. *Territory* is a type of ordering and quantification of space based on the idea of a homogeneous area (Elden 2010); it is simultaneously "juridical, political, economic, social and cultural, and even affective" (November 2002, 17,

cited in Elden 2010). Although territory is considerably more than just "bounded space,"[1] the concept raises questions of the production, defense, management, and contestation of politically bounded areas (Sassen 2000; Elden 2010). The concept of territory also provides a sociospatial framework through which to examine the state's governance of capital, population, and resource movement (Foucault 2007). *Place* refers to "situatedness" of everyday human action and interaction (Agnew 1996). Agnew's (1987) tripartite definition of place as *locale*, *location*, and *sense of place* indicates that places are the spaces we see, smell, taste, touch, hear, and bond to (or not) as we inhabit, visit, and pass through them. Places are the spaces of social practice and meaning: We can consume, produce, or transgress place (Massey 1994; Cresswell 1996). *Scale* has multiple definitions but generally denotes "a set of territorially nested, malleable relationships among territorially embedded or constituted agents and institutions, shaping their responsibilities, capacities, opportunities, and constraints through territory-specific rule regimes, resources, and identities" (Miller 2009, 62). Scale discourses and practices are frequently "inserted into processes of identification and place-making" (MacKinnon 2011, 27) and can manifest in strategies such as the downloading or uploading of state functions, scale-shifting of identity by political movements, and other nested restructuring approaches related to political struggles over resources and scopes of management (Keil and Mahon 2009). *Networks* likewise have a diverse intellectual and analytical history but are commonly characterized by nodes and connections and associated with a "space of flows." Deleuze and Guattari (1987) presented a metaphorical "rhizomatic" notion of dynamic networks that contrast with, and resist, the fixity associated with common understandings of territory. For others (e.g., Latour 2005), concern with networks is primarily a methodological issue relating to the formation of connections among actants, whereas for still others (e.g., Castells 2000; Taylor 2003), networks are the focus of structural-empirical analysis.

To consider the variety of ways in which territory, place, scale, and networks, as spatial technologies of power, enable and constrain mobility in everyday life, we turn to Cresswell's "constellations of mobility," which bears a strong resemblance to Lefebvre's (1991, 38–46) schema of *spatial practice* (perceived space), *representations of space* (conceived space), and *representational spaces* (lived space). Cresswell's framework offers the advantage of specifically addressing mobility, while retaining the insights of Lefebvre. For Cresswell, *patterns of movement* are the physical processes of movement that lend themselves to empirical observation and quantitative analysis; they are the stuff of motion studies and transportation planning. *Representations of mobility* address the "profound array of meanings from conformity to rebellion" (Creswell 2010, 19) found in diverse narratives and portrayals of mobility. *Practices of mobility* concern the everyday "embodied and habitualized" experiences of the mobile subject. These three aspects of mobility stand in relationship to each other, sometimes in a mutually reinforcing manner—such as when driving an expensive automobile reinforces notions of (high) social status and sometimes in a dissonant manner—as when greenhouse gas emissions associated with driving provoke censure for inadequate environmental stewardship.

Considering territory, place, scale, and networks in relationship to patterns of movement, representations of mobility and practices of mobility (Figure 1) provide us with a broad framework for thinking through the ways in which a variety of spatialities are implicated in the production of mobility and how mobility is simultaneously physical, representational, and embodied or experiential practice. Although Figure 1 is by no means comprehensive, it begins to illustrate ways in which territory, place, scale, and networks are employed as spatial technologies of power, enabling or constraining the production of mobility. Conversely, mobility, as a spatial technology of power, has, in turn, implications for territorial, place-based, scalar, and networked relations.

Mobility should be understood as constituted through territory, place, scale, and networks, each with physical, representational, and embodied or experiential dimensions. The territorial dimension of mobility is perhaps most apparent when a mobile object or body tries to cross a border. We should also consider the governance techniques and rationalities that seek to govern mobile practices and subjectivities throughout a territory, however, as well as the effects territorial representations have on mobile practices. Mobilities are frequently thought to occur between places, but as Cresswell (2006) demonstrated in his analysis of bricklayers and ballroom dancers, mobility also occurs in places and, indeed, practices of mobility are central to the constitution of places. Movements that produce, conform to, and transgress place-based norms illustrate the mutual constitution of place and mobility. Networks can be thought of as the connections or channels through which people, objects, and information move. Networks are neither ubiquitous nor absent power relations. Rather, they are social, cultural, political, and economic achievements, with all the social and geographical inequities such achievements imply. The scalar structure of the state significantly affects the resources available for investment in infrastructure that facilitates mobility, as well as the geographical evenness of that investment. Moreover, mobile practices are shaped, motivated, and constrained by governance techniques enacted at different scales of government and by scalar practices of representation and identification that might influence desire to travel to particular destinations. Clearly, actions that shape the characteristics of territory, place, scale, or networks are likely to affect mobile practices and, in some cases, render what was once mobile immobile or what was once immobile mobile.

Complex Automobility and Sociospatial Relations

For many, no technology embodies the ideal of mobility more than the automobile. Certainly no technology is more literally named: The automobile derives its name from the Greek *autos* (self) and the French *mobile* (moving). The automobile has become a symbol of freedom and status in much of the world and it is not uncommon to hear automobile-associated interest groups speak glowingly of our "love affair with the automobile"—a phrase originally coined by Groucho Marx as part of a 1961 promotional campaign for automobiles, sponsored by Dupont and General Motors (Badger 2015). The phrase implied not only free choice but a passion for automobility—a framing integral to the construction of automobility as a hegemonic regime (Walks 2015). Yet automobility, like all forms of mobility, cannot be understood apart from the conditions that enable it. As Böhm et al. (2006) explained:

> Automobility is ultimately impossible *in its own terms*. Its impossibility is contained in the very combination of autonomy and mobility. At the point at which a subject attempts to move, the specifics of that movement—the technologies deployed, the spaces which need to be made available, the consequences of the form and place of movement, and so on—require a set of external interventions to render it possible. (11)

Spatiality	Patterns of Physical Movement	Representations of Mobility	Practices of Mobility
Territory	How does physical infrastructure create opportunities and barriers to move within boundaries? How do infrastructures of movement serve to define territory?	How is mobility through or across territories depicted? How do these depictions reinforce or challenge affective attachments and governance techniques?	How are mobile subjectivities, e.g., citizens' rights to mobility, formed throughout a territory? How are they are they reinforced or challenged in daily life?
Place	How do the physical features of place, e.g., the site-specific built environment and situation relative to other places, shape movement?	What is the normative order: what mobile practices are appropriate where? How is movement in a place represented as appropriate or inappropriate?	How are mobile subjectivities formed in 'place-appropriate' ways? How does mobility in place affect memory, significance, bonding, and attachments?
Networks	How are the channels through which people, objects and information move physically connected? What connections are lacking? How does the structure and capacity of connections shape the speed and nature of movement?	How is mobile social activity portrayed? How do people recognize the availability or possibility of connection and interaction? How do representations and knowledge affect flows of people, objects and information?	How do people come to know themselves as connected to or "motile" in a network? How does connectivity affect mobile activity, opportunities, and identity construction? How are the un-connected understood and viewed?
Scale	How does the scalar structure of the state affect the availability of resources for, and evenness of, investment in roads, rails, airports, communications technology, etc.? How does this unevenness affect movement decisions, routes, velocities, etc.?	How does the scalar framing of issues affect policy debates? How do scalar frameworks of governance shape understandings of appropriate avenues for policy transfer, e.g., city to city or central state to central state? Might scale be strategically re-constructed to affect movement?	How do mobile subjects come to identify with, and know themselves as, citizens of multiple scalar jurisdictions, e.g., city, region, nation-state? How do mobile practices make, re-make, or challenge constructions of scale?

Figure 1. Spatialities and the production of mobility: Some initial questions.

The interventions to which Böhm et al. (2006) referred are themselves spatially constituted. In other words, automobility must be understood not as a free-floating spatial practice but as a matrix, or assemblage, of intertwined sociospatial relations. Consider automobility as it relates to networks, place, territory, and scale. The automobile represents the promise and capacity to move freely among networked places, connected through extensive physical and regulatory infrastructure. The practices of automobility include not only the material networks of roads, freeways, homes, and places of employment but also networked social relations. Lofland (1973, 136) observed that the automobile links "widely dispersed urban spaces to such a degree that, for all practical purposes, they are spatially contiguous." Dispersed urban villages, as she called them, frequently form networks of ideologically consistent places. Henderson's (2006) examination of automobility in Atlanta demonstrates how trips among predominately white places represent "a distinctive politics of secessionist automobility, couched in a racialized, anti-urban, anti-density, anti-transit set of ideologies and values" (298). As he explained, this is an example of how "automobility embodies deeper social conflicts" (304). Although these examples highlight the ideological and spatial qualities of movement through networks, it is also important to consider the production of mobile subjectivities within and across networks. Position within networks is critical not only to the capacity to move oneself and one's things but also to self-understanding, social standing, and identity.

The automobile is not placeless. On the contrary, the automobile has a place, is a place, and transports drivers and passengers between places. The automobile can be a place for doing office work (Laurier 2004), a place for sexual relations (Flink 1990), and a place for private conversations. As Lofland (1973) put it, the automobile "allows its passengers to move through … [the city] encased in a cocoon of private space" (136). The automobile, a mobile place in its own right, also has its place. Road markings denote the right place for parked cars, moving cars, high-occupancy cars, and cars with handicapped drivers. There are places for going fast (freeways) and places for going slow (school zones). The mobile practices conducted in these kinds of velocity-separated places are governed through techniques produced in complex epistemological networks. Forstorp (2006), for example, described the quantitative production of knowledge that drives the governance of motorists in Sweden. Before these place-specific mobile practices can be governed, however, automobile-friendly (or automobile-exclusive) places must be produced. In her study of the governance of motorists in Adelaide, Bonham (2006) described a shift from "the street as a place for conducting (or accessing) economic activities" to "the street as a site for facilitating the economical conduct of movement between activities" (60) in the early twentieth century. Automobility fundamentally alters urban form, meanings, and social relations where it operates (Walks 2015).

Motorists frequently cross territorial borders. When they do, the territorial considerations can dominate the practices of automobility. Territory might be informally defined (like a gang's turf) but, more often, motorists move (or attempt to move) across and through the established and recognized territorial domains of states. Of course, the polysemic nature of borders (Bauder 2011) means that

travelers (Neumayer 2006), immigrants (Fassin 2011), and border guards (Côté-Boucher, Infantino, and Salter 2014) see, perform, and experience territorial mobility in different ways. Territory, as a relatively homogeneous ideological space, gives rise to feelings of safety or fear, comfort or discomfort, attraction or repulsion as one travels by automobile. Territory, as a space of relatively homogeneous regulation, sets rules of conduct and controls infrastructure for activities affecting automobility—everything from speed limits, to road design, to the priority given to different modes of transportation, to regulations shaping the built environment, all, in turn, affecting the physical opportunities for, and constraints on, automobility.

Issues of mobility inherently entail questions of policy, investment, finance, and, ultimately, the state. The production of U.S. automobility has, for the better part of a century, involved ongoing struggles among governments at different scales of the state. Since the early twentieth century, federal government policies have set the overall direction and framework for the promotion of automobility, whereas subnational (state and local) governments have struggled for control of routing and minimal financial commitment (Jackson 1985; Goddard 1994). A partnership of the federal and subnational state governments was responsible for planning and financing more than 200,000 miles of roads in the first half of the twentieth century. By the early 1950s the rise of the automobile industry and related lobby groups, combined with President Eisenhower's direct observation of the military advantages of Germany's autobahn during World War II (an extraordinarily ironic example of policy mobility), created pressure for a massive expansion of the highway network. In 1953 the Eisenhower administration proposed the largest public works project in U.S. history—a nationwide interstate highway system to be coordinated by a "National Highway Authority [that would act as] a command central, dictating where and how expressways would be built and financing them with tolls" (Goddard 1994, 181–82). Political battles with the states and the American Municipal Association, however, resulted in the devolution of routing decisions to the states and the introduction of a new policy of building highways through the centers of cities, as well as the elimination of tolls as a primary financing mechanism. The net result was a radically different configuration of automobility from what had been originally envisioned: Instead of toll roads that would encircle, but not enter, major cities, the United States built public highways with no user fees, extending into the centers of cities, in turn facilitating racialized and class-based "secessionist automobility" (Henderson 2006). Furthermore, the construction of an oil-dependent mobility system set the United States on a geopolitical course tied to securing global access to oil, as well as a pattern of resource consumption directly driving rising greenhouse gas emissions and climate change.

There is nothing "auto" about automobility. A host of sociospatial relations come into play in constructing practices of automobility, and automobile practice—a dominant spatial practice in many regions of the world—in turn influences a variety of other sociospatial relationships. Indeed, this recursive relationship exists for all forms of mobility. Although the sociospatial polymorphism of mobility is seldom highlighted, the sociospatial practice of mobility depends on, and shapes, the sociospatial qualities of territory, place, scale, and networks.

Trying to Pin Down a Conclusion

There has been considerable debate around the ontological status of mobility: Some scholars assert that the world is always mobile, including not only objects, bodies, and information but the very concepts and categories through which we comprehend the world. Such anti-foundationalist positions stand in opposition to notions of a rigidly ordered sedentary world. Although such debates are beyond what can be addressed here, we agree with McCann and Ward's (2011b, xvi) assertion that we need to move past "the sorts of easy analytical dichotomies—fixity/mobility, global/local" to consider how mobilities and immobilities are relationally and differentially produced. Regardless of whether mobility is an ontological baseline or simply the provisional opposite of fixity, it is clear that mobility, as sociospatial practice, is inextricably bound up in myriad polymorphic sociospatial relations including territory, place, scale, and networks. These sociospatial relations, as spatial technologies of power, are employed and manipulated to impose—if only temporarily—a sedentary order or to enable mobile practices that counter such an order. Neither mobility nor fixity can be assumed. Mobility is, rather, a social, cultural, and political achievement, inherently power-laden and recursively bound up with other spatialities.

Note

1. Elden (2010) argued that "boundaries only become possible in their modern sense through a [territorial] notion of space, rather than the other way round" (13).

References

Adey, P. 2006. If mobility is everything then it is nothing: Towards a relational politics of (im)mobilities. *Mobilities* 1 (1): 75–94.

———. 2010. *Mobility*. London and New York: Routledge.

Agnew, J. 1987. *Place and politics*. London: Allen and Unwin.

———. 1996. Mapping politics: How context counts in electoral geography. *Political Geography* 15 (2): 129–46.

Badger, E. 2015. The myth of the American love affair with cars. *The Washington Post* 27 January 2015. http://www.washingtonpost.com/blogs/wonkblog/wp/2015/01/27//debunking-the-myth-of-the-american-love-affair-with-cars/ (last accessed 7 October 2015).

Bauder, H. 2011. Toward a critical geography of the border: Engaging the dialectic of practice and meaning. *Annals of the Association of American Geographers* 101 (5): 1126–39.

Böhm, S., C. Jones, C. Land, and M. Paterson. 2006. Introduction: Impossibilities of automobility. *The Sociological Review* 54 (S1): 1–16.

Bonham, J. 2006. Transport: Disciplining the body that travels. *The Sociological Review* 54 (S1): 55–74.

Castells, M. 2000. *The rise of the network society*. Malden, MA: Blackwell.

Côté-Boucher, K., F. Infantino, and M. Salter. 2014. Border security as practice: An agenda for research. *Security Dialogue* 45 (3): 195–208.

Cresswell, T. 1996. *In place/out of place: Geography, ideology, and transgression*. Minneapolis: University of Minnesota Press.

———. 2006. *On the move: Mobility in the modern western world*. London and New York: Routledge.

———. 2010. Towards a politics of mobility. *Environment and Planning D: Society and Space* 28 (1): 17–31.

de Certeau, M. 1984. *The practice of everyday life*. Berkeley: University of California Press.

Deleuze, G., and F. Guattari. 1987. *A thousand plateaus: Capitalism and schizophrenia*. Minneapolis: University of Minnesota Press.

Elden, S. 2010. Land, terrain, territory. *Progress in Human Geography* 34 (6): 1–19.

Fassin, D. 2011. Policing borders, producing boundaries: The governmentality of immigration in dark times. *Annual Review of Anthropology* 40:213–26.

Flink, J. 1990. *The automobile age*. Cambridge, MA: MIT Press.

Forstorp, P. 2006. Quantifying automobility: Speed, "zero tolerance" and democracy. *Sociological Review* 54 (S1): 93–112.

Foucault, M. 2007. *Security, territory, population: Lectures at the College de France 1977–1978*. New York: Picador.

Goddard, S. 1994. *Getting there: The epic struggle between road and rail in the American century*. Chicago: University of Chicago Press.

Hannam, K., M. Sheller, and J. Urry. 2006. Editorial: Mobilities, immobilities and moorings. *Mobilities* 1 (1): 1–22.

Harvey, D. [1982] 2007. *The limits to capital.* New York: Verso.

Henderson, J. 2006. Secessionist automobility: Racism, anti-urbanism, and the politics of automobility in Atlanta, Georgia. *International Journal of Urban and Regional Research* 30 (2): 293–307.

Jackson, K. T. 1985. *Crabgrass frontier: The suburbanization of the United States.* New York: Oxford University Press.

Jensen, A. 2011. Mobility, space and power: On the multiplicities of seeing mobility. *Mobilities* 6 (2): 255–71.

Jessop, B., N. Brenner, and M. Jones. 2008. Theorizing sociospatial relations. *Environment and Planning D: Society and Space* 26 (3): 389–401.

Kaufmann, V., M. Bergman, and D. Joye. 2004. Motility: Mobility as capital. *International Journal of Urban and Regional Research* 28 (4): 745–56.

Keil, R., and R. Mahon. 2009. *Leviathan undone? Towards a political economy of scale.* Vancouver, Canada: UBC Press.

Latour, B. 2005. *Reassembling the social: An introduction to actor-network theory.* New York: Oxford University Press.

Laurier, E. 2004. Doing office work on the motorway. *Theory, Culture & Society* 21 (4–5): 261–77.

Lefebvre, H. 1991. *The production of space.* Oxford, UK: Wiley-Blackwell.

Leitner, H., and B. Miller. 2007. Scale and the limitations of ontological debate: A commentary on Marston, Jones and Woodward. *Transactions of the Institute of British Geographers* 32 (1): 116–25.

Leitner, H., E. Sheppard, and K. M. Sziarto. 2008. The spatialities of contentious politics. *Transactions of the Institute of British Geographers* 33 (2): 157–72.

Lofland, L. 1973. *A world of strangers: Order and action in urban public space.* New York: Basic Books.

MacKinnon, D. 2011. Reconstructing scale: Towards a new scalar politics. *Progress in Human Geography* 35 (1): 21–36.

Massey, D. 1994. *Space, place, and gender.* Minneapolis: University of Minnesota Press.

McCann, E. 2010. Urban policy mobilities and global circuits of knowledge: Toward a research agenda. *Annals of the Association of American Geographers* 101 (1): 107–30.

McCann, E., and K. Ward, eds. 2011a. *Mobile urbanism: Cities and policymaking in the global age.* Minneapolis: University of Minnesota Press.

———. 2011b. Urban assemblages: Territories, relations, practices, power. In *Mobile urbanism: Cities and policymaking in the global age,* ed. E. McCann and K. Ward, xiii–xxxv. Minneapolis: University of Minnesota Press.

Miller, B. 2009. Is scale a chaotic concept? Notes on processes of scale production. In *Leviathan undone? Towards a political economy of scale,* ed. R. Mahon and R. Keil, 51–66. Vancouver, Canada: UBC Press.

———. 2013. Spatialities of mobilization: Building and breaking relationships. In *Spaces of contention: Spatialities of social movements,* ed. W. Nicholls, B. Miller, and J. Beaumont, 285–98. Aldershot, UK: Ashgate.

Neumayer, E. 2006. Unequal access to foreign spaces: How states use visa restrictions to regulate mobility in a globalized world. *Transactions of the Institute of British Geographers* 31:72–84.

November, V. 2002. *Les territoires risqué: Le risqué comme objet de reflexion geographique* [Territories of risk: Risk as an object of geographic reflection]. Bern, Switzerland: Peter Lang.

Sassen, S. 2000. Territory and territoriality in the global economy. *International Sociology* 15:372–93.

Sennett, R. 1970. *The uses of disorder.* New York: Norton.

Sheller, M., and J. Urry. 2006. The new mobilities paradigm. *Environment and Planning A* 38 (2): 207–26.

Sibley, D. 1995. *Geographies of exclusion.* London and New York: Routledge.

Smith, N. 1992. Contours of a spatialized politics: Homeless vehicles and the production of geographical scale. *Social Text* 33:54–81.

Taylor, P. 2003. *World city network: A global urban analysis.* London and New York: Routledge.

Urry, J. 2000. *Sociology beyond societies: Mobilities for the twenty-first century.* London and New York: Routledge.

———. 2007. *Mobilities.* Cambridge, UK: Polity.

Walks, A. 2015. Driving cities: Automobility, neoliberalism, and urban transformation. In *The urban political economy and ecology of automobility,* ed. A. Walks, 3–20. London and New York: Routledge.

Algorithmic Geographies: Big Data, Algorithmic Uncertainty, and the Production of Geographic Knowledge

Mei-Po Kwan

Department of Geography and Geographic Information Science, University of Illinois at Urbana–Champaign

Drawing on examples from human mobility research, I argue in this article that the advent of big data has significantly increased the role of algorithms in mediating the geographic knowledge production process. This increased centrality of algorithmic mediation introduces much more uncertainty to the geographic knowledge generated when compared to traditional modes of geographic inquiry. This article reflects on important changes in the geographic knowledge production process associated with the shift from using traditional "small data" to using big data and explores how computerized algorithms could considerably influence research results. I call into question the much touted notion of data-driven geography, which ignores the potentially significant influence of algorithms on research results, and the fact that knowledge about the world generated with big data might be more an artifact of the algorithms used than the data itself. As the production of geographic knowledge is now far more dependent on computerized algorithms than before, this article asserts that it is more appropriate to refer to this new kind of geographic inquiry as algorithm-driven geographies (or algorithmic geographies) rather than data-driven geography. The notion of algorithmic geographies also foregrounds the need to pay attention to the effects of algorithms on the content, reliability, and social implications of the geographic knowledge these algorithms help generate. The article highlights the need for geographers to remain attentive to the omissions, exclusions, and marginalizing power of big data. It stresses the importance of practicing critical reflexivity with respect to both the knowledge production process and the data and algorithms used in the process.

我运用人类能动性研究的案例, 于本文中主张, 大数据的出现, 已显着地增加了演算法在中介地理知识生产过程中的角色。相较于传统的地理探问模式而言, 演算法中介的中心性之强化, 为地理知识生产带来了更多的不确定性。本文反思从运用传统的 "小数据" 转而运用大数据的地理知识生产过程中的重要改变, 并探讨电脑化的演算法如何能够大幅影响研究结果。我质问 "数据驱动的地理" 此一备受吹捧之概念, 该概念忽略了演算法对于研究结果所具有的潜在显着影响, 以及透过大数据生产的世界知识, 或许较数据本身而言更像演算法的人工物之事实。随着当今地理知识的生产较以往更为依赖电脑化的演算法, 本文主张, 将此般新形式的地理探问指称为 "演算法驱动的地理" (或演算地理), 相较于 "数据驱动的地理" 之指称更为合适。演算地理的概念, 同时凸显出必须关注演算法对于其所促成的地理知识的内容、可信度与社会意涵的效应。本文强调地理学者必须持续留意大数据的遗漏、排除与边缘化的力量, 并着重对于知识生产过程以及该过程中使用的数据和演算法, 进行批判性反思的重要性。

Con base en ejemplos de investigación sobre movilidad humana, sostengo en este artículo que el advenimiento de los *big data* ha incrementado significativamente el papel de los algoritmos para mediar el proceso de producción de conocimiento geográfico. Esta creciente centralidad de la mediación algorítmica introduce un mayor grado de incertidumbre en el conocimiento geográfico generado cuando se la compara con los modos tradicionales de pesquisa geográfica. Este artículo reflexiona sobre las transformaciones importantes del proceso de producción de conocimiento geográfico asociados con el cambio de usar "datos pequeños" tradicionales por el uso de datos mayores, y explora la manera como los algoritmos computarizados podrían influenciar considerablemente los resultados de la investigación. Cuestiono la noción muy publicitada de la geografía orientada por datos, que ignora la influencia potencialmente significativa de los algoritmos en los resultados de la investigación, y el hecho de que el conocimiento acerca del mundo generado con *big data* podría ser más un artefacto de los algoritmos usados que los propios datos. Como la producción de conocimiento geográfico es ahora mucho más dependiente de algoritmos computarizados que antes, este artículo afirma que es mucho más apropiado referirnos a este nuevo tipo de pesquisa geográfica como geografía de orientación algorítmica (o geografías algorítmicas) que geografía orientada por datos. La noción de geografías algorítmicas también pone en primer plano la necesidad de dar atención a los efectos de los algoritmos en el contenido, confiabilidad e implicaciones del conocimiento geográfico que estos algoritmos ayudan a generar. El artículo resalta la necesidad que tienen los geógrafos de permanecer atentos a las omisiones, exclusiones y poder marginador de los *big data*. Enfatiza la importancia de practicar la reflexividad crítica con respecto del proceso de producción de conocimiento y de los datos y algoritmos utilizados en ese proceso.

Although much has been written about the advent of big data, the implications of using big data for the generation of geographic knowledge are still far from clear. One of the most important elements missing in this discussion to date, with a few exceptions (e.g., Kitchin 2014b), is the role of algorithms in generating, processing, and analyzing big data in the process of geographic knowledge production. Although algorithms have been used to handle and analyze geographic data for decades, there are indications that the process of geographic knowledge production is increasingly mediated by computerized algorithms with the emergence of big data. Such an increase in algorithmic mediation introduces much more uncertainty to the geographic knowledge generated when compared to traditional modes of geographic inquiry. Drawing on examples from research on human mobility and activity-travel patterns and with a focus on how geoprocessing algorithms could influence research results, this article discusses various sources of algorithmic uncertainty. It reflects on important changes in the geographic knowledge production process associated with the shift from using traditional "small data" to using big data. I call into question the much touted notion of data-driven geography, which neglects the potentially significant influence of algorithms on research results and the fact that knowledge about the world generated with big data might be more an artifact of the algorithms used than the data itself. As the production of geographic knowledge is now far more dependent on computerized algorithms than before, this article asserts that it is more appropriate to refer to this new kind of geographic inquiry as algorithm-driven geographies (or algorithmic geographies) rather than data-driven geography. Through the notion of algorithmic geographies, the article also foregrounds the need to pay attention to the effects of algorithms on the content, reliability, and social implications of the geographic knowledge these algorithms help generate. I highlight the need for geographers to remain attentive to the omissions, exclusions, and marginalizing power of big data. I stress the importance of practicing critical reflexivity with respect to both the knowledge production process and the data and algorithms used in the process.

The Algorithmic Mediation of Geographic Knowledge Production

In the long chains of events that happen before results in geographic research are obtained, many sets of procedures need to be implemented to collect, process, and analyze relevant data, which could be qualitative or quantitative data. Some of these procedures are or can be performed manually (e.g., transferring data from survey instruments to digital data files), whereas many are implemented as computerized procedures because it would be extremely tedious and time consuming to perform them manually on most empirical geographic data sets. This article refers to these sets of procedures for collecting, processing, and analyzing data in geographic research as algorithms, which are well-defined sequences of steps for solving problems or performing specific tasks with or without computerized implementations.

It should be noted that implementation as computerized procedures (in the form of computer code or software) is not a necessary condition for a set of procedures to be considered an algorithm, contrary to the definition used in computer science or in some recent works by geographers (e.g., Graham and Shelton 2013; Kitchin 2014b). For instance, the shortest path between a source node and a destination node in a small network can be found using Dijkstra's algorithm without implementing any computerized procedures. It should also be noted that although many analytical methods might be considered conceptually separate from and can be described without referring to the algorithms that implement them (e.g., the Moran's I or accessibility measures can be expressed as mathematical equations), some other methods can only be expressed in the form of specific algorithms (e.g., Dijkstra's algorithm and evolutionary algorithms; Kwan, Xiao, and Ding 2014). It is thus often impossible to maintain a clear conceptual distinction among methods, procedures, techniques, and algorithms.

Further, because geographic data concern a variety of geographic entities and relations, algorithms are used to perform not only mathematical computation but also complex geoprocessing operations on georeferenced data (e.g., line simplification, spatial search, spatial interpolation, surface modeling, or identifying the topological relationships between two geographic objects; Shi 2010; Xiao 2016). In addition, geoprocessing algorithms are often necessary for representing geographic data at suitable spatial and temporal scales using appropriate data models before any analysis can be performed (e.g., addressing geocoding or digital elevation models). The data used in much of geographic research are thus the product of prior processing using a wide variety of algorithms. It is therefore important to recognize that algorithms are an essential and integral element of geographic data.

Because no results from geographic research involving data can be generated without using algorithms, the production of geographic knowledge derived from data is necessarily mediated by algorithms even before the widespread use of computerized procedures. This is a fundamental reality of the geographic knowledge production process. When algorithms or procedures are applied to generate, process, and analyze geographic data, some uncertainty or error might be introduced and research findings might differ slightly, or even considerably, depending on the specific algorithms used. For example, an algorithm identified the wrong street segments for about 20 percent of the trips in a Global Positioning System (GPS) data set (Gong et al. 2012). Further, different algorithms that implement the same analysis or different implementations of the same algorithm could lead to different results, and the differences in research findings might vary considerably. As cogently demonstrated by Fisher's (1993) study, there can be more than 50 percent difference in the results obtained with different viewshed analysis algorithms. Even the computer languages, compilers, and computational platforms used might introduce some differences to the results generated with the same algorithm due to different processor precision and methods of handling interim values. Algorithmic uncertainty is thus an essential element in the geographic knowledge production process due to the use of different sets of procedures, implementations, data environments (data model and data structure), or computational platforms. These uncertainties are often magnified in big data research.

Algorithms and Traditional Data in Human Mobility Research

For decades, human mobility studies conducted by geographers and transport researchers collected the needed data largely through activity-travel diary surveys. These traditional data sets were obtained with custom-designed survey instruments. They were created to answer specific questions about human activity-travel patterns based on some prior theoretical understanding of these patterns. These data sets tended to be small to moderate in size (with several hundred to several thousand participants) and were often collected with specific sample designs that seek to obtain representative samples of a population. Although activity-travel diary data sets are costly and time consuming to collect, they contain highly detailed information about participants' sociodemographic attributes and activity-travel behavior (e.g., activity purpose and location, start and end time of activities, travel mode, and travel route) that enables rich description and analysis of their mobility patterns.

Because these data sets are not huge, computerized procedures were typically not necessary and were often not used in the data generation and preparation phases. For instance, data tend to be transferred from survey instruments to digital data files manually. Little algorithmic uncertainty is introduced by computerized data transformation operations because definite mathematical relationships govern how new variables are generated. In addition, it is still possible to examine particular data records or items to check and clean the data as well as to address specific data quality issues (e.g., missing data or anomalies such as outliers) via researchers' experience and expertise. For instance, in past research I have contacted research participants to rectify missing or inconsistent data in their surveys and corrected errors in data spreadsheets with hundreds of records. Further, even when computerized algorithms are implemented to analyze traditional "small data," it is still quite feasible to examine the effects of different algorithms on research results. This is because it is not prohibitively costly or time consuming to rerun statistical tests or analyses using different algorithms and to use researchers' experience to identify and correct errors in analytical procedures or results.

Thus the algorithmic mediation of the geographic knowledge production process is limited when using traditional "small" data sets (especially in the data generation and preparation phases, as most of the tasks during these phases can be performed manually). Meanwhile, computerized procedures or algorithms are mainly used in the data analysis and modeling phase when using these traditional data sets (except when analyzing most qualitative data). It is important to note that when data are or can be handled manually, researchers and data interact more directly, and both the data and procedures are more tangible and visible in the research process. This is especially true for procedures that handle or analyze qualitative data, as researchers or their associates often need to perform these tasks themselves (e.g., coding or interpreting interview transcripts) instead of relegating them to computerized procedures (note that this is true even when computer-aided qualitative data analysis software is used).

The Advent of Big Data

In recent years, the widespread use of location-aware technologies, mobile devices, and social media has made it possible to assemble huge amounts of data about people's location and movement from various sources (e.g., cell phone companies, public transit and taxi companies, real-time GPS/geographic information system (GIS) functionalities in mobile devices, Internet search engines, and social media providers). The rapid increase in the volume, diversity, and intensity of data from these sources has led to the emergence of big data and stimulated new developments in human mobility research. Big data are not just massive in volume (e.g., about 10 million geotagged tweets are generated every day); it is generated continuously at high velocity and high space–time resolution (Richardson et al. 2013).

Although big data seem promising for advancing human mobility research, how algorithms might influence the data that researchers obtain and the research findings they report have received very little attention to date. A fundamental fact about big data should be noted, however: No big data can be generated without using some computerized procedures or algorithms (e.g., searching, selecting, or ranking using specific parameters, such as the PageRank algorithm used by Google to generate search results). Thus, no big data or research results obtained using big data are unmediated by algorithms. An instructive example of how big data are the result of prior processing before reaching the public or research community is provided by Fischer (2014). In the process of developing a software tool for mapping geotagged tweets from Twitter, he observed a banding phenomenon: The original tweet locations tend to align with the closest latitude or longitude. This suggests that tweet locations might have been fuzzed by Twitter through snapping them to the closest latitude or longitude to prevent people's exact locations being disclosed. Further, the study observed a strange phenomenon of missing data at the Prime Meridian when zooming in on London and suggested that Twitter might have filtered out the tweets on the

Prime Meridian for some reason. Two algorithms were then implemented to address these issues and to make the tweet maps look more natural. As a result, every map generated by the mapping tool is mediated by Fischer's own algorithms, Twitter's privacy-protection algorithms, or both.

An important change in the geographic knowledge production process associated with the increasing use of big data is the greatly expanded use of computerized algorithms, which have become necessary even in the data collection and generation phase. As algorithms are increasingly implemented as computerized procedures, many of which are now relegated to computer programmers, they become increasingly detached from and less visible to researchers who use them. One important reason for this trend is the limited informational content of big data. For instance, many variables needed for addressing specific questions about human mobility (e.g., home and workplace location, travel route, travel mode, gender, income, and race) are often not available in popular big data sets. Algorithms are thus needed to infer their values indirectly from the big data used (e.g., the filtering and clustering algorithms used in Widhalm et al. [2015] to infer trip sequences from cell phone data). In addition, algorithmic uncertainty has increased because checking raw data with researchers' experience and expertise, rerunning analyses, or comparing the effects of different algorithms on research results are often infeasible or prohibitively costly, given the huge volume, complexity, and dynamic nature of big data.

Another important factor contributing to increased algorithmic uncertainty when using big data is that search algorithms used to generate data could change over time (algorithmic dynamics). Lazer et al. (2014) presented a highly instructive example that illustrates the unstable and changing nature of the algorithms used in the generation of big data as well as the significant influence of algorithms on the knowledge produced. In early 2013, Google Flu Trends (GFT), the flu tracking system created by Google, made significant prediction error about influenza-related doctor visits in the United States. Lazer et al. (2014) explained that Google search algorithms are constantly being modified by its engineers to improve its service and that "GFT was an unstable reflection of the prevalence of the flu because of algorithm dynamics affecting Google's search algorithm" (1204). The authors further highlighted the fact that because the algorithms underlying Google, Twitter, and Facebook are always being modified and constantly changing, it is far from clear "whether studies conducted even a year ago on data collected from these platforms can be replicated in later or earlier periods" (Lazer et al. 2014, 1204). Thus, the production of geographic knowledge in the big data era is far more uncertain and affected by algorithms than before. The section that follows examines issues concerning the use of computerized algorithms in human mobility research in greater detail.

Algorithms and Human Mobility Research Using Big Mobile Phone Data Sets

Popular sources of big data for human mobility research include GPS tracks of vehicles, public transit smart card data, GPS data from bicycle sharing programs, and social media data (e.g., Ma et al. 2013; Corcoran and Li 2014; Hawelka et al. 2014). Among them, passive cell phone data, which are recorded automatically by cell phone companies without implementing additional data collection procedures such as GPS tracking, is perhaps the most popular because of its advantages over conventional surveys (e.g., González, Hidalgo, and Barabasi 2008; Ahas et al. 2010; Bayir, Demirbas, and Eagle 2010; Silm and Ahas 2014; Widhalm et al. 2015). For instance, when phone companies are willing to provide these data at reasonable prices, researchers can have cost-effective access to very large numbers of communication records (e.g., over 1 billion) from a large number of users (e.g., several million users) that cover a high proportion of the population over large areas and long periods of time (e.g., six months or one year; Song et al. 2010; Wang, Chen, and Ma 2014). Researchers also do not need to worry about the problem of sample attrition over time because most data provided by phone companies are kept for a long period of time. In light of these advantages, big cell phone data sets have become a major data source in recent human mobility research.

There has been little discussion to date, however, about how algorithms might affect the data or research results when big cell phone data sets are used. Algorithms for processing and analyzing big mobile phone data sets are necessary largely because of their limited informational content. For instance, the actual location of users is not recorded and thus is unknown in passive cell phone data. Instead, the recorded location is the geographic location of the serving cell towers that handled users' communication activities (e.g., a call or a text message). Further, unless complemented by data obtained directly from individuals through surveys

(e.g., Licoppe et al. 2008), people's activity location, activity duration, activity purpose, travel route, travel mode, and other trip characteristics (e.g., trip distance) are unknown and can only be inferred using algorithms. For instance, whether an individual is staying at a location instead of moving around needs to be inferred from the data based on spatial and temporal constraints that identify a sequence of consecutive cell phone records as an activity location. A spatial constraint sets the roaming distance within which a user is considered staying at a location. A temporal constraint sets the minimum duration a user needs to spend at a location for it to be considered an activity location (instead of moving around). For instance, Jiang et al. (2013) used a 300-m roaming distance as the spatial constraint and a temporal constraint of ten minutes to identify certain stay points as activity locations. Because different constraints and parameters can be used in these inferential algorithms, slightly or even considerably different patterns of activity location could be observed, depending on the exact algorithms and constraints used.

Similarly, both trip distance and activity need to be inferred from the data using algorithms and specific parameters. When trip distance is estimated using inferred activity locations, considerable positional error can occur. For instance, for one respondent in Ahas et al. (2010), the home location inferred from cell phone data is 830 m from the real home location, and the inferred workplace location is 300 m from the real workplace location. Although this does not seem to be a lot of error, the home–work distance derived from these two inferred locations is about 1 km longer than the real home–work distance. Given that the original home–work distance is about 2 km (estimated based on a visual examination of Figure 4 in Ahas et al. 2010), this amounts to a 50 percent error. The potential uncertainty introduced by algorithms used to process big cell phone data sets could thus be considerable. The difficulty is, unlike with traditional data sets, estimating and correcting this kind of positional error are prohibitively costly and often infeasible when there are millions of records in the data set.

Information about the activity being performed when data are recorded is available only for certain types of data sets (e.g., people are riding taxis in taxi-tracking data sets). For big cell phone data sets, we know almost nothing about what people are actually doing when they performed various communication events. Some studies have used algorithms to infer people's most likely activities at different locations based on the space–time characteristics of the activity (e.g., a long stay at a location from the evening to early morning as staying home and a long stay at a location during the day as work). Some studies used algorithms based on land use data to infer the activity being performed when people's communication events took place. Even if a person is located in a particular type of land use (e.g., commercial or recreational land use), though, we still do not know and cannot verify what the person is actually doing because many different activities can be performed in a given type of land use (e.g., shopping, dining, and running errands can happen in commercial land use). This is particularly true as people can now perform a wide range of activities in the same type of land use or at the same place with the greatly increased use of information and communication technologies (Kwan 2007; Couclelis 2009).

Further, although human mobility research using big cell phone data sets often represents people's movement trajectories as if they can be directly observed from the data, these data sets do not record people's actual location or movement over space and time. The location associated with a particular communication event in passive mobile phone records is actually the location of the cell tower that handled that event. The movement trajectories derived from big cell phone data sets are thus not the actual movement trajectories of phone users but the paths connecting consecutive cell towers that handled users' communication activities. To reconstruct these paths, algorithms are used to infer whether a phone user was moving or not and the likely routes traversed. Because the routes of movement are inferred and are just arbitrary lines connecting consecutive serving cell towers, trip distance cannot be accurately estimated and such use of movement inference algorithms introduces an unknown amount of uncertainty into analytic results.

Several issues arise when using algorithms to infer the movement trajectories of cell phone users in big data mobility research. For example, movement is detected only when the current serving cell tower of a phone shifts to another one. Short trips that do not lead to such a cell tower shift (i.e., when the actual movement is not long enough to shift the serving cell tower to another one), however, will not be recorded in the data set. Because most human trips are over a short distance, considerable measurement error can occur when using big cell phone data sets. In fact, recent studies have observed that mobility measures derived from big cell phone data sets using cell tower for location tend to underestimate people's daily travel distance when compared to those obtained from GPS data (Wang, Chen, and Ma 2014).

Other issues arise when cell phone users make short trips across cell tower service areas or when they are located in the overlapping service areas of two or more cell towers. In the former case, a short trip might be recorded as a much longer one because the recorded movement distance is the distance between the serving cell towers instead of the actual trip distance. In the latter case, when a phone is located in the overlapping service areas of two or more cell towers, a cell phone might switch or oscillate between neighboring cell towers for load balancing or signal strength optimization (Ahas et al. 2010; Wang, Chen, and Ma 2014). Referred to as the *ping-pong effect*, this kind of oscillation was also observed in Wi-Fi networks (Bayir, Demirbas, and Eagle 2010). When this happens, a phone user, although not moving, could appear to have traveled back and forth for several kilometers in just a few seconds. Algorithms are thus necessary to assign a phone user to the most likely cell tower, but it remains unclear which cell tower better approximates the actual location of the phone user.

To highlight the algorithmic uncertainty involved due to the need to use algorithms to infer unavailable information, I provide two examples in what follows to illustrate how the precise movement trajectories obtained from passive big cell phone data can be affected by the inferential algorithms and their interactions with particular data environments. In Figures 1 and 2, the actual movement trajectory of an individual is represented by the solid black arrow that runs from

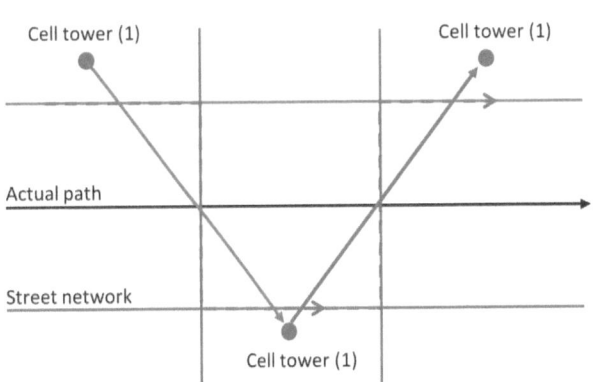

Figure 1. The effects of different trajectory inference algorithms on the inferred distance and movement trajectory with cell tower Set 1 (red dots). The actual movement trajectory is represented by the solid black arrow that runs from left to right across the center of the figure. This movement took place along one of the streets of the local street network, which is represented by a grid of light blue lines. (Color figure available online.)

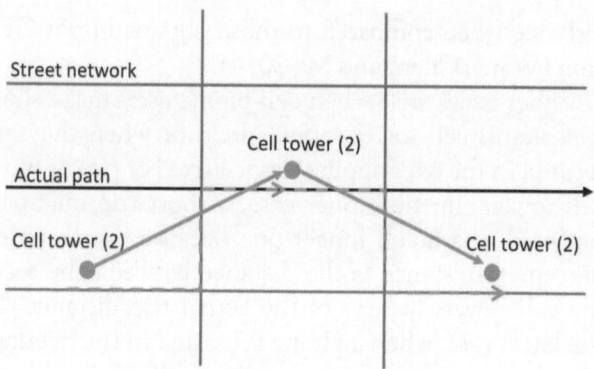

Figure 2. The effects of different trajectory inference algorithms on the inferred distance and movement trajectory with cell tower Set 2 (green dots). The actual movement trajectory is represented by the solid black arrow that runs from left to right across the center of the figure. This movement took place along one of the streets of the local street network, which is represented by a grid of light blue lines. (Color figure available online.)

left to right across the center of the figures. This movement took place along one of the streets of the local street network, which is represented by a grid of light blue lines. Two possible geographic distributions of cell towers are represented: cell tower Set 1 in Figure 1 is represented as red dots, and cell tower Set 2 in Figure 2 is represented as green dots.

Let us first consider Figure 1. If the trajectory inference algorithm uses cell towers as a proxy of the phone user's actual location, then the inferred movement trajectory obtained will be the solid red line. If the algorithm attempts to take into account actual movement possibilities in the area and snaps the movement path to the street network, then the inferred movement trajectory obtained will be the red dashed line. Note that both of these lines are different from and longer than the actual movement trajectory represented by the solid black arrow that runs from left to right across the center of Figure 1. This means that the inferred travel distance is longer than the actual travel distance, no matter which trajectory inference algorithm is used. Note that even when the algorithm uses the real street network to better approximate people's actual movement paths, this might be helpful only under specific conditions—for example, when the street network is not dense or when there is only one possible transport route in the study area, as illustrated by Gong et al. (2012), where wrong street segments were identified for about 20 percent of trips due to the dense street network of the study area.

A comparison of Figures 1 and 2 highlights how the location of cell towers in the study area could affect both the inferred movement trajectory and inferred

travel distance even when using the same algorithm. Although the actual movement is the same (depicted by the solid black arrow that runs from left to right across the center of both Figures 1 and 2), either the red or green solid lines are inferred as the movement trajectories, depending on which set of cell towers is present in the study area. Note that the inferred travel distance based on the red cell towers is longer than that obtained with the green cell towers, and both of these inferred distances are longer that the actual movement distance represented by the solid black arrow, independent of whether the trajectory inference algorithm takes the street network into account or not. These examples show how algorithms and their interactions with particular data environments (cell tower location) could introduce a considerable amount of uncertainty to the inferred movement trajectories and inferred travel distance when using big cell phone data sets.

Toward Reflexive Algorithmic Geographies

No geographic knowledge derived from data can be created without using algorithms, which are sets of procedures for collecting, generating, processing, and analyzing data. All knowledge created with any data is thus necessarily mediated by algorithms, which might or might not be implemented as computerized procedures and are thus more than the computer code or software that implements them. Understanding algorithms this way renders them amenable even to research with qualitative methods and to the practice of critical reflexivity discussed later.

Using computerized algorithms in geographic research is not new. They have been used in studies that collected and used traditional data sets of small to moderate size. The advent of big data, however, significantly increases the role of algorithms in the geographic knowledge production process, especially in the data generation and preparation phases. Drawing from examples from human mobility research with big cell phone data sets, this article shows that many more procedures in the geographic knowledge production process are now performed by computerized algorithms, frequently due to the need to infer variables that are not directly available in big data sets. An important consequence of this trend is a considerable potential increase in the algorithmic uncertainty in the knowledge created, as many algorithms introduce uncertainty that will propagate in the process, leading to slightly or even significantly different findings. Further, as the algorithmic mediation of the knowledge production process increases, the precise ways in which data are generated, processed, and analyzed tend to become increasingly invisible to and detached from researchers.

Many have argued that the advent of big data is leading to the rise of a new paradigm of scientific inquiry (the fourth paradigm) and a new mode of data-driven knowledge discovery, which entails a shift from deductive forms of inquiry (based on the sequence of hypothesis formulation, data collection, and analysis) toward more inductive and emergent forms of analysis that allow the data to speak for itself (Kitchin 2014a). Following this line of thinking, some geographers have begun to advocate the data-driven generation of geographic knowledge based on the latest advances in big data geographic information science, spatial data mining, and visual analytics. Yet, the notion of data-driven geography is misleading and untenable. It ignores the potentially significant influence of algorithms on research results and the fact that knowledge about the world generated with big data might be more an artifact of the algorithms used than the data itself (Lazer et al. 2014). As the examples of this article illuminate, the existence of pristine and pure big data is largely a myth because the generation of big data itself necessitates the use of computerized algorithms, not to mention further processing and analysis. Thus, no big data can reach researchers or the public untainted by some algorithmic uncertainty. Further, as the production of geographic knowledge is now far more dependent on and affected by the algorithms used in the process, it seems appropriate to refer to this new kind of geographic inquiry as algorithm-driven geographies (or algorithmic geographies) rather than data-driven geography.

The notion of algorithmic geographies foregrounds the need for geographers to pay attention to the effects of algorithms on the content, reliability, and social implications of the geographic knowledge these algorithms help generate. It alerts us to the perils in elevating the promise of big data at the expense of ignoring critical issues concerning the scientific and social consequences of the knowledge generated with such data. It also highlights the need for geographers to mitigate the tendency and consequences of becoming increasingly detached from both algorithms and data and to remain attentive to the omissions, exclusions, and marginalizing power they entail. It stresses the imperative for us to practice critical reflexivity with respect to both the knowledge production process and the data used in the

process. Geographers need to recognize that algorithms might have played an important role in determining their research results. We also need to be aware of how using big data could lead us to address questions that are less central to pressing societal concerns when compared to using traditional data. For instance, past studies using traditional data are often interested in questions concerning how social difference such as gender or race affects people's mobility experience, but these kinds of questions cannot be addressed by big data due to their lack of detailed sociodemographic information.

To address the scientific and social consequences of algorithmic uncertainty when using big data in geographic research, several steps can be undertaken to practice critical reflexivity: (1) evaluate how different algorithms and their interactions with data might lead to different results, omissions, and exclusions; (2) assess the amount of algorithmic uncertainty involved, how much confidence about the research findings is warranted, and whether it is acceptable with respect to the research questions and geographic scale of the study (e.g., error of 1 km or less may be tolerated for studies on long-distance intercity travel but might not be acceptable for research on intraurban travel where people make a lot of short trips); (3) examine or validate the algorithms using smaller subsets of the data that have been enriched with additional information; (4) complement big data by traditional data, especially with regard to information that is not available in big data sets but is often obtained directly from research participants; (5) evaluate whether the algorithms are capable of revealing the effects of interpersonal and social difference such as gender, race, and class (e.g., some accessibility measures just mimic the spatial patterns of urban opportunities and are not capable of capturing interpersonal differences in individual accessibility; see Kwan 1998); and (6) assess the stability of the algorithms used to generate the data and its implications for the replicability of the findings.

In the final analysis, geographers need to proceed with great caution when using big data in their research. It is important to bear in mind that "big data sets do not, by virtue of their size, inherently possess answers to interesting questions" (Reich 2015, 34). Using data sets of enormous size "does not mean that one can ignore foundational issues of measurement. ... The core challenge is that most big data that have received popular attention are not the output of instruments designed to produce valid and reliable data amenable for scientific analysis" (Lazer et al. 2014, 1203).

Acknowledgments

I thank Tim Schwanen for handling the blind review process and making editorial decisions for this article (including the abstract). His helpful suggestions and the thoughtful comments of three anonymous reviewers have helped improve the article considerably.

Funding

This article was written while I was supported by grants NSF IIS-1354329, NSF BCS-1244691, and NSFC-41529101.

References

Ahas, R., S. Silm, O. Järv, E. Saluveer, and M. Tiru. 2010. Using mobile positioning data to model locations meaningful to users of mobile phones. *Journal of Urban Technology* 17 (1): 3–27.

Bayir, M. A., M. Demirbas, and N. Eagle. 2010. Mobility profiler: A framework for discovering mobility profiles of cell phone users. *Pervasive and Mobile Computing* 6 (4): 534–54.

Corcoran, J., and T. Li. 2014. Spatial analytical approaches in public bicycle sharing programs. *Journal of Transport Geography* 41:268–71.

Couclelis, H. 2009. Rethinking time geography in the information age. *Environment and Planning A* 41 (7): 1556–75.

Fischer, E. 2014. Making the most detailed tweet map ever. https://www.mapbox.com/blog/twitter-map-every-tweet/ (last accessed 8 October 2015).

Fisher, P. F. 1993. Algorithm and implementation uncertainty in viewshed analysis. *International Journal of Geographical Information Science* 7 (4): 331–47.

Gong, H., C. Chen, E. Bialostozky, and C. T. Lawson. 2012. A GPS/GIS method for travel mode detection in New York City. *Computers, Environment and Urban Systems* 36 (2): 131–39.

González, M. C., C. A. Hidalgo, and A. L. Barabasi. 2008. Understanding individual human mobility patterns. *Nature* 453 (7196): 779–82.

Graham, M., and T. Shelton. 2013. Geography and the future of big data, big data and the future of geography. *Dialogues in Human Geography* 3 (3): 255–61.

Hawelka, B., I. Sitko, E. Beinat, S. Sobolevsky, P. Kazakopoulos, and C. Ratti. 2014. Geo-located Twitter as a proxy for global mobility patterns. *Cartography and Geographic Information Science* 41 (3): 260–71.

Jiang, S., G. A. Fiore, Y. Yang, J. Ferreira, E. Frazzoli, and M. C. González. 2013. A review of urban computing for mobile phone traces: Current methods, challenges and opportunities. Paper presented at the ACM SIGKDD International Workshop on Urban Computing, Chicago.

Kitchin, R. 2014a. Big data, new epistemologies and paradigm shifts. *Big Data and Society* 1 (1): 1–12.

———. 2014b. Thinking critically about and researching algorithms. Article presented at the Programmable City Working, National University of Ireland, Maynooth, Ireland. http://articles.ssrn.com/sol3/articles.cfm?abstract_id=2515786 (last accessed 8 October 2015).

Kwan, M.-P. 1998. Space–time and integral measures of individual accessibility: A comparative analysis using a point-based framework. *Geographical Analysis* 30 (3): 191–216.

———. 2007. Mobile communications, social networks, and urban travel: Hypertext as a new metaphor for conceptualizing spatial interaction. *The Professional Geographer* 59 (4): 434–46.

Kwan, M.-P., N. Xiao, and G. Ding. 2014. Assessing activity pattern similarity with multidimensional sequence alignment based on a multiobjective optimization evolutionary algorithm. *Geographical Analysis* 46 (3): 297–320.

Lazer, D., R. Kennedy, G. King, and A. Vespignani. 2014. The parable of Google Flu: Traps in big data analysis. *Science* 343:1203–05.

Licoppe, C., D. Diminescu, Z. Smoreda, and C. Ziemlicki. 2008. Using mobile phone geolocalisation for "socio-geographic" analysis of co-ordination, urban mobilities, and social integration patterns. *Tijdschrift voor Economische en Sociale Geografie* 99 (5): 584–601.

Ma, X., Y.-J. Wu, Y. Wang, F. Chen, and J. Liu. 2013. Mining smart card data for transit riders' travel patterns. *Transportation Research Part C* 36:1–12.

Reich, J. 2015. Rebooting MOOC research. *Science* 347 (6217): 34–35.

Richardson, D. B., N. D. Volkow, M.-P. Kwan, R. M. Kaplan, M. F. Goodchild, and R. T. Croyle. 2013. Spatial turn in health research. *Science* 339 (6126): 1390–92.

Shi, J. 2010. *Principles of modeling uncertainties in spatial data and spatial analyses.* Boca Raton, FL: CRC.

Silm, S., and R. Ahas. 2014. Ethnic differences in activity spaces: A study of out-of-home nonemployment activities with mobile phone data. *Annals of the Association of American Geographers* 104 (3): 542–59.

Song, C., Z. Qu, N. Blumm, and A.-L. Barabási. 2010. Limits of predictability in human mobility. *Science* 327 (5968): 1018–21.

Wang, T., C. Chen, and J. Ma. 2014. Mobile phone data as an alternative data source for travel behavior studies. Paper presented at the Transportation Research Board 93rd annual meeting, Washington, DC.

Widhalm, P., Y. Yang, M. Ulm, S. Athavale, and M.C. González. 2015. Discovering urban activity patterns in cell phone data. *Transportation* 42:597–623.

Xiao, N. 2016. *GIS algorithms.* London: Sage.

Mobility Research in the Age of the Smartphone

Amit Birenboim and Noam Shoval

Department of Geography, The Hebrew University of Jerusalem

At the end of 2014, over 93 percent of the world's population owned cellular phones, with penetration rates that exceeded 100 percent in most developed countries. Mobile phones have become an integral part of our lives and have had a significant impact on society—including on individuals' daily movement and mobility patterns. Cellular phones have also been used for research and have been employed to collect time–space data about the mobility of relatively large populations. In this article, we provide a comprehensive review of the potential of advanced mobile phones—known as smartphones—in the investigation of the geographies of mobility. We discuss how these devices can be employed in research, tracking individuals in time and space and functioning as location-aware survey tools in real time, among other things. We also engage in a debate over the advantages, disadvantages, and limitations of smartphones in this context and highlight new research trends that are beginning to appear following the introduction of smartphones.

2014年底，全世界有超过百分之九十三的人口拥有移动电话，在大多数的已开发国家中，渗透率更超过百分之百。移动电话已成为我们生活的一部分，并已对社会产生显着的影响——包含对个人的每日移动和行动模式。移动电话同时被用来进行研究，并用来搜集相对而言大型人口移动的时空数据。我们在本文中，综合回顾高阶移动电话——亦即智能手机——之于探讨能动性地理的潜能。我们探讨这些机器如何能够被应用至研究当中，追踪个人所在的时空，并作为即时感知定位的调查工具及其他应用。我们同时于此脉络中，涉入有关智能手机的优势、劣势与限制之辩论，并强调自智能手机问世之后开始浮现的崭新研究趋势。

A finales de 2014, más del 93 por ciento de la población del mundo tenía teléfonos celulares, con tasas de adopción que excedían más del 100 por ciento en la mayoría de los países desarrollados. Los teléfonos móviles se han convertido en una parte integral de nuestras vidas y han tenido un impacto significativo en la sociedad—lo cual incluye los movimientos diarios del individuo y sus patrones de movilidad. Los teléfonos celulares también han sido usados para investigación y se han empleado para recoger datos de tiempo-espacio acerca de la movilidad de poblaciones relativamente grandes. En este artículo presentamos una revisión amplia del potencial de teléfonos móviles avanzados—conocidos como teléfonos inteligentes—en la investigación de las geografías de la movilidad. Presentamos la discusión sobre cómo pueden emplearse estos aparatos en investigación, siguiendo a los individuos en tiempo y espacio y sirviendo como herramientas para el estudio consciente de la localización en tiempo real, entre otras cosas. Nos comprometemos también en un debate sobre las ventajas, desventajas y limitaciones de los teléfonos inteligentes en este contexto, y destacamos nuevas tendencias de investigación que han estado surgiendo desde la aparición de los teléfonos inteligentes.

The development of mobile technologies that can be used for tracking—such as Global Positioning System (GPS) receivers and mobile phones—has been very dynamic. These devices represent a boon to researchers, giving them the opportunity to collect continuous and intensive high-resolution data in time (seconds) and space (meters) for long periods of time—something that has never before been possible in geographic and social science research.

The use of mobile tracking technologies for the study of human mobility began at the end of the 1990s. The first studies conducted, experiential in nature, established data collection practices (Wolf, Guensler, and Bachman 2001; Shoval and Isaacson 2006) and data analysis procedures for the new high-resolution, tempo-spatial information gathered (Schuessler and Axhausen 2009).

Mobile tracking technologies hold clear advantages over traditional methods such as surveys and self-report diaries; they are more accurate, reduce the burden on participants, and are not dependent on respondents' enthusiasm and memory (Isaacson et al. 2014). As a result, these technologies have become a central data collection tool in human mobility studies. Today, tracking technologies are applied in large- and small-scale surveys in a wide variety of fields such as transportation (Bohte and Maat 2009), tourism (Shoval and Isaacson 2010), and medicine (Richardson et al. 2013). This includes several important works by geographers (Ahas et al. 2010; Silm and Ahas 2014) and others (Calabrese and Ratti 2006) that analyzed passive cellular data collected by cellular mobile providers.

One of the most prominent developments in the field of mobile tracking technologies in recent years is the introduction of advanced mobile phones, known as smartphones. These devices are highly programmable, and they incorporate strong computation power, intuitive and convenient user interfaces, advanced tracking and telecommunication technologies, and built-in sensors—all of which can be utilized to enhance geographical and mobility studies. The use of smartphones in empirical studies is still in its infancy, but they appear to bear great potential as a valuable tool in geographical studies about human spatial behavior.

The aim of this article is twofold: (1) to review the growing usage of smartphones in empirical studies, including the development of unique sampling procedures and approaches and (2) to assess the advantages, disadvantages, and potential of smartphones to serve as data collection tools in geography and mobility studies. Before we begin our review of the use of smartphones, we take a brief look at the smartphone—its place in today's market, its capabilities, and its uses.

The Emergence of the Smartphone

The Cellular Revolution

Although the commercial use of cellular communications began as far back as 1983, it was limited primarily to business purposes due to the high price of both the service and the devices. Cellular phone prices began to drop drastically in the mid-1990s; today, in the developed world, cell phones are owned by people of all ages, professions, and income levels. Cell-phone penetration in the developed world recently crossed the 100 percent mark (World Bank 2014). The saturation of this form of communications technology has accelerated in many parts of the developing world as well; at the time of the writing of this article, more than 93 percent of the world's population owns a cellular phone (World Bank 2014). We can therefore refer to modern human society as a whole as a cellular society (Shoval 2007).

The rapid adoption of mobile phones demonstrates that our social interactions, economic engagement, and individual existence—including personal life activities—are now being shaped by the use of these technologies (Schwanen and Kwan 2008; Dijst 2009).

Smartphones and Sensors

Modern smartphones began to appear on the market in the mid-2000s. The launch of the first-generation iPhone on 29 June 2007 and the widespread adoption of the Android operating system one year later, however, opened a new phase in the history of mobile phones and mobile computing. In February 2012, more than 50 percent of the phones in the United States were smartphones (Nielsen 2012); today, the numbers are significantly higher—and not only in the most developed countries.

Smartphones and other mobile devices, such as tablet computers, incorporate a set of embedded sensors that include various technologies: GPS, cell tower identification and Wi-Fi positioning, proximity technologies such as Bluetooth and radio frequency identification (RFID), accelerometer, gyroscope, magnetometer (compass), light sensor, microphone, and camera. In addition, sensors such as the barometer, ambient thermometer, humidity sensor, and pedometer, available in some higher end smartphones today, might become the standard for many smartphones in the near future, furnishing them with even more sophisticated sensing abilities. Table 1 summarizes the most common tracking technologies that are incorporated in current smartphones, many of which could be useful in mobility studies. Table 1 only includes technologies that can be used in studies without employing additional external hardware. Although differences in quality exist between the hardware of various phone brands and models, the capability and resolution levels that are indicated in Table 1 should be found even in low-end devices.

The array of sensors that most current smartphones are equipped with enhance the phones' ability to collect location and movement parameters in several

Table 1. Handset-based time–space and movement measurement technologies

Technology	Description	Spatial resolution[a]	Sampling frequency[b]	Indoor (I) or outdoor (O)	Remarks
GPS (A-GPS[c])	GPS technology can acquire high-resolution time–space information based on the trilateration of radio signals that are transmitted from satellites.	A few meters	Seconds (1 Hz and lower)	O	The American NAVSTAR is the most commonly used GPS system. Some devices also take advantage of the Russian GPS system (GLONASS) to enhance the functionality of satellite-based tracking.
Cell ID[d]	Location is determined based on the location of the cell tower to which the mobile phone is assigned.	50 m–5 km+ (dependent on tower density)	Seconds (1 Hz and lower)	I + O	Cell ID positioning can be easily obtained in most devices. Accuracy is better in urban environments where tower density is higher. In rural areas, accuracy might fall to as low as several kilometers.
Wi-Fi positioning[d]	The location of mobile devices can be inferred from wireless access points based on the received signal strength.	10–50 m	Seconds (1 Hz and lower)	I + O	Wi-Fi positioning can be easily obtained in most devices. It requires that the mobile user keep his or her Wi-Fi on. Wi-Fi positioning is not possible in places without wireless access points (e.g., rural areas).
Bluetooth	Bluetooth technology is based on short radio wavelength. It can be used to determine the proximity of one smartphone device to another (relative position).	A few meters	Seconds (1 Hz and lower)	I + O	Setting external Bluetooth transmitters might allow researchers to identify the absolute location of smartphones.
Accelerometer[e]	This detects the acceleration of the device. It is used to infer the mode of transportation, intensity of activity, and type of activity (e.g., walking, running, and driving).		Milliseconds (~100 Hz)	I + O	Since the introduction of smartphones, accelerometers are becoming more popular in mobility studies.
Magnetometer[e]	The magnetometer recognizes the direction of movement.		Milliseconds (~100 Hz)	I + O	Measures direction and not location.
Gyroscope[e]	Allows the phone to determine its orientation.		Milliseconds (~100 Hz)	I + O	Gyroscope information is still not widely used.
Barometer	Measures barometric pressure. This information can be used to detect vertical movement and other contextual information.		Seconds (1 Hz and lower)	I + O	Barometer is still not very common in standard smartphones.

Note: Radio frequency identification (RFID) technology could also be used to detect locations but it requires the deployment of external infrastructure. GPS = Global Positioning System.
[a]Expected spatial resolution in optimal conditions.
[b]These are the highest sampling frequencies that can be expected in current smartphones. Such high sampling frequencies could exhaust the battery and it is therefore not usually feasible to conduct a continuous high-frequency sampling for long periods of time.
[c]Assisted GPS (A-GPS) technology makes use of external network information to augment GPS functionality and especially its time to first fix. It is becoming a standard in smartphones.
[d]External information about the network and infrastructure is required to determine the device location. Most current smartphone devices and operating systems supply network positioning information in a straightforward manner. Moreover, network positioning is usually used as a default replacement in smartphones whenever GPS location is unavailable.
[e]The use of magnetometer, accelerometer, and gyroscope in combination might make it possible to determine the relative position of a device in indoor environments for short distances.

ways relative to the more traditional stand-alone devices.

1. The sensors allow both indoor and outdoor tracking options; dedicated tracking technologies such as GPS receivers can only track people outdoors.
2. The combination of location sensors and technologies engenders a compensation mechanism: One technology can assist or replace another. For example, when GPS location is not available, the phone can use Wi-Fi positioning; when a Wi-Fi signal cannot be found, cell ID location can be extracted.
3. Some of the sensors garner additional information about spatial activity. Accelerometers, for instance, allow researchers to record indoor movement parameters and assist in detecting mode of transportation (Miluzzo et al. 2008).

Recent Research Using Mobile Phone Sensing

The introduction of smartphones and tablet computers, with their cheap embedded sensors, gave rise to a new research paradigm in computer science that could be called mobile-phone sensing (Lane et al. 2010), human sensing (Shoval 2007), or people-centric sensing (Campbell et al. 2008). The main notion behind this paradigm is that one can use advanced mobile phones to sense and monitor mobile user activity and behavior. It is possible to collect information about the immediate environment of the users as well. What sparked the imagination of computer scientists regarding smartphones was not just the cheap embedded sensors and the computing and communication capabilities of these devices. Researchers also understood that the rapid penetration rates of smartphones in developed countries, and the fact that people willingly carry their phones everywhere, make these devices the ultimate apparatus for the development of new ubiquitous data collection practices (Cuff, Hansen, and Kang 2008; Raento, Oulasvirta, and Eagle 2009).

Three main types of research evolved simultaneously with smartphone development. The first type, and probably the least explored, is the analysis of large data sets that are generated by smartphones independent of research projects. Online social networks such as Twitter, Flickr, and Foursquare, which usually include location-based capabilities, are a major data source for these analyses (Cheng et al. 2011; Noulas et al. 2011; Roick and Heuser 2013). Most of the studies in this category explore mass behavior patterns of large populations (Noulas et al. 2011).

The second type of research includes pioneering or proof-of-concept studies that demonstrate the feasibility of using smartphones to collect information about human behavior (Xiao et al. 2012). In some of these studies the focus is on the utilization and analysis of the data generated by smartphones rather than on the data collection procedure itself.

Finally, computer science researchers have published several studies that describe the development and implementation of new mobile applications and systems that take advantage of smartphone capabilities—especially their sensors. In these studies, researchers usually use their applications in small pilot studies to test their systems (Froehlich et al. 2007; Miluzzo et al. 2008; Hicks et al. 2010).

These current research trends indicate that the use of smartphones in academic studies is still in its early stages. Next we present the primary data collection practices that use smartphones, developed in recent years. This is followed by a review of geographical and mobility studies that make use of such methods.

Using Smartphones in Human Behavior Studies

Participatory Versus Opportunistic Sensing

Mobile sensing includes two types of sensing approaches—the participatory approach and the opportunistic sensing approach (Lane et al. 2008). *Participatory sensing* requires the active participation of users. For example, users might be requested to send a report or take a photograph. Due to the active role of mobile users in participatory sensing, this sensing procedure is highly dependent on participants' enthusiasm. With *opportunistic sensing,* in contrast, the user is passive. Background processes that run on the phone can automatically record sensor data, such as GPS location and sounds; they can also document other types of information—for example, phone usage logs. The passive nature of opportunistic sensing reduces the burden on participants. Hence, one can assume that it will generate more systematic information and allow longer periods of sensing.

Creative implementation of passive sensing can supply valuable information about the behavior of mobile users. The CenceMe application (Miluzzo et al. 2008), for example, uses accelerometers to detect activity type (sitting, standing, walking, running) and Bluetooth sensors to identify nearby "friendly" phones. The application also makes use of the phone microphone to sense whether the user is having a conversation and whether he or she is at a loud party. On the other hand, participatory sensing can supply valuable information about subjective experiences and participants' motivations as well as additional information that cannot be collected through passive sensing.

Although Lane et al. (2008) argued that there is an advantage to using opportunistic sensing in research—especially when working with a large number of mobile users—there is no conclusive agreement about which is the superior approach. In fact, many of the studies that use smartphones combine both approaches (e.g., passive location recording with active reports that participants are asked to send). In this regard, it is important to match the sensing approach used to the objectives of the research.

Reports, Surveys, and the Experience Sampling Method

The ubiquity of smartphone devices allows researchers to communicate with their participants throughout the day while asking them to complete various surveys and send various feedback and reports—all in an ecological manner, in situ, and in real time. One application of such reporting is the gathering of information about the immediate environment of mobile users. For example, participants can be asked to self-report about what they see, hear, or smell in a specific moment (e.g., is the place crowded? noisy?). In this context, human senses can be thought of as an extension of the phone's sensing capabilities.

Self-reporting techniques could also be used to collect information about subjective feelings, thoughts, and moods of participants. The experience sampling method (ESM) is one such technique; it is commonly applied in studies that use smartphones to collect such information. In ESM—sometimes referred to as the ecological momentary assessment (EMA)—participants are asked to self-report about their behaviors, subjective experiences, and moods repeatedly as they occur in real time and in the daily environment

(Csikszentmihalyi and Larson 1987). In the past, researchers used various technologies, such as pagers and stopwatches, to signal participants to send their feedback. Advanced mobile devices such as smartphones allow researchers to generate a sophisticated set of triggers that prompt surveys based on various contextual parameters—the location of the phone or the activity of the mobile user, for example (Froehlich et al. 2007; Hicks et al. 2011).

Phone Logs

Smartphones make it possible for researchers to log both automatic background processes that run on the phone (e.g., detection of cell towers) and mobile user actions (e.g., use of applications). This type of information could be highly valuable in social science research. For example, logs can be used to evaluate social networks by analyzing the number of unique incoming and outgoing phone calls and nearby Bluetooth devices (Raento, Oulasvirta, and Eagle 2009). The use of these data is highly sensitive in terms of privacy, as we discuss in the final section of this article.

Smartphones and Geographical Research

Increasing numbers of researchers from disciplines other than computer sciences are beginning to exploit smartphones for their research. In most cases, researchers' efforts focus on the development and validation of new mobile data collection techniques. We are nearing the point, however, when smartphones will be adopted as legitimate data collection tools. The majority of smartphone studies published thus far utilized the devices' location technologies. Most of these studies, therefore, tend to be highly similar to geographical works on the subject of spatial behavior and mobility. In the field of health science, Doherty, Lemieux, and Canally (2014) explored the relationship between well-being and outdoor activity in natural environments. Similarly, the work of Palmer et al. (2013) is considered a demographic study. Their research, however, used an open-source Android application to explore the mobility of people in various places in the world; it can be easily classified as a mobility study.

In this section we review studies conducted by geographers and researchers in related disciplines in which some sort of mobile phone sensing technique was implemented. The number of these works is limited; as with studies from other disciplines, most of them are pioneering or pilot studies that demonstrate the feasibility of a suggested data collection technique.

Three noteworthy studies examined the feasibility of using smartphones as tracking devices. Wiehe et al. (2008) used Blackberry phones to track the spatial activity of adolescents. A similar assessment was conducted by Wan et al. (2013), who measured the spatial ranges of activity—or "life-spaces"—of older adults using smartphones. De Nazelle et al. (2013) examined the feasibility of using smartphones for measuring the exposure of residents in Barcelona to air pollution. In this last study, the researchers measured the activity of participants using the smartphones' location technologies and accelerometer; they then combined the information with external pollution measurements. In all of these studies, the researchers concluded that smartphones—although inferior in their GPS capabilities to dedicated devices (both in accuracy and sampling rate)—are very practical for collecting tempo-spatial information about human behavior and mobility.

Geographers have also used smartphones as self-reporting devices. To explore the subjective experiences of students during walks, Ettema and Smajic (2014) used smartphones that recorded *in situ* self-reported experiences of students. They found that the affect level of students was higher in lively places and in less busy places. Chen and Bishop (2011) developed an iPhone application that they used to track visitors in the Royal Botanic Garden Cranbourne in Australia. Aside from tracking and recording participants' locations, the application could also trigger surveys based on the phone's location and store photos taken by the visitors. They used the data in an agent-based model that predicted the movement patterns of visitors.

Another study that implemented geotagged smartphone surveys was that of MacKerron and Mourato (2013). This is one of the more impressive geographical studies in terms of sample size, with more than 20,000 participants throughout the United Kingdom continuously reporting about their happiness level using an iPhone application developed by MacKerron and Mourato. In addition to happiness level, participants were asked to report on other parameters of their behavior—the type of activity in which they were engaged and the people they were with, for example. The data indicated that people are significantly happier in natural environments.

Smartphones have also been used to enhance more qualitative studies. Jones, Drury, and McBeath (2011), for example, used a simple Windows Mobile application that stored the coordinates of photographs taken by participants. The photos were used as part of a study examining daily positive and negative experiences in the area of Birmingham, England.

Adding Contextual Information to Geographical and Mobility Studies

As we saw, smartphones are valuable in that they allow researchers to collect contextual information beyond the where and when in a more straightforward way (Froehlich et al. 2007). These data might include subjective reports of behavior as well as more objective variables about a social and physical environment.

Transportation and mobility studies are probably the most immediate applications to benefit from new mobile sensing technology, but many other geographical research fields can and should take advantage of these technological advances—geographies of health and well-being, spatial behavior and decision making, and urban geography. When it comes to research applications, the potential of smartphones is almost endless. There is still considerable work to be done, though, to turn this data collection approach into something that is both viable and reliable. We discuss this further in the concluding section of this article.

Theoretical and Methodological Reflections

Earlier, we reviewed data collection practices and smartphones' potential to contribute to the investigation of geographies of mobility. Smartphones provide several prominent advantages, such as accurate time–space data recording capabilities and the option for real-time data collection, allowing researchers to investigate the human agent's motivation (or "why" dimension). On the other hand, the use of smartphones for research is fraught with disadvantages.

The relatively short battery life of phones, at least at the time of writing this article, limits the use of the devices for continuous sensing purposes (Raento, Oulasvirta, and Eagle 2009; Lane et al. 2010). New selective sampling algorithms and new batteries recently introduced might be able to tackle this shortcoming, however.

Another drawback has to do with the validity of the samples; researchers who use smartphones in their studies and want to use a representative sample group might encounter difficulties in doing so. To date, not all of the population owns a smartphone, which might lead to sample biases, especially with regard to some specific segments, such as the elderly. This situation is expected to change in the near future, with the increase of smartphone penetration rates. Moreover, when the study involves a small sample group, this bias can be overcome by using adequate sampling schemes.

A third disadvantage that could hinder the smooth utilization of smartphones as a research tool is that the development of mobile applications requires knowledge in programming. Nevertheless, there has been progress in the development of several new generic sensing applications and systems that can be easily implemented for research purposes without any prior knowledge of computers.

Researchers also report inferior GPS performance in smartphone devices relative to dedicated GPS loggers; however, it seems that this quality gap is narrowing with the introduction of new smartphone models.

In addition, with the use of smartphones, old questions receive new context—the geoprivacy question, violation of data confidentiality, and human subject protection when personal and especially location data are collected or used without being regulated by strict protective procedures (e.g., the use of unprotected social media data). Questions of privacy have significant implications for research practice and methodology in geography and beyond (Burke et al. 2006; Shoval et al. 2014). These concerns are greatly reduced when participants are drafted directly by the researchers (i.e., when they must download a dedicated application or sign an informed consent form). In such cases, researchers are usually obligated to follow and implement the ethical procedures of their institution, which tend to protect the privacy of participants.

Finally, it is important to note that smartphones run on several different operating systems, most notably Android and iOS, which are frequently being upgraded. This requires that researchers who use their own sensing systems adjust their applications occasionally to meet the specifications of as many operating systems and mobile devices as possible.

Key Questions for Future Research

One open question that is relevant to all types of tracking studies is this: Do people, once they know they are being followed, change their activity? If so, how? These questions should be further explored in empirical studies. Even the more traditional methods for gathering information like time–space activities,

however, raise concerns as to whether reports are inaccurate (in questionnaires and time budgets) and whether people change their patterns of activity due to participation in a study, feeling that they must behave in a certain way. The advantage of smartphones is that, in contrast with GPS devices, people are used to carrying their phones on a daily basis regardless of the research intervention.

A further direction of inquiry that has yet to be explored is the integration of tempo-spatial information with additional digital sources of data. One interesting prospect in this regard is the employment of external sensors that are compatible with smartphones. These include biometric sensors that measure excitement and physical effort, as well as sensors that measure characteristics of the environment such as noise and pollution levels. Externally worn devices, such as heart rate sensors, have already been used successfully (Gaggioli et al. 2011). It is important to note, though, that the incorporation of worn sensors is usually logistically inconvenient and does not allow long-term, continuous sampling.

Another question remaining relates to the large amount of data that is accumulated when using these methods, which requires the development of more algorithms to enable automatic scripts to analyze the data in a fast and practical way. It would be beneficial if the software being developed today by different research teams and private companies around the globe on an ad hoc basis could be standardized at some point so that common measures for data analysis could be developed. More important, this could lead to an increase in the number of researchers in the field, because the current challenge of analyzing the data no doubt limits the number of prospective users of these methods.

Smartphones open exciting new horizons, which might lead to new research trends in mobility and geography. First, the growing availability of geotagged social information generated through smartphones might increase the number of studies that rely on volunteered geographic information. Second, because smartphones improve our ability to collect contextual information, researchers can now focus on new aspects of mobility such as subjective experiences and the effect of external environmental factors. Third, smartphones' communication technologies allow researchers to obtain live information that can be analyzed and presented automatically. Therefore, we expect to witness a growing number of applicable real-time mobility and transportation studies in the near future. Finally, indoor location technologies and sensors such as accelerometers allow researchers to investigate the activity of people in closed environments. The increased interest of the private sector in commercial applications of indoor navigation is very likely to further enhance this trend in the future.

A recent study by Shoval et al. (2014) indicates that geographers are not the central group to adopt such mobile sensing and tracking technologies. The study found that less than 13 percent of the total articles that made use of GPS and other tracking technologies were published in geographic journals, whereas transportation and health studies were the two largest fields, with more than half of all related publications. These findings were rather surprising; one would expect that geographers, who were among the most prominent in the development and application of geographic information systems and remote sensing, would also be leaders in the use of GPS and other tracking technologies. For some reason, however, the deployment of tracking technologies in geographic research appears to be neglected. It will be interesting to see whether and how geographers make use of advanced mobile phones in their research.

Funding

This project received funding from the European Union's Seventh Framework Programme for research, technological development, and demonstration under grant agreement no. 261652.

References

Ahas, R., A. Aasa, S. Silm, and M. Tiru. 2010. Daily rhythms of suburban commuters' movements in the Tallinn metropolitan area: Case study with mobile positioning data. *Transportation Research Part C: Emerging Technologies* 18 (1): 45–54.

Bohte, W., and K. Maat. 2009. Deriving and validating trip purposes and travel modes for multi-day GPS-based travel surveys: A large-scale application in The Netherlands. *Transportation Research Part C: Emerging Technologies* 17 (3): 285–97.

Burke, J. A., D. Estrin, M. Hansen, A. Parker, N. Ramanathan, S. Reddy, and M. B. Srivastava. 2006. *Participatory sensing*. Los Angeles: UCLA Center for Embedded Network Sensing.

Calabrese, F., and C. Ratti. 2006. Real time Rome. *Networks and Communication Studies* 20 (3–4): 247–58.

Campbell, A. T., S. B. Eisenman, N. D. Lane, E. Miluzzo, R. A. Peterson, H. Lu, X. Zheng, M. Musolesi, K. Fodor, and G.-S. Ahn. 2008. The rise of people-centric sensing. *Internet Computing, IEEE* 12 (4): 12–21.

Chen, Y., and I. D. Bishop. 2011. iSurv: Using iPhone to collect visitor data for behaviour simulation system. Paper presented at the conference on Computers in Urban Planning and Urban Management, Lake Louise, AB, Canada.

Cheng, Z., J. Caverlee, K. Lee, and D. Sui. 2011. Exploring millions of footprints in location sharing services. In *Proceedings of the Fifth International AAAI Conference on Weblogs and Social Media*, 81–88. Palo Alto, CA: AAAI Press.

Csikszentmihalyi, M., and R. Larson. 1987. Validity and reliability of the experience-sampling method. *The Journal of Nervous and Mental Disease* 175 (9): 526–36.

Cuff, D., M. Hansen, and J. Kang. 2008. Urban sensing: Out of the woods. *Communications of the ACM* 51 (3): 24–33.

De Nazelle, A., E. Seto, D. Donaire-Gonzalez, M. Mendez, J. Matamala, M. J. Nieuwenhuijsen, and M. Jerrett. 2013. Improving estimates of air pollution exposure through ubiquitous sensing technologies. *Environmental Pollution* 176:92–99.

Dijst, M. 2009. ICT and social networks: Towards a situational perspective on the interaction between corporeal and connected presence. In *The expanding sphere of travel behaviour research*, ed. R. Kitamura, T. Yoshii, and T. Yamamoto, 45–75. Bingley, UK: Emerald Group.

Doherty, S. T., C. J. Lemieux, and C. Canally. 2014. Tracking human activity and well-being in natural environments using wearable sensors and experience sampling. *Social Science & Medicine* 106:83–92.

Ettema, D., and I. Smajic. 2014. Walking, places and well-being. *The Geographical Journal* 181 (2): 102–09.

Froehlich, J., M. Y. Chen, S. Consolvo, B. Harrison, and J. A. Landay. 2007. MyExperience: A system for in situ tracing and capturing of user feedback on mobile phones. In *Proceedings of the 5th International Conference on Mobile Systems, Applications and Services—MobiSys '07*, 57–70. New York: ACM.

Gaggioli, A., G. Pioggia, G. Tartarisco, G. Baldus, D. Corda, P. Cipresso, and G. Riva. 2011. A mobile data collection platform for mental health research. *Personal and Ubiquitous Computing* 17 (2): 241–51.

Hicks, J., N. Ramanathan, H. Falaki, B. Longstaff, K. Parameswaran, M. Monibi, D. H. Kim, J. Selsky, J. Jenkins, H. Tangmu, and D. Estrin. 2011. *Ohmage: An open mobile system for activity and experience sampling*. Los Angeles: UCLA Center for Embedded Network Sensing.

Hicks, J., N. Ramanathan, D. Kim, M. Monibi, J. Selsky, M. Hansen, and D. Estrin. 2010. AndWellness: An open mobile system for activity and experience sampling. In *Proceedings of Wireless Health 2010 (WH '10)*, 34–43. New York: ACM.

Isaacson, M., N. Shoval, H.-W. Wahl, F. Oswald, and G. Auslander. 2014. Compliance and data quality in GPS-based studies. *Transportation*. Advance online publication.

Jones, P., R. Drury, and J. McBeath. 2011. Using GPS-enabled mobile computing to augment qualitative interviewing: Two case studies. *Field Methods* 23 (2): 173–87.

Lane, N., S. B. Eisenman, M. Musolesi, E. Miluzzo, and A. T. Campbell. 2008. Urban sensing systems: Opportunistic or participatory? In *Proceedings of the 9th Workshop on Mobile Computing Systems and Applications—HotMobile '08*, 11–16. New York: ACM.

Lane, N., E. Miluzzo, H. Lu, D. Peebles, T. Choudhury, and A. Campbell. 2010. A survey of mobile phone sensing. *IEEE Communications Magazine* 48 (9): 140–50.

MacKerron, G., and S. Mourato. 2013. Happiness is greater in natural environments. *Global Environmental Change* 23 (5): 992–1000.

Miluzzo, E., N. D. Lane, K. Fodor, R. Peterson, H. Lu, M. Musolesi, S. B. Eisenman, X. Zheng, and A. T. Campbell. 2008. Sensing meets mobile social networks: The design, implementation and evaluation of the CenceMe application. In *Proceedings of the 6th ACM Conference on Embedded Network Sensor Systems—SenSys '08*, 337–50. New York: ACM.

Nielsen. 2012. Smartphones account for half of all mobile phones, dominate new phone purchases in the US. http://www.nielsen.com/us/en/insights/news/2012/smartphones-account-for-half-of-all-mobile-phones-dominate-new-phone-purchases-in-the-us.html (last accessed 24 November 2014).

Noulas, A., S. Scellato, C. Mascolo, and M. Pontil. 2011. An empirical study of geographic user activity patterns in foursquare. In *Proceedings of the Fifth International AAAI Conference on Weblogs and Social Media*, 570–73. Palo Alto, CA: AAAI Press.

Palmer, J. R. B., T. J. Espenshade, F. Bartumeus, C. Y. Chung, N. E. Ozgencil, and K. Li. 2013. New approaches to human mobility: Using mobile phones for demographic research. *Demography* 50 (3): 1105–28.

Raento, M., A. Oulasvirta, and N. Eagle. 2009. Smartphones: An emerging tool for social scientists. *Sociological Methods & Research* 37 (3): 426–54.

Richardson, D. B., N. D. Volkow, M.-P. Kwan, R. M. Kaplan, M. F. Goodchild, and R. T. Croyle. 2013. Spatial turn in health research. *Science* 339 (6126): 1390–92.

Roick, O., and S. Heuser. 2013. Location based social networks—Definition, current state of the art and research agenda. *Transactions in GIS* 17 (5): 763–84.

Schuessler, N., and K. Axhausen. 2009. Processing raw data from global positioning systems without additional information. *Transportation Research Record: Journal of the Transportation Research Board* 2105:28–36.

Schwanen, T., and M.-P. Kwan. 2008. The Internet, mobile phone and space–time constraints. *Geoforum* 39 (3): 1362–77.

Shoval, N. 2007. Sensing human society. *Environment and Planning B: Planning and Design* 34 (2): 191–95.

Shoval, N., and M. Isaacson. 2006. Application of tracking technologies to the study of pedestrian spatial behavior. *The Professional Geographer* 58 (2): 172–83.

———. 2010. *Tourist mobility and advanced tracking technologies*. London and New York: Routledge.

Shoval, N., M.-P. Kwan, K. H. Reinau, and H. Harder. 2014. The shoemaker's son always goes barefoot: Implementations of GPS and other tracking technologies for geographic research. *Geoforum* 51:1–5.

Silm, S., and R. Ahas. 2014. Ethnic differences in activity spaces: A study of out-of-home nonemployment activities with mobile phone data. *Annals of the Association of American Geographers* 104 (3): 542–59.

Wan, N., W. Qu, J. Whittington, B. C. Witbrodt, M. P. Henderson, E. H. Goulding, A. K. Schenk, S. J. Bonasera, and G. Lin. 2013. Assessing smart phones for generating life-space indicators. *Environment and Planning B: Planning and Design* 40 (2): 350–61.

Wiehe, S. E., A. E. Carroll, G. C. Liu, K. L. Haberkorn, S. C. Hoch, J. S. Wilson, and J. D. Fortenberry. 2008. Using GPS-enabled cell phones to track the travel patterns of adolescents. *International Journal of Health Geographics* 7 (1): 22.

Wolf, J., R. Guensler, and W. Bachman. 2001. Elimination of the travel diary: Experiment to derive trip purpose from global positioning system travel data. *Transportation Research Record: Journal of the Transportation Research Board* 1768:125–34.

World Bank. 2014. World development indicators 2014. http://data.worldbank.org/products/wdi (last accessed 24 November 2014).

Xiao, Y., D. Low, T. Bandara, P. Pathak, H. B. Lim, D. Goyal, J. Santos, C. Cottrill, F. Pereira, C. Zegras, and M. Ben-Akiva. 2012. Transportation activity analysis using smartphones. In *Proceedings of Consumer Communications and Networking Conference (CCNC), 2012 IEEE*, 60–61. New York: IEEE.

Rosa Parks Redux: Racial Mobility Projects on the Journey to Work

Virginia Parks

Department of Urban and Environmental Policy, Occidental College

The iconic image of Rosa Parks sitting at the front of a bus documents the most famous commute in history. Rosa Parks was traveling home from work when she refused to give her seat to a white passenger in 1955, an act of civil disobedience that set the Montgomery bus boycott in motion and propelled civil rights onto the national stage. Sixty years later, cities in the putatively postracial era continue to generate profound racial inequalities. Drawing on Rosa Parks's defiant commute as a framing device, I situate the journey to work as a racial mobility project that extends from historic urban processes of racial discrimination, reveals lived experiences of intersectional inequality, and generates future racial disparities. I define commuting as a racial mobility project that organizes, redistributes, and mobilizes resources along racial lines in conjunction with the movement of bodies across space. This framework links the discourses of race and mobility, both of which highlight the dynamics of politics and power. By positioning the journey to work as a racial mobility project, this article seeks to resituate the commute for geographers—conceptually, empirically, and politically—at the nexus of geography, mobility, and the struggle for racial justice in the city.

罗莎. 帕克斯坐在公交车前半部的代表性意象, 记载了历史上最富盛名的通勤。1955 年, 罗莎. 帕克斯在下班回家的途中, 拒绝让位给一位白人乘客, 这是一个公民不服从的行动, 不仅引发了蒙哥马利公交车抵制行动, 更将公民权搬上了全国舞台。六十年后, 在推定为后种族时代中的城市, 仍然持续生产深刻的种族不平等。我运用罗莎. 帕克斯的违抗通勤作为框架的工具, 将上班旅程置放作为一个种族能动性之计画, 该计画延伸自种族歧视的历史性城市过程, 揭露多元交织的不平等经验, 并生产未来的种族差异。我将通勤定义为结合跨越空间的身体运动, 寻着种族的界限, 组织、再分配并动员资源的种族能动性计画。此般架构连结种族与能动性的论述, 而两者皆强调政治与权力的动态。本文透过将上班的旅程置放作为种族能动性计画, 企图为地理学者在概念上、经验上和政治上, 将通勤重新置于地理、能动性和争取城市中的种族平等的连结之中。

La imagen icónica de Rosa Parks sentada en la parte frontal de un bus documenta el más célebre viaje al trabajo de la historia. Rosa Parks estaba yendo a casa desde el trabajo cuando rehusó ceder su puesto a un pasajero blanco en 1955, un acto de desobediencia civil que desencadenó el sabotaje a los buses en Montgomery e impulsó el tema de los derechos civiles a la palestra nacional. Sesenta años después, las ciudades de lo que es putativamente considerado la era posracial siguen generando profundas desigualdades raciales. A partir del desafiante viaje de trabajo de Rosa Parks, a título de artilugio ilustrativo, sitúo el viaje al trabajo como un proyecto de movilidad racial que se desprende de procesos urbanos históricos de discriminación racial, revela vívidas experiencias de desigualdad interseccional y genera futuras disparidades raciales. Defino el viaje al trabajo como un proyecto de movilidad racial que organiza, redistribuye y moviliza recursos a lo largo de líneas raciales en conjunción con el movimiento de cuerpos a través del espacio. Este marco enlaza los discursos de raza y movilidad, los cuales destacan la dinámica de la política y el poder. Al posicionar el viaje al trabajo como un proyecto de movilidad social, este artículo busca resituar para los geógrafos el viaje al trabajo—conceptual, empírica y políticamente—en la conexión de la geografía, la movilidad y la lucha por justicia racial en la ciudad.

The iconic image of Rosa Parks sitting at the front of a bus documents the most famous commute in history. Rosa Parks, a seamstress, was traveling home from work on 1 December 1955, when she refused to give her seat to a white passenger, an act of civil disobedience that set the Montgomery bus boycott in motion and propelled civil rights onto the national stage. Her refusal crystallized the insidious nature of segregation in the South and laid bare its brutal banality. Sixty years later, cities in the putatively postracial era continue to generate profound racial inequalities, and commuting continues to embody, reveal, and sometimes contest the twenty-first-century city as a generator of racial inequality.

Drawing on Rosa Parks's defiant commute as a framing device, I situate the journey to work as a racial mobility project that extends from historic urban processes of racial discrimination, reveals lived experiences of intersectional inequality, and generates future racial disparities. The journey to work sits at the nexus of multiple spatial and temporal processes, providing a window onto past, present, and future trajectories of mobility. Yet, as a racial project, the journey to work is fundamentally political. Urban political economies emphatically shape commutes, even as actors leverage the commute as a political site of engagement to reveal, contest, and remake these urban political economies. By positioning the journey to work as a racial mobility project, I seek to reinvigorate the commute as a strategic research site for investigations of urban inequality and contentious politics.

In this article, I sketch the conceptual, empirical, and political dimensions of the commute that place it at the nexus of geography, mobility, and racial justice in the city. Geographers have long engaged questions of urban mobility and commuting. The brief nature of this article disallows a full accounting of this extensive body of geographic literature, although I offer signposts to direct readers' exploration. Instead, my primary aim is to highlight the distinctly racial experience of commuting in U.S. cities throughout most of the twentieth century that continues today and its significance as a site of political contestation. Tied to a white supremacist racial order (King and Smith 2005) and the logic of the capitalist city (Harvey 1989; Farmer and Noonan 2014), commuting is a sociospatial and political–economic phenomenon I describe, first and foremost, as a racial mobility project.

Commuting as a Racial Mobility Project

Racial disparities in commuting emerge first and foremost from the racialization of urban space in the capitalist city. Under the racial urban order of the twentieth century, black labor was fixed in its reproductive space (home) while simultaneously mobilized for capitalist production (work). The two spaces of home and work historically conjoined only under the most exploitive conditions: the plantation. Throughout most of the twentieth century, the multifaceted practice of racial residential segregation defined the dominant racial order of urban space (Massey and Denton 1993). Yet although African Americans were restricted in their choice of neighborhoods, they were employed in jobs across the space economy (Ellis, Wright, and Parks 2004). The storied spatial mismatch between the spheres of segregated neighborhoods and the economic

demand for black workers has generated a peculiar and pronounced experience of commuting among African Americans that differs strikingly from that of other racial and ethnic groups in a variety of ways (e.g., speed, duration, mode, safety, ease, dignity).

I argue that the experience of commuting in the twentieth and early twenty-first century in U.S. cities is best understood as a racial mobility project. This theoretical framework builds directly on Omi and Winant's (1994) explanation of the role that "racial projects" play in the process of racial formation. A racial project is "simultaneously an interpretation, representation, or explanation of racial dynamics, and an effort to reorganize and redistribute resources along particular racial lines" (Omi and Winant 1994, 56). This approach emphasizes the symbolic and representational work of racial projects as well as their material consequences. A racial mobility project organizes, redistributes, and mobilizes resources along racial lines in conjunction with the movement of bodies across space. This framework links the discourses of race and mobility, both of which highlight the dynamics of politics and power.

Racial projects are political contests that produce, mediate, and disrupt racial inequality. In their work on the role of race in U.S. political development, King and Smith (2005) conceived of *racial institutional orders*, "in which political actors have adopted (and often adapted) racial concepts, commitments and aims in order to help bind together their coalitions and structure governing institutions that express and serve the interests of their architects" (75). Notably, King and Smith focused on the interactional dynamic of two primary racial institutional orders in the United States: a white supremacist order and an egalitarian transformative order. Racial projects, then, involve multiple participants and can be driven by perpetrators of racial injustices as well as by those who struggle against these injustices. Racial projects are neither static nor totalizing; instead, they are constantly in flux, in motion, and mobile.

Commuting sits at the nexus of three primary urban racial projects: labor markets, housing, and transportation. I discuss each briefly in turn, before focusing on the distinct aspects of commuting as a racial mobility project. Racial residential segregation stands out as the most visible and well-documented urban racial project, literally reorganizing urban space across the United States by constraining the housing options of African

Americans (Massey and Denton 1993). Through the allocation and redistribution of housing resources along racial lines, the practice of racial residential segregation mediated the availability, flow, and quality of a host of other territorial resources attached to the neighborhood, such as schools, crime, health, recreation, food, pollution, and jobs.

Labor markets also function as racial projects, distributing employment opportunities and outcomes along racial lines (Parks 2012). Racial disparities in employment abound: African Americans experience higher rates of unemployment than other racial and ethnic groups (Western and Beckett 1999), the black–white wage gap persists (Western and Pettit 2005), and multiple experiences of labor market precarity—low-wage work, unpredictable scheduling, insufficient hours—affect black workers disproportionately (Luce and Weinbaum 2008). Labor market processes overlap and embed with patterns of residential segregation, amplifying the spatial impact of residential segregation practices just described (Ellis, Wright, and Parks 2004).

Urban transportation systems enable the movement of workers between the two spheres of home and work yet contribute unique racial disparities. Transportation resources, including mode type, service levels, capital expenditures, fare subsidies, and even safety, have exhibited sharp racial disparities across U.S. cities and continue to do so (Bullard and Johnson 1997; Farmer and Noonan 2014). As Hodge (1990) pithily stated, "Embedded in urban transportation systems are relations that are intimately and inescapably tied to the most fundamental, and often contentious, elements of cities and societies" (97). The fundamental and contentious effects of race in the United States translate into racially unequal transportation systems, a phenomenon Bullard (2004) called *transportation racism*.

Commuting emerges as a discrete interaction of the racial and spatial dynamics of labor markets, housing, and transportation systems. I more narrowly define commuting as a racial mobility project to emphasize the organization, redistribution, and movement of resources along racial lines that is directly connected to the movement of bodies across space. I build on other geographers who have made similar insights, particularly with regard to the concept of mobility as movement embedded within power structures (Cresswell 2006; Alderman, Kingsbury, and Dwyer 2013; Nagel et al. 2015). Critically, commuting as a racial mobility project centers on the movement of bodies, particularly bodies of color, through racially defined

space. Analytically, the racial project framework challenges us to explicate the racial politics and power structure that informs, regulates, and sometimes contests the movement of racially defined bodies along the journey to work.

I focus on three aspects of commuting as a racial mobility project: lived experience, economic process, and site of political transgression and mobilization. First, commuting constitutes a first-order experience of the movement of a body over space between an origin (home or work) and a destination (work or home). At the urban scale, commuting illustrates different patterns of geographic mobility among urban workers, produced by and constitutive of sociospatial processes that link home and work. At the scale of the body, commuting comprises a lived experience of movement and travel. The physical, social, and psychological conditions of this movement and travel are integral to understanding the immediate and visceral effects of racial inequality.

These conditions have long formed the basis of demands by people of color for fair treatment, racial dignity, and equality. The physical and verbal abuse, sometimes fatal, exacted by white bus drivers and passengers against African American bus riders in the Jim Crow South exemplify the racially unjust conditions of travel experienced historically by black commuters on the journey to work (Williams 2006; McGuire 2010). Other conditions include, but are not limited to, safety, comfort, ease, accommodation, sanitation, speed, duration, and frequency. Is a commuter free from the threat of physical violence or harassment while traveling or waiting to board a bus or train? Does a commuter have access to adequate space for his or her body and possessions while traveling? Is the mode of travel comfortably climate-controlled? Does the bus stop or train platform provide protection against the elements? Is the bus or train safely operated and in sound condition? Are the roads and rails properly maintained? These conditions of travel emerge from and reflect the racial politics that structure the larger transportation system, such as lower levels of infrastructure investment and service in neighborhoods of color (Bullard and Johnson 1997).

Second, commutes manifest as indicators of economic accessibility and mobility. Geographic accessibility to employment opportunities is a key determinant of labor market outcomes as well as future economic mobility. The relative ability of workers to physically travel from their place of residence to a place of work is illustrated through aggregate commute times. When workers are constrained in their housing options, as African Americans were through discriminatory housing markets, commute times reflect conditions of constraint rather than of choice. The preponderance of empirical research testing the spatial mismatch hypothesis shows a significant correlation between lower levels of spatial accessibility to jobs, higher rates of unemployment, and longer commutes among African American workers throughout much of the twentieth century (e.g., Kain 1968; Holzer 1991; Mouw 2000).

Longer commutes are costly in both time and money, especially for lower paid workers. Longer commutes are not always an indicator of relative economic hardship, however. Commutes are contextual—for example, highly paid male commuters have the luxury of trading off the costs of a longer commute for other amenities, such as housing. On average, women have shorter commutes than men and suffer economic consequences as a result. Dubbed the spatial entrapment hypothesis by feminist geographers, women work closer to home to accommodate their dual responsibilities at home and at work (Hanson and Pratt 1988). Shorter commutes, a function of more spatially delimited job search areas, partly explain occupational sex segregation, the gender pay gap, and underemployment (Madden and Chiu 1990; Hanson and Pratt 1991).

Women of color are particularly disadvantaged. Several geographic studies have found that Latinas and African American women experience longer commutes than their white counterparts (Johnston-Anumonwo 1995; McLafferty and Preston 1997). African American women experience the longest commutes, reflecting their heavy reliance on public transit and their poor spatial accessibility to jobs (McLafferty and Preston 1992; Sultana 2003; Parks 2004). Parks (2014) found that among working poor women in Chicago in 2011, African American women experienced the longest commutes. These women commuted eight minutes longer to their low-wage jobs than their white female counterparts, a total of eighty minutes more per week.

The effect of the accumulation of these disparities over the life course is nontrivial, generating negative economic returns not only for the worker but for her children as well. New evidence demonstrates the relationship between geographic accessibility and intergenerational economic mobility. Chetty et al. (2014) found that children from lower income families were more likely to rise into the ranks of the middle class in cities with lower levels of residential segregation and greater levels of spatial accessibility to employment. Commuting as a contemporary mobility project generates future intergenerational inequalities, bestowing a huge multiplier effect onto each minute captured in the daily commute disparity.

Third, commuting serves as a site of political contestation and mobilization. The mobility of African Americans en route to work stems from processes of racial subordination and directs these discriminatory practices in space. Black bodies on the move literally become a target of subjugation. They are also agents of resistance. When commuters such as Rosa Parks defy racial subjugation on their journey to work, they mobilize political opposition through their immediate mobility for the purposes of an egalitarian agenda. In the next section, I focus on the refusal of Rosa Parks, elucidating how the spatiality of commuting shapes the contentious politics of racial resistance (Leitner, Sheppard, and Sziarto 2008). Parks's refusal is one node within a more extensive web of political oppression and resistance of African Americans in relation to transportation, movement, and mobility (Alderman, Kingsbury, and Dwyer 2013). Yet her action bears reexamination because of its historical familiarity, its strategic connection to the Montgomery bus boycott, and its direct relevance to ongoing racially disparate experiences of commuting.

Commuting as Political Contestation

The contradictions of racial residential segregation and the capitalist imperative of production render commuting inherently transgressive as a spatial practice: Black workers need to move out of their home spaces to access their work spaces, crossing racially divided urban space en route. Jim Crow laws in Southern cities belied Southern whites' recognition of this inherent paradox. Thus, segregation practices extended from home to work and on the route between. No space was left unregulated or unpatrolled, an insidiousness exemplified by segregated buses. Unwilling to pay the full costs of segregation, bus companies in the South did not operate "separate but equal" buses. Instead, they offered separate and unequal bus service. The demand for segregation was met by segregating space within the buses themselves, designating separate seating areas for whites and blacks.

City ordinances dictated bus segregation, but drivers were given wide discretion in interpreting and enforcing the law. Rosa Parks, for example, was not sitting at the front of the bus when the driver ordered her to give up her seat. Parks was sitting in a middle row, an area in which blacks were allowed to sit, but they were expected to relinquish seats for white riders. These rows represented the flux and mutability of the boundaries of segregation—boundaries that, because of their instability, focused and mobilized activities of defense and contestation. On 1 December 1955, these boundaries were further unsettled by contextual factors at the scale of the bus itself—it was full. All available seats were occupied, in both the black and white sections of the bus. Blacks in Montgomery had won an earlier provision to the city's segregation code stipulating that they did not have to give up their seats if the bus was full—a law Parks knew well. Yet the driver, acting as a frontline defender of segregation, exercised his discretionary powers to make Parks move, ultimately calling on the police to arrest her (Theoharis 2013). As liminal space, Parks's middle-row seat provided a purchase for resistance and skirmish ground in the fight against segregation.

The significance of Parks's act of resistance stems not only from its location on the bus but also its quotidian timing. Parks refused to stand during her commute home. For black workers such as Rosa Parks, the commute was the spatiotemporal link between the racialized spaces of home and work. Fundamentally, the commute signified access to economic opportunity. Most jobs available to blacks in the Jim Crow economy were of poor quality, but they provided a means of economic support and, for some black workers, upward mobility. Thus, the economic necessity of commuting raised its political significance and sharpened the economic paradox of segregation. The South's racialized political economy depended on black labor, yet the burden of accommodating the irrationality of segregation was foisted on black workers through long, costly commutes under servile conditions. Parks herself emphasized this paradox. Theoharis (2013) wrote:

> Parks's frustration came also from how she was expected to act at work, tailoring clothes in the men's department at Montgomery Fair, and how she was treated in public life. "You spend your whole lifetime in your occupation, actually making life clever, easy and convenient for white people. But when you have to get transportation home, you are denied an equal accommodation. Our existence was for the white man's comfort and well-being; we had to accept being deprived of just being human." (65)

The timing of Parks's refusal was also significant in collective terms, both historically and contemporaneously. Parks's act of defiance was not unique; it extended from a tradition of black protest on mass transit (Alderman, Kingsbury, and Dwyer 2013). Black passengers in the South, women in particular, used a range of tactics to resist segregation and incivility on the bus, including sitting in white seats and arguing with drivers and white passengers (Washington 1995; R. D. G. Kelley 1996; B. Kelley 2010). Further, Parks's refusal was rooted in her own political activism, especially her concurrent efforts through the National Association for the Advancement of Colored People to organize a campaign around the arrest of Claudette Colvin, the teenager arrested eight months earlier for also refusing to give up her seat on a Montgomery bus. Although Parks's act was spontaneous, the groundwork had been laid through months, even years, of collective action to take up her individual case as a cause for mobilization in the struggle for civil rights (Theoharis 2013).

The Montgomery bus boycott of 1955 and 1956 that followed Parks's arrest did more than pry open a space of contestation in the fight against Jim Crow: As Alderman, Kingsbury, and Dwyer (2013) aptly described, the boycott created "Montgomery's alternative transportation system" (176). This black-led, alternative system of movement was an intricate collective endeavor, mobilizing social networks and resources within the African American community (most notably developing a free carpool system) and, in turn, fostering solidarity (Alderman, Kingsbury, and Dwyer 2013). As a black-led movement combining dissent and visionary claims-making (calls to end segregation were accompanied by demands to improve service and hire more black drivers), the boycott was an emancipatory racial mobility project: It redistributed black bodies through space as the fundamental resource of justice. By moving African Americans out of the buses and away from spaces of oppression, it repositioned them onto the streets where they reoccupied urban space in an expression of their social power. African Americans denied the bus companies their bodies and their fares; they denied Jim Crow their bodies and their acquiescence.

Parks's site of protest—on the bus, in the middle seats, on her way home from work—was located at an interstitial space in the larger terrain of segregation, revealing the transgression inherent in any form of movement that traverses racialized space. In the case of the Jim Crow South, capitalism propelled black bodies into motion, rendering the project of segregation ad hoc and incomplete. Although movement of African Americans largely represented experiences of constraint rather than freedom, movement opened spaces of contestation (Leitner, Sheppard, and Sziarto 2008). The African American experience of rebellion and protest is replete with acts and symbols of movement: the Underground Railroad, the Great Migration, the Montgomery bus boycott, the Freedom Rides. Yet Rosa Parks's act captures the subversion embedded within daily acts of movement, even those as seemingly routine and mundane as the journey to work.

Commuting equity is once again on the urban agenda, framed as transit justice by residents and activists in urban communities of color. Transit justice functions as a collective action frame that negotiates a shared understanding of the problem of transportation inequities, assigns blame, imagines an alternative solution, and urges people to act (Benford and Snow 2000). Transit justice frames extend from an analysis of racial justice that explicitly links class and racial oppression, a grassroots mobilization of intersectionality that locates the plight of poor and working-class communities of color at the nexus of runaway economic inequality and austerity politics.

The Bus Riders' Union/Sindicato de Pasejeros (BRU) in Los Angeles provides a robust example of political mobilization around transit justice, in both the organization's framing activities as well as its organizing tactics. The BRU identifies the redistribution of transit resources away from the system's poorest riders of color to its wealthiest, mostly white, riders as the key pattern of transportation inequality (Mann 2001). Bus riders in Los Angeles, as in most cities, comprise the poorest transit riders who receive the lowest fare subsidies relative to riders on other modes, especially commuter rail. In the early 1990s, 57 percent of all bus riders on the Metropolitan Transportation Authority (MTA) system earned less than $15,000 annually, compared to 20 percent of all county residents. Riders on commuter rail, by contrast, earned $65,000 (Garrett and Taylor 1999). The bus system received the lowest levels of capital investment, resulting in overcrowding on poorly maintained, older buses. Spatial inequalities abound: The most decrepit buses, often gross emitters, have tended to service the heaviest routes that travel through Los Angeles's lowest income neighborhoods, concentrating the environmental externalities of the geographic home–work divide on the residential side of the equation.

The BRU, founded on socialist principles, identifies class inequality as the basis of transportation inequality. In the contemporary United States, class inequality imbricates with racial inequality, widening the frame for the BRU as a strongly class-inflected racial justice approach. In its call to combat "transit racism," the BRU draws together the economic, environmental, and racial dimensions of transportation inequality (Mann 2001). This framing gave rise to an innovative tactic: In 1994, the BRU sued the Los Angeles MTA for operating a racially discriminatory separate but unequal transit system in violation of Title VI of the 1964 Civil Rights Act. Unable to sue on the basis of class discrimination, civil rights law nonetheless provided an avenue of redress for Los Angeles's poorest transit riders because they were largely people of color. The suit resulted in a consent decree that reinstated the unlimited ride monthly bus and rail pass and guaranteed its existence for ten years and created the first national standard to restrict overcrowding on buses. Later victories included committing the MTA to expand its clean-fuel fleet and phase out diesel buses (Mann 2001).

The BRU builds a vast membership base of bus riders by organizing them where they directly engage the transit system: on the buses. BRU organizers board buses daily, reaching out to bus riders and enlisting them directly in the fight for transit justice (Pulido, Barraclough, and Cheng 2012). These on-the-bus organizers claim bus space as a site of politics and protest. In 1997, tens of thousands of bus riders refused to pay their fares as part of the BRU's "No Seat No Fare" campaign (Mann 2001). Today the BRU has helped build a coalition of locally based community organizations called Transit Riders for Public Transportation to fight for racial equity in transit at the federal level (The Labor/Community Strategy Center n.d.). These organizations, ranging from the Atlanta Transit Riders Union to Communities United for Transportation Equity in New York, continue the fight for civil rights, like Rosa Parks did nearly a half-century ago, on the bus.

Conclusion

1 December 2015 marks the sixtieth anniversary of Rosa Parks's defiant act. Her resistance reveals the possibility of an emancipatory racial mobility project in the urban United States. Racial disparities in commuting are not simply a technical problem to be fixed but

a political project that engages larger questions about justice and the urban experience. Commuting as a racial mobility project connects multiple racial projects in the city—in housing, labor markets, urban transportation systems—and reveals the varied ways in which mobility through the city confers advantage and disadvantage. Empirical analysis remains necessary to focusing and sustaining attention on these racial patterns of inequality. Out on the streets, the case of Rosa Parks illustrates how mobility confers subjective power through which political claims can be mobilized, a lesson exercised daily on the buses of Los Angeles by the BRU and by other transportation activists fighting for racial justice on the journey to work.

References

Alderman, D. H., P. Kingsbury, and O. J. Dwyer. 2013. Reexamining the Montgomery bus boycott: Toward an empathetic pedagogy of the Civil Rights Movement. *The Professional Geographer* 65:171–86.

Benford, R. D., and D. A. Snow. 2000. Framing processes and social movements: An overview and assessment. *Annual Review of Sociology* 26:611–39.

Bullard, R. 2004. The anatomy of transportation racism. In *Highway robbery: Transportation racism & new routes to equity*, ed. R. Bullard, G. Johnson, and A. Torres, 15–31. Cambridge, MA: South End Press.

Bullard, R., and G. S. Johnson, eds. 1997. *Just transportation: Dismantling race and class barriers to mobility*. Stony Creek, CT: New Society.

Chetty, R., N. Hendren, P. Kline, and E. Saez. 2014. Where is the land of opportunity? The geography of intergenerational mobility in the United States. http://obs.rc.fas.harvard.edu/chetty/mobility_geo.pdf (last accessed 31 August 2015).

Cresswell, T. 2006. *On the move: Mobility in the western world*. London and New York: Routledge.

Ellis, M., R. Wright, and V. Parks. 2004. Work together, live apart? Geographies of racial and ethnic segregation at home and at work. *Annals of the Association of American Geographers* 94:620–37.

Farmer, S., and S. Noonan. 2014. The contradictions of capital and mass transit: Chicago, USA. *Science and Society* 78:61–87.

Garrett, M., and B. Taylor. 1999. Reconsidering social equity in public transit. *Berkeley Planning Journal* 13:6–27.

Hanson, S., and G. Pratt. 1988. Spatial dimensions of the gender division of labor in the local labor market. *Urban Studies* 9:180–202.

———. 1991. Job search and the occupational segregation of women. *Annals of the Association of American Geographers* 81:229–53.

Harvey, D. 1989. *The urban experience*. Oxford, UK: Blackwell.

Hodge, D. C. 1990. Geography and the political economy of urban transportation. *Urban Geography* 11:87–100.

Holzer, H. J. 1991. The spatial mismatch hypothesis: What has the evidence shown? *Urban Studies* 28:105–22.

Johnston-Anumonwo, I. 1995. Racial differences in the commuting behavior of women in Buffalo, 1980–1990. *Urban Geography* 16:23–45.

Kain, J. F. 1968. Housing segregation, Negro employment, and metropolitan decentralization. *Quarterly Journal of Economics* 82:175–97.

Kelley, B. 2010. *Right to ride: Streetcar boycotts and African American citizenship in the era of Plessy v. Ferguson*. Chapel Hill: University of North Carolina Press.

Kelley, R. D. G. 1996. *Race rebels: Culture, politics, and the black working class*. New York: Free Press.

King, D. S., and R. M. Smith. 2005. Racial orders in American political development. *American Political Science Review* 99:75–92.

The Labor/Community Strategy Center. n.d. Transit Riders for Public Transportation. http://www.thestrategycenter.org/project/transit-riders-public-transportation (last accessed 31 August 2015).

Leitner, H., E. Sheppard, and K. M. Sziarto. 2008. The spatialities of contentious politics. *Transactions of the Institute of British Geographers* 33:157–72.

Luce, S., and E. Weinbaum. 2008. Low-wage women workers: A profile. *New Labor Forum* 17:20–31.

Madden, J. F., and L. C. Chiu. 1990. The wage effects of residential location and commuting constraints on employed married women. *Urban Studies* 27:353–69.

Mann, E. 2001. "A race struggle, a class struggle, a women's struggle all at once": Organizing on the buses of L.A. *Socialist Register* 37:259–73.

Massey, D. S., and N. A. Denton. 1993. *American apartheid*. Cambridge, MA: Harvard University Press.

McGuire, D. L. 2010. *At the dark end of the street: Black women, rape, and resistance—A new history of the civil rights movement from Rosa Parks to the rise of black power*. New York: Knopf.

McLafferty, S., and V. Preston. 1992. Spatial mismatch and labor market segmentation for African-American and Latina women. *Economic Geography* 68:277–88.

———. 1997. Gender, race, and the determinants of commuting: New York in 1990. *Urban Geography* 18:192–212.

Mouw, T. 2000. Job relocation and the racial gap in unemployment in Detroit and Chicago, 1980 to 1990. *American Sociological Review* 65:730–53.

Nagel, C., J. Inwood, D. Alderman, U. Aggarwal, C. Bolton, S. Holloway, R. Wright, et al. 2015. The legacies of the U.S. Civil Rights Act, fifty years on. *Political Geography* 48:159–68.

Omi, M., and H. Winant. 1994. *Racial formation in the United States: From the 1960s to the 1990s*. London and New York: Routledge.

Parks, V. 2004. Access to work: The effects of spatial and social accessibility on unemployment for native-born black and immigrant women in Los Angeles. *Economic Geography* 80:141–72.

———. 2012. The uneven geography of racial and ethnic wage inequality: Specifying local labor market effects. *Annals of the Association of American Geographers* 102:700–25.

———. 2014. Density for all: Linking urban form to social equity. Paper presented at the annual meeting of the American Association of the Advancement of Science, Chicago.

Pulido, L., L. Barraclough, and W. Cheng. 2012. *A people's guide to Los Angeles*. Oakland: University of California Press.

Sultana, S. 2003. Commuting constraints of black female workers in Atlanta: An examination of the spatial mismatch hypothesis in married-couple, dual-earner households. *Southeastern Geographer* 43:249–59.

Theoharis, J. 2013. *The rebellious life of Mrs. Rosa Parks*. Boston: Beacon Press.

Washington, R. 1995. You don't have to ride Jim Crow. http://www.robinwashington.com/jimcrow/home.html (last accessed 31 August 2015).

Western, B., and K. Beckett. 1999. How unregulated is the U.S. labor market? The penal system as a labor market institution. *American Journal of Sociology* 104:1030–60.

Western, B., and B. Pettit. 2005. Black–white wage inequality, employment rates, and incarceration. *American Journal of Sociology* 111:553–78.

Williams, D., and W. Greenshaw. 2006. *The thunder of angels: The Montgomery bus boycott and the people who broke the back of Jim Crow*. Chicago: Lawrence Hill.

Revisiting Gender, Race, and Commuting in New York

Valerie Preston* and Sara McLafferty[†]

*Department of Geography, York University
[†]Department of Geography and Geographic Information Science, University of Illinois at Urbana–Champaign

In the 1990s, many women commuted shorter distances and less time than men, and research underscored the pernicious effects of racial and ethnic segregation and access to transportation on minority women's commuting. Since then, growing income inequality and the bifurcation of employment between well-paid and secure jobs and a growing number of insecure and poorly paid jobs have been accompanied by the concentration of jobs at central and suburban locations and the transformation of women's roles in the labor market. We investigate some of the geographical implications of these trends by analyzing commuting in the New York metropolitan region. In 2010, gender and race differences in commuting varied across the metropolitan area. Regression analysis demonstrates that the impacts of wages and household composition on commuting differ between the highly valued center that has benefited from private and public investment, the suburbs where traditional gender roles persist, and the deteriorating inner ring where minority women still commute long times on slow public transit. The findings highlight racial and gender disparities in geographical access to employment within the metropolitan region.

1990 年代时，许多女性的通勤路程和时间皆较男性为短，而研究则强调阶级与族裔隔离及运输管道对少数族裔女性通勤的有害影响。此后，逐渐增大的所得差距，以及高薪且稳固的工作和数量日渐增加的不稳定且低薪工作之间的分歧，伴随着工作集中于市中心与城郊的地点，以及女性在劳动市场中的角色转变。我们透过分析纽约大都会区域的通勤，探讨这些趋势的若干地理意涵。2010 年时，通勤中的性别与种族差异，在大都会各地有所不同。迴归分析证实，薪资与家户组成对于通勤的影响，在受惠于私人及公共投资的高价市中心、传统性别角色仍然续存的城郊，以及少数族裔女性仍然耗费长时间在缓慢的公共运输上的衰败内环区之间具有差异。研究结果凸显出大都会区域中，就业地理管道中的种族与性别差异。

Durante los años 1990, muchas mujeres hacían viajes pendulares a distancias más cortas y de menos tiempo que los hombres, y la investigación enfatizó los efectos perniciosos de la segregación racial y étnica y acceso al transporte en el desplazamiento pendular de las mujeres de las minorías. Desde entonces, la creciente desigualdad en el ingreso y la bifurcación del empleo en trabajos seguros y bien remunerados y un creciente número de empleos inseguros y mal remunerados, han estado acompañados de una concentración de los puestos de trabajo en localidades centrales y suburbanas y la transformación de los papeles de las mujeres en el mercado laboral. Nosotros investigamos algunas de las implicaciones geográficas de estas tendencias analizando el viaje pendular en la región metropolitana de Nueva York. En el 2010, las diferencias de género y raza en el desplazamiento pendular variaron a través del área metropolitana. El análisis de regresión demuestra que los impactos de los salarios y la composición del hogar sobre el viaje pendular difieren entre las del centro altamente valorado, que se ha beneficiado de la inversión privada y pública, la de los suburbios donde persisten los papeles tradicionales de género, y el anillo más interior en deterioro donde las mujeres de las minorías todavía viajan por tiempo más largo en tránsito público lento. Los descubrimientos destacan disparidades raciales y de género en el acceso geográfico al empleo dentro de la región metropolitana.

I n the 1990s, many women commuted shorter distances and less time than men, and research underscored the pernicious effects of racial and ethnic segregation and access to transportation on minority women's commuting. Since then, income inequality and the bifurcation of employment between "good" and "bad" jobs have worsened. An increasing proportion of women are the principal breadwinners in their households (W. Wang, Parker, and Taylor 2013), compelled to seek remunerative rather than convenient jobs. Gentrification and suburbanization have continued, concentrating jobs and residences near the city center and in distant suburbs. We investigate the geographical implications of these trends for gendered and racialized differences in commuting by analyzing commuting in the New York metropolitan region in 2010. The analysis compares travel times among three distinctive urban zones—the center, inner ring, and suburbs—and examines the impacts of wages, household type, and mode of transportation on commuting time in each urban zone. The findings highlight the advantages of central residential locations where good jobs are readily accessible by rapid transit for white and Asian men and women and underscore the lengthy work trips that persist for blacks living in the New York metropolitan area.

What Do We Know?

In 2000, women in the United States still commuted shorter distances and times than men, although the gender gap was diminishing (Rosenbloom 2006). Women's short commuting distances, like so many forms of mobility (Cresswell 2010), have multiple interpretations (Hanson 2010). For women who have remunerative jobs near their residential locations, short work trips are beneficial. For others, short commutes are a sign of spatial entrapment that forces women to settle for nearby jobs that might pay less or offer fewer opportunities for advancement than jobs available farther away. Gender differences in commuting have often been attributed to women's efforts to accommodate competing responsibilities for employment and family by finding jobs near home (Preston and McLafferty 1999; Hanson 2010). This explanation has always been partial. It does not account for the long commutes of well-educated women working in professional occupations (Hanson and Pratt 1995) and it has little relevance for many minority women. Housework and child care responsibilities have little impact on the work trips of minority women (Preston, McLafferty, and Hamilton 1993; Joassart-Marcelli 2009) and their travel time is also not compensated by high wages. In the 1990s, research uncovered little or no gender gap in commuting for minority workers, particularly blacks (McLafferty and Preston 1991, 1992; Mauch and Taylor 1997; Crane 2007; Johnston-Anumonwo 2014). Racial and ethnic residential segregation and limited access to transportation strongly affected minority women's labor market outcomes and commuting (Parks 2004; Q. Wang 2010).

We also know that the effects of race and gender vary depending on the relative locations of places of residence and workplaces. For example, after analyzing commuting in several U.S. cities, Johnston-Anumonwo (2014) concluded that race did not affect the travel times of women commuting to central work locations, but black women who reverse commuted to suburban workplaces traveled farther than white women. Similarly, a study of gender and race disparities in commuting in Atlanta observed differences between workers living in central and suburban locations (Sultana 2005).

Since the 1990s, the residential and employment geographies of many U.S. cities have changed. The spatial distribution of jobs has grown more uneven with concentrations in central locations along with the rise of suburban employment nodes (Frey 2012). Highly valued central locations, well served by transit, have attracted well-educated and affluent professional men and women. Metropolitan residents have also moved to affordable housing in distant suburbs, accepting long commutes as the price of homeownership. In this shifting geographical context, residential income segregation has increased steadily (Taylor and Fry 2012). Low-income households live near other low-income households and high-income households also concentrate. Racial segregation has also persisted (Logan and Stults 2011) with minority residents, particularly blacks, concentrated in less desirable residential areas, near relatively few jobs.

These trends are readily apparent in the New York metropolitan area. Between 2000 and 2009, employment increased in two zones: the central core, specifically, Manhattan and adjacent districts of Brooklyn and Queens, and suburbs on the periphery, such as Suffolk and Orange counties (Moss, Qing, and Kaufman 2012). At the same time, the New Jersey counties adjacent to Manhattan lost jobs. The relocation of employment has exacerbated residential income segregation. In 2010, income segregation in the New York metropolitan area was fifth-highest among the thirty largest metropolitan areas in the United States (Taylor and Fry 2012). The metropolitan area also ranks consistently among the most racially segregated in the United States, with values for the Index of Dissimilarity ranging from 79.1 for blacks to 63.1 and 49.5 for Hispanics and Asians, respectively (Logan and Stults 2011; Population Studies Center 2015). Although levels of segregation have hardly changed, the locations of racial groups have shifted. Whites predominate in southern Manhattan and in western Brooklyn, Staten Island, and the peripheral suburbs, whereas blacks are concentrated in eastern Brooklyn, southern Queens, New Jersey counties adjacent to

Manhattan, and the Bronx. Hispanics also live in the New Jersey counties and throughout northern Manhattan, the Bronx, Queens, and Nassau County.

The impacts of these shifting geographies on gender differences in commuting are not well understood. The decentralization of jobs in the 1980s and 1990s has been linked to persistent spatial mismatch for minority residents of many U.S. metropolitan areas (Liu and Painter 2012; Jang and Yao 2014). Long work trips in the suburbs have also been attributed to rapid urban development and they are expected to diminish as suburbs mature (Sultana and Weber 2014). These studies did not consider gender differences in commuting, however. We address this question by an empirical analysis that emphasizes how the intersecting effects of race and gender on travel time vary across the New York metropolitan area.

Trends in New York Consolidated Statistical Area

To analyze commuting times by gender and race in the New York metropolitan area, we extracted data for employed persons from the Public Use Microdata Sample (PUMS) of the American Community Survey (ACS) for the five-year period between 2008 and 2012. We designate these data as 2010 PUMS, although it is important to keep in mind that they cover a five-year period. With New York City at its center, the study area is a highly urbanized, twenty-four-county region that includes parts of New York, New Jersey, and Connecticut. To maintain comparability with previous work, we did not include data for Pike County in Pennsylvania, which was added to the metropolitan area in the 1990s. The PUMS file contains data records for approximately 5 percent of the region's residents.[1] Because the ACS PUMS represents a weighted sample, population weights were used in generating various statistics.

The main variable of interest is commuting time, the self-reported one-way time spent traveling to work. To examine geographic variations in gendered and racialized differences in commuting, we analyzed three zones within the metropolitan area: the center, the inner ring, and the suburbs. The center encompasses Manhattan, the most gentrified part of the metropolitan area, where a dense network of subways and buses serves the area's largest employment concentration. The inner ring includes the densely populated area neighboring Manhattan including the Bronx, Brooklyn, and Queens, and Essex, Hudson, Passaic, and Union counties in New Jersey. Many residents commute to Manhattan via an extensive, but congested, bus, subway, and rail system. In some neighborhoods, the quality of public services is deteriorating as public facilities age and public funds are channeled to private service providers, such as charter schools (Zukin 2010). In the suburbs, low densities, strictly separated land uses, and limited public transportation foster dependence on the automobile and commuter rail.

The 2010 PUMS data reveal the uneven geographical concentration of jobs and residences among these three zones. The center is a huge employment concentration, accounting for 25 percent of the jobs held by the region's resident workforce. The center has 2.74 jobs per resident compared with less than one job per resident in the suburbs and inner ring (Table 1). The center is a job magnet for all gender and race groups,

Table 1. Places of work and residence by zone, New York, 2010

Race/gender group	Percentage working in zone[a]			Percentage living in zone			Ratio (%working/%living)		
	Center	Inner ring	Suburbs	Center	Inner ring	Suburbs	Center	Inner ring	Suburbs
Asian women	35.3	30.1	33	12	48.9	39.1	2.94	0.62	0.84
Asian men	33.2	32.1	32.3	9	50.7	40.3	3.69	0.63	0.80
Black women	25.7	44.1	29.6	6	65	29	4.28	0.68	1.02
Black men	25.7	43.4	29.7	6.5	63.3	30.2	3.95	0.69	0.98
Hispanic women	24.3	37.8	37.1	9	55.5	35.5	2.70	0.68	1.05
Hispanic men	24	36.6	38.4	7.8	55.1	37	3.08	0.66	1.04
White women	20.5	17.9	59.9	10	24.2	65.8	2.05	0.74	0.91
White men	24.1	20.6	53.1	9.4	24.7	65.8	2.56	0.83	0.81
Total	24.5	28.1	45.8	8.9	39.8	51.2	2.74	0.71	0.89

[a]Percentages for working in zone do not sum to 100 because 1.60 percent of the resident workforce is employed outside the New York metropolitan area.
Source: Public Use Microdata Sample 2010. Calculations by authors.

with the percentage of each group employed in the center exceeding the percentage of the group living there by a wide margin. Residential segregation is evident in the inner ring and suburbs. The majority from every minority group lives in the inner ring, whereas the majority of white men and women live in the suburbs. As a result, access to employment varies geographically and among racial groups. In the suburbs, the ratio comparing the percentages of jobs and residents[2] ranges between 0.80 and 1.05. Even though a small percentage of blacks are employed in the suburbs, the ratio hovers close to 1.0 because so few black men and women live there. The ratio of jobs to residents is lowest in the inner ring that is home to large minority populations. The paucity of jobs in the inner ring is particularly isolating for minority men and women who commute in large numbers to jobs at the center (Table 1).

Gender and Racial Disparities in Commuting Time

In 2010, gender differentials in commuting time in the New York metropolitan area were similar to those in the 1980s and 1990s (McLafferty and Preston 1991). Average commuting time in 2010 is 34.8 minutes; however, the mean masks substantial variations between men and women and across racial groups (Figure 1). In three racial groups—white, Hispanic, and Asian[3]—men commute longer times on average than women; however, the gender differential varies from 5.8 minutes for whites to 0.9 minutes for Hispanics.[4] Blacks of both sexes commute approximately the same time, with black men reporting average travel times that are 0.7 minutes less than those reported by black women. The small gender differences for minority

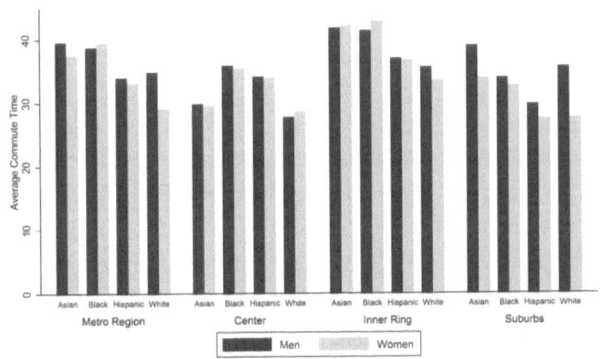

Figure 1. Commuting time by gender, race, and zone. *Source*: Public Use Microdata Sample 2010. Calculations by authors. (Color figure available online.)

men and women suggest that minority women are unable to adjust their work trips to accommodate household responsibilities and this is particularly the case for black women. This conclusion is reinforced by the persistent evidence that white women have shorter work trips than any other female workers. In the New York metropolitan area, white women commute 29.1 minutes one way on average, compared with 33.1 minutes, 37.5 minutes, and 39.5 minutes for Hispanic, Asian, and black women.

The Impacts of Residential Location

Commuting varies across the three zones, with the expected gender differences emerging mainly in the suburbs (Figure 1). In the center, black and Hispanic residents commute longer than Asians and whites, with average travel times ranging from 27.8 minutes to 35.9 minutes for white and black men, respectively. There are virtually no gender differences in commuting time. White women commute slightly longer than white men; however, the average difference in travel time is so small (0.8 minutes) as to be meaningless.

In the inner ring that encircles Manhattan, racial differences in travel time are evident, with blacks and Asians commuting longer than whites or Hispanics of both sexes. Once again, gender differences appear only for whites. As expected, white men commute longer than white women. For Hispanics and Asians, the travel times of men and women are approximately equal. Black women commute slightly longer than black men, but the difference of 1.5 minutes is small.[5]

The expected gender differences in travel time are most apparent in the suburbs, where the effects of race and gender are intertwined. White and Asian men commute the longest, spending 35.7 and 39.0 minutes, respectively, traveling to work. As a result of men's long travel times, there are pronounced gender differences in commuting between white and Asian men and women. On average, white women commute 8.1 minutes less than white men and the equivalent difference in travel time for Asian men and women is 6.0 minutes. The gender gap is relatively small for Hispanics—only 2.3 minutes—and even smaller (1.5 minutes) for blacks.

Explaining Commuting Time

Whereas gender and race have little impact on commuting times in the center, their effects persist in the inner ring and suburbs. These geographically vary-ing disparities are associated with a complex mix of economic, social, and transportation factors that affect workers' residential and employment locations and the efficiency of travel between them. In the remainder of this article, we investigate how these factors influence gender and racial differences in commuting time in the three zones.

Multiple regression models were estimated with commuting time as the dependent variable. Indepen-dent variables included hourly wage (log), an indicator of the economic return to commuting, mass transit use (subway, bus, rail, and ferry), and a set of variables measuring domestic responsibilities based on marital status and the presence of children under age eighteen in the household. Because household responsibilities are expected to have differing effects on commuting for men and women, we included effect-coded interac-tion variables for gender with marriage and presence of children (Wendorf 2004). With effect coding, model coefficients measure the differential between the group mean and the grand mean, with all other variables held constant. Gender disparities that are independent of wages, transit use, and household vari-ables are represented via a dummy variable equal to one for women. Models were estimated separately for the three metropolitan zones and by racial group to reveal geographic and racial disparities in factors asso-ciated with the time spent traveling to work.[6] Although our discussion emphasizes overall gender and racial disparities in commuting, the relatively low R^2 values (<0.30) reveal substantial unexplained vari-ation within and among groups that we acknowledge when interpreting the findings.

Wages

Commuting is often seen as tangible evidence of a rational trade-off between wages and travel time (Madden 1981). Commuters extend their work trips to obtain higher wages and reduce travel times when wages fall. Women who receive lower wages on average in the labor market spend less time commuting than men. The same relationship should emerge for minority workers who earn less on average than white workers. Racial and gender disparities in wages are readily apparent in the New York metropolitan area. Average hourly wages are highest for white and Asian men, at $39.81 and $33.60, respectively, and lowest for Hispanic men and women. Men's average wages are higher than

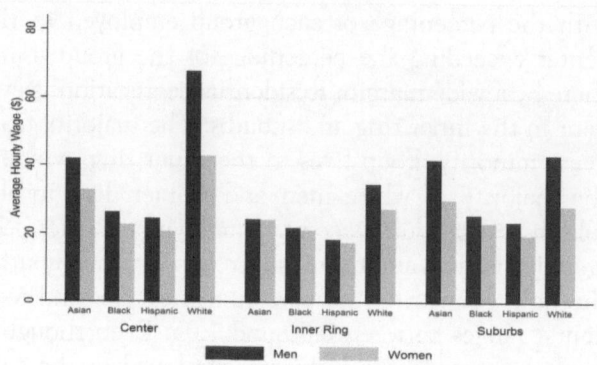

Figure 2. Average hourly wages by gender, race, and zone. *Source:* Public Use Microdata Sample 2010. Calculations by authors. (Color figure available online.)

women's for all racial groups; however, the magni-tude of the gender gap varies. White women earn approximately 70 percent of white men's average hourly wage, whereas black women earn 93 percent of the average black male hourly wage.

Residential location also affects wages. Wages are higher in the center even though gender and racial wage gaps persist. Average wages are exceptionally high for white men who live in Manhattan ($58.63), followed by Asian men ($39.40) and white women ($38.88; Figure 2). In contrast, wages for blacks and Hispanics living in the center are much lower. Earning just 35 percent of white men's average hourly wages, black and Hispanic women are economically disadvantaged, perhaps because of the persistent racial and gender gap in wages (Goldin 2014). Residents of the inner ring earn less than workers living in either the center or the sub-urbs, and gender and racial disparities in wages are small. Residents of the suburbs are in an intermedi-ate position, earning less than central residents but more than residents of the inner ring.

In the New York metropolitan area, where average wages are highly differentiated by location, wages have a significant influence on commuting time for every racial group in each metropolitan zone (Table 2). The regression coefficients underscore the advantages of central locations where commuting time declines sig-nificantly as hourly wages increase. Contrary to the norm of longer commutes for higher wages, the best paid workers living in the center commute least, and this relationship is found for all racial groups, despite substantial differences in average wages between whites and minorities.

Table 2. Regression coefficients for travel time by zone and race, New York, 2010

Variable	Center				Inner ring				Suburbs			
	White	Black	Hispanic	Asian	White	Black	Hispanic	Asian	White	Black	Hispanic	Asian
Log hourly wage	-1.20**	-1.51**	-1.29**	-1.15**	2.66**	1.43**	2.38**	0.918**	5.01**	6.77**	7.10**	5.37**
	(0.173)	(0.491)	(0.393)	(0.382)	(0.117)	(0.186)	(1.69)	(0.205)	(0.075)	(0.260)	(0.198)	(0.224)
Transit	10.62**	16.04**	17.70**	13.69**	21.78**	25.26**	22.77**	21.82**	42.06**	32.96**	29.08**	39.46**
	(0.306)	(0.864)	(0.682)	(0.726)	(0.186)	(0.260)	(0.233)	(0.336)	(0.206)	(0.495)	(0.406)	(0.492)
Woman	ns	ns	ns	ns	-.608	ns	ns	ns	-1.94**	ns	ns	ns
					(0.278)				(0.228)			
Married, men	1.36**	ns	ns	ns	1.67**	0.94**	1.05**	2.21**	1.79**	2.14**	2.02**	1.38**
	(0.377)				(0.208)	(0.299)	(0.245)	(0.398)	(0.149)	(0.441)	(0.306*)	(0.514)
Married, women	ns	ns	ns	ns	-.965**	ns	ns	ns	-1.30**	ns	-.67*	ns
					(0.205)				(0.142)		(0.321)	
Kids, men	ns	2.21**	1.68**	ns	ns	0.763**	0.579**	ns	1.06**	1.60**	ns	1.58**
		(0.984)	(0.716)			(0.298)	(0.244)		(0.132)	(0.435)		(0.398)
Kids, women	ns	ns	ns	ns	-0.420**	-0.985**	-0.850**	ns	-0.963**	-1.21**	-0.794*	-1.55**
					(0.222)	(0.266)	(0.256)		(0.134)	(0.398)	(0.324)	(0.415)
Constant	26.01**	31.01**	26.83**	24.84**	18.44**	25.15**	20.02**	28.07**	13.05**	8.25**	5.91**	13.33**
	(0.767)	(1.733)	(1.291)	(1.43)	(0.412)	(0.602)	(0.507)	(0.758)	(0.276)	(0.849)	(0.607)	(0.856)
N	14,460	3,087	4,290	3,123	54,177	37,214	35,904	20,484	152,226	16,185	23,695	17,678
F	187.71**	53.66**	101.44**	52.28**	2,149.24**	1,362.52**	1,389.14**	607.67**	8,198.75**	741.50**	741.50**	1,150.03**
Adj. R²	0.08	0.11	0.14	0.10	0.22	0.20	0.21	0.17	0.27	0.24	0.22	0.31

Note: Unstandardized regression coefficient (standard error).
*$p \leq 0.05$.
**$p \leq 0.01$.
$ns = p > 0.05$.
Source: Public Use Microdata Sample 2010. Calculations by authors.

In the inner ring and suburbs, higher wages are associated with longer work trips, as expected. The relationship between wages and commuting time also strengthens from the center to the suburbs, with coefficients increasing from a low of 0.918 for Asians in the inner ring to a high of 7.10 for Hispanics in the suburbs. In each zone, the coefficients for wages are low for Asians (Table 2). Their work trips are influenced less by wages than are those of other racial groups, raising the possibility that factors other than those considered here play an important role in their insertion into the labor market.

Transportation Mode

Commuting times also depend tremendously on transportation mode. Typically, the car is the quickest and most flexible form of transportation, and various forms of mass transit, particularly buses, result in longer travel times (Shen 2001). One explanation for gender differences in commuting has focused on access to various modes of transportation, recognizing that women are often more reliant on public transportation, which lengthens commuting time (Rosenbloom 2006). The New York metropolitan area stands out because of its dense subway, bus, and commuter rail systems, which carry more than 4 million passengers each day. The availability of mass transit, slow driving times, and high parking costs in central Manhattan mean that workers in the New York metropolitan area are more likely than any others in the United States to commute on public transportation (Moss, Qing, and Kaufman 2012).

Despite the widespread availability of mass transit, the transportation modes used for work trips vary among racial groups in the New York metropolitan area. White men and women are more likely to travel by car, whereas minority workers rely heavily on public transportation (Table 3). Almost three quarters of whites commute by car, compared with approximately half of minority workers. There are no gender differences in reliance on the car for whites, whereas slight gender differences emerge for minority workers. Minority men are more likely than women to commute by car and minority women, particularly blacks and Hispanics, use buses more than others. The racialized differences in transportation mode suggest that white men and women have more access to employment than minorities.

The transportation mode used for commuting varies between Manhattan and the suburbs. In Manhattan,

Table 3. Transportation mode by metropolitan zone, race, and gender (percentage)

	Car	Public transit	Other[a]	Work at home
Center				
Asian women	4.9	59.4	31.4	4.4
Asian men	10.8	54.0	32.3	2.8
Black women	8.9	70.9	15.4	4.8
Black men	14.7	64.5	15.4	5.4
Hispanic women	7.8	65.3	21.8	5.2
Hispanic men	14.5	59.4	20.6	5.5
White women	6.8	55.6	29.8	7.9
White men	10.0	53.2	29.2	7.6
Inner ring				
Asian women	31.1	56.4	10.0	2.5
Asian men	43.2	46.1	8.2	2.6
Black women	37.8	53.8	6.3	1.9
Black men	46.2	55.4	6.0	2.4
Hispanic women	34.5	50.5	12.1	2.9
Hispanic men	41.8	44.6	11.4	2.3
White women	48.6	37.3	9.8	4.4
White men	52.4	34.7	8.1	4.9
Suburbs				
Asian women	77.5	14.7	3.9	3.9
Asian men	76.1	16.7	3.8	3.4
Black women	74.2	18.1	5.6	2.1
Black men	76.5	15.6	5.4	2.5
Hispanic women	76.9	13.1	7.2	2.7
Hispanic men	76.1	12.7	8.7	2.4
White women	85.9	6.6	2.5	5.1
White men	81.5	10.9	2.7	5.0

[a]Includes bicycle, walking, and other modes.
Source: Public Use Microdata Sample 2010. Calculations by authors.

more than half of each race and gender group commutes by transit and at least 15 percent walks to work. Whites and Asians are more likely to walk than Hispanics and blacks, suggesting that the latter live far from their jobs. Gender differences are apparent only for commuting by car and bus. Men in each group are more likely to commute by car, and women are more likely to take the bus. Black women and Hispanics are most likely to travel by bus, a slow mode of transport even in Manhattan. Reliance on cars increases steadily from the center to the inner ring and suburbs as use of mass transit declines. In each setting, though, minority women's reliance on mass transit far exceeds that of their male counterparts. At the same time that gender differences in transportation mode increase in the inner ring and suburbs for minorities, increasing reliance on the car causes the transportation modes used by white men and women to converge.

As expected, the regression analyses confirm that transit use increases travel time, even in Manhattan

(Table 2). Reflecting the high frequency of transit service and the proximity of job opportunities, the regression coefficients for transit use are smaller in the center than in the inner ring and suburbs. The impact of transit use on commuting time increases steadily from the inner ring to the suburbs, with regression coefficients increasing from a minimum of 21.78 and 21.82 for whites and Asians in the inner ring to 42.06 for whites in the suburbs (Table 2). The steady increase in coefficients for transportation mode from the center of the metropolitan area to the periphery reflects the advantages of driving in the suburbs, as contrasted with poorly developed bus networks and long travel times by commuter rail from the suburbs to Manhattan.

At each location, the regression coefficients for transportation mode underscore the advantages of commuting via mass transit for whites and Asians living in the center and inner ring (Table 2). In both zones, transit use has less impact on the commuting times of whites and Asians than on those of blacks and Hispanics. In the suburbs, transit use increases the work trips of black and Hispanic workers less than those of whites and Asians, whose transit use mainly involves commuter rail.

Household Type

Women are expected to commute shorter distances and less time than men to accommodate the demands of child care and other reproductive tasks (Hanson 2010). Overall, in New York, there is some evidence that women with household responsibilities work closer to home than comparable men. Indeed, on average, married women travel approximately six minutes less than married men. Similarly, women with children at home have commutes that are five minutes less on average than those of men with children (Table 4).

No gender differences exist among men and women who do not have these household responsibilities, indicating that marriage and child care influence men's as well as women's commuting.

When we examine the effects of marital status and the presence of children for men and women from the four main racial groups, the findings are more nuanced. White and Asian mothers with children at home commute shorter times, on average, than their male counterparts, and similar but slightly smaller gender differences exist for whites and Asians based on marital status. In contrast, black and Hispanic women who are married and those whose households include dependent children do not make the same adjustments for family circumstances, traveling approximately the same time as their male counterparts and as long as white and Asian married men who have dependent children.

The regression coefficients suggest that when wages and transit use are controlled, the direct effects of family responsibilities vary across the metropolitan region (Table 2). In the center, marital status and presence of children have little impact on men's and women's commutes. In Manhattan, women's commutes are not associated with household responsibilities. For men in the center, although marriage lengthens travel time for white men and presence of children lengthens trips for black and Hispanic men, the bulk of coefficients are nonsignificant, suggesting that household responsibilities have little impact on their commutes.

Expected gender variations begin to emerge in the inner ring. Married men consistently commute longer than their single counterparts, perhaps because the residential decisions of households with two wage earners are often more constrained geographically than those of their single counterparts. The presence of dependent children influences both men's and women's commuting times in the inner ring, reducing women's commuting times and increasing men's, except among

Table 4. Average commuting time (minutes) by marital status, presence of dependent children, gender, and race (standard deviation)

	All workers		Asian		Black		Hispanic		White	
	Men	Women	Men	Women	Men	Women	Men	Women	Men	Women
Married	37.1 (27.6)	31.5 (25.3)	41.1 (28.5)	37.3 (27.2)	39.0 (27.6)	38.8 (28.0)	34.8 (25.4)	32.3 (24.7)	36.5 (27.9)	28.3 (23.8)
Not married	34.6 (25.0)	34.5 (25.5)	36.9 (25.0)	38.0 (25.3)	39.1 (26.4)	40.2 (28.2)	34.5 (24.1)	34.6 (24.9)	32.3 (24.7)	30.7 (23.4)
Dependent children	37.3 (27.5)	32.3 (25.5)	41.7 (28.5)	37.1 (27.0)	39.1 (27.4)	38.7 (27.6)	34.8 (25.0)	32.5 (24.6)	37.3 (28.3)	27.8 (23.6)
No dependent children	35.2 (26.0)	33.4 (25.4)	38.6 (26.9)	38.3 (26.2)	39.0 (26.7)	40.8 (28.6)	34.5 (24.5)	34.9 (25.1)	33.9 (26.1)	30.0 (23.7)

Source: Public Use Microdata Sample 2010. Calculations by authors.

Asians. The end result is a widening gender gap in commuting time reflecting varying responses by men and women to household circumstances.

Gender disparities in commuting time linked to household responsibilities peak in the suburbs. Marriage and the presence of dependent children consistently increase suburban men's commuting times and reduce work trips for suburban women. Notably, white men and women are able to adjust their work trips to accommodate household responsibilities, whereas among minority workers, the adjustment is made typically by one gender or the other. Still, the coefficients are remarkably consistent among all groups, indicating that in low-density suburban landscapes, gendered responsibilities associated with marriage and child care affect men's and women's commuting.

Once we control for wages, transportation mode, and household composition, the independent effects of gender on commuting time diminish, except among the most privileged workers (Table 2). The persistence of significant ($p \leq 0.05$) coefficients for sex for whites in the inner ring and suburbs suggests that gender remains an important organizing construct for the labor market participation of white residents in these contexts. In both cases, women commute less time than men, suggesting that white women can find work near home irrespective of wages, transit use, and household characteristics. These advantages do not extend to minority women, whose commuting times are similar to those of men in similar circumstances.

Conclusions

In the New York metropolitan area, gender and racial inequalities in commuting time persist. White women travel substantially shorter times to work than white men, there is a moderate gender gap for Asians, and there is no gender disparity for other minority workers. These patterns that are identical to those observed in the 1980s and 1990s indicate that gender- and racially segmented labor market processes endure. Although these findings focus solely on the New York metropolitan area with its unique geographies of employment and population, dense rapid transit, and long history of racial segregation, the region's size and complexity and its position at the forefront of urban economic and social transformations make it a compelling case study.

Although commuting time is still influenced by wages, transportation mode, and household composition as it was in the past, the impacts of these factors vary across the metropolitan area. In the center, whites and Asians have short work trips to well-paid jobs, whereas elsewhere commuting time increases as wages rise. The advantages of central residential locations are reinforced by rapid transit that speeds central residents to work. In the inner ring and suburbs, commuters who use public transportation report long work trips. Even the effects of household responsibilities vary, having the greatest impact on commuting times in the suburbs, where men and women adjust their commutes to accommodate the responsibilities of marriage and children, and marriage still increases men's travel times.

Manhattan, where gender differences in commuting time are muted, offers insight into the built and social characteristics that promote more equitable access to employment. These include well-developed and frequent rapid transit and a built environment in which residential locations are nearby dense employment opportunities. Despite the diminution of gender differentials in commuting in the center, racial differences persist, with blacks of both sexes commuting longer than whites and earning substantially less. Racialized differences persist, in part because wages for minorities, particularly for Hispanics and blacks of both sexes, are lower than those paid to white workers throughout the metropolitan area. One interpretation of our findings is that privileged white men and women enjoy better access to employment that reinforces and maintains their privilege. To undo this white privilege requires research about pockets of poverty and wealth within each of the three rings and the diverse, lived experiences of racialized inequality in each. Additional research is also needed to assess the unexplained variation in commuting times and how well findings from the New York metropolitan area apply elsewhere. Our conclusions also need to be tested by asking men and women from different racial backgrounds in each zone how they view their work trips and how these everyday experiences affect their well-being.

Acknowledgments

We acknowledge with gratitude helpful comments from three reviewers and the editor, the exceptional research assistance of Jamie Fishman, and informative discussions with Silvia D'Addario. The authors have contributed equally to this article.

Notes

1. The data are organized by place of residence, so unless otherwise stated, zones refer to place of residence.
2. For each race and gender group, the ratio compares the percentage of the group population working in a zone with the percentage of the group's population living there.
3. We use the racial categories reported in the ACS recognizing that notions of race are social constructions with material consequences for everyday life (Wilson 2009).
4. Due to the very large sample sizes, small differences in average travel time (0.1–0.2 minutes) would be deemed statistically significant. Therefore, we focus on the magnitude of differences, rather than their statistical significance.
5. The concentration of blacks in the inner ring and the preponderance of women among black workers mean that this gender difference dominates metropolitan-wide. For blacks, there are more women than men in the labor force, but for all other racial groups, the labor force is majority male.
6. We tested other model formulations, including models disaggregated by gender, and the results were highly consistent with those discussed here.

References

Crane, R. 2007. Is there a quiet revolution in women's travel? Revisiting the gender gap in commuting. *Journal of the American Planning Association* 73 (3): 298–316.

Cresswell, T. 2010. Mobilities I: Catching up. *Progress in Human Geography* 35 (4): 550–58.

Frey, W. H. 2012. *Population growth in metro America since 1980.* Washington, DC: The Brookings Institution. http://www.brookings.edu/~/media/research/files/papers/2012/3/20-population-frey/0320_population_frey.pdf (last accessed 24 August 2015).

Goldin, C. 2014. A grand gender convergence: Its last chapter. *American Economic Review* 104 (4): 1091–1119.

Hanson, S. 2010. Gender and mobility: New approaches for informing sustainability. *Gender, Place and Culture* 17 (1): 5–23.

Hanson, S., and G. Pratt. 1995. *Gender, work, and space.* London and New York: Routledge.

Jang, W., and X. Yao. 2014. Tracking ethnically divided commuting patterns over time: A case study of Atlanta. *The Professional Geographer* 66 (2): 274–83.

Joassart-Marcelli, P. 2009. The spatial determinants of wage inequality: Evidence from recent Latina immigrants in Southern California. *Feminist Economics* 15 (2): 33–72.

Johnston-Anumonwo, I. 2014. Women's work trips and multifaceted oppression. In *Diversity, social justice, and inclusive excellence,* ed. S. N. Asumah, and M. Nagel, 93–112. Albany: State University of New York Press.

Liu, C. Y., and G. Painter. 2012. Immigrant settlement and employment suburbanisation in the US: Is there a spatial mismatch? *Urban Studies* 49 (5): 979–1002.

Logan, J. R., and B. Stults. 2011. The persistence of segregation in the metropolis: New findings from the 2010 census. Census Brief prepared for Project US 2010. http://www.s4.brown.edu/us2010 (last accessed 24 August 2014).

Madden, J. F. 1981. Why women work closer to home. *Urban Studies* 18 (2): 181–94.

Mauch, M., and B. D. Taylor. 1997. Gender, race, and travel behavior: Analysis of household-serving travel and commuting in San Francisco Bay Area. *Transportation Research Record: Journal of the Transportation Research Board* 1607 (1): 147–53.

McLafferty, S., and V. Preston. 1991. Gender, race, and commuting among service sector workers. *The Professional Geographer* 43 (1): 1–15.

———. 1992. Spatial mismatch and labor market segmentation for Black and Latina women. *Economic Geography* 68 (4): 406–31.

Moss, M. L., C. Y. Qing, and S. Kaufman. 2012. *Commuting to Manhattan.* New York: Rudin Center for Transportation Policy, New York University.

Parks, V. 2004. Access to work: The effects of spatial and social accessibility on unemployment for native-born black and immigrant women in Los Angeles. *Economic Geography* 80 (2): 141–72.

Population Studies Center. 2015. *Race segregation for largest metro areas.* Ann Arbor: Institute for Social Research, University of Michigan.

Preston, V. S., and S. McLafferty. 1999. Spatial mismatch research in the 1990s: Progress and potential. *Papers in Regional Science* 78 (4): 387–402.

Preston, V., S. McLafferty, and E. Hamilton. 1993. The impact of family status on black, white, and Hispanic women's commuting. *Urban Geography* 14 (3): 228–50.

Rosenbloom, S. 2006. Understanding women's and men's travel patterns: The research challenge. In *Research on women's issues in transportation: Report of a conference. Vol. 1: Conference overview and plenary papers,* ed. N. Kassabian, 7–27. Washington, DC: Transportation Research Board of the National Academies.

Shen, Q. 2001. A spatial analysis of job openings and access in a US metropolitan area. *Journal of the American Planning Association* 67 (1): 53–68.

Sultana, S. 2005. Racial variations in males' commuting times in Atlanta: What does the evidence suggest? *The Professional Geographer* 57 (1): 66–82.

Sultana, S., and J. Weber. 2014. The nature of urban growth and the commuting transition: Endless sprawl or a growth wave? *Urban Studies* 51 (3): 544–76.

Taylor, P., and R. Fry. 2012. *The rise of residential segregation by income.* Washington, DC: Pew Research Center.

Wang, Q. 2010. How does geography matter in the ethnic labor market segmentation process? A case study of Chinese immigrants in the San Francisco CMSA. *Annals of the Association of American Geographers* 100 (1): 182–201.

Wang, W., K. Parker, and P. Taylor. 2013. *Breadwinner moms.* http://www.pewsocialtrends.org/2013/05/29/breadwinner-moms/ (last accessed 24 August 2015).

Wendorf, C. A. 2004. Primer on multiple regression coding: Common forms and the additional case of repeated contrasts. *Understanding Statistics* 3 (1): 47–57.

Wilson, D. 2009. Introduction: Racialized poverty in US cities: Toward a refined racial economy perspective. *The Professional Geographer* 61 (2): 139–49.

Zukin, S. 2010. *Naked city: The death and life of authentic urban places.* New York: Oxford University Press.

Mobility, Communication, and Place: Navigating the Landscapes of Suburban U.S. Teens

Meghan Cope* and Brian H. Y. Lee[†]

*Department of Geography, University of Vermont
[†]School of Engineering and the Transportation Research Center, University of Vermont

In the context of sprawl and car dependence in U.S. metropolitan areas, young people—especially teens in middle-class suburbs—create new mobility practices with near-universal adoption of cellphones and high levels of access to automobiles. The growth in the use of handheld mobile devices for communication and information might enhance independent mobility and accessibility for higher socioeconomic segments of the youth population. In a project with teens in two high schools near Burlington, Vermont, representing somewhat different land-use contexts, we examined how often and in what ways teens use information and communication technologies (ICTs) to arrange transportation, what travel needs are being met and which transportation modes are used, and how household situations contextualize the use of ICTs for mobility. We explore the ways in which access to cellphones and cars affects how high school teens organize and enact their daily lives in suburban and rural contexts. We employ a conceptual framework that connects mobility, communication, and place based on the notion that contemporary teens generate new intersections among the built, digital, and social landscapes.

在美国大都会地区的蔓延和车辆依赖之脉络中, 年轻人——特别是中产阶级郊区中的青少年——随着近乎全球普及的手机使用和高度的汽车可及性, 创造出崭新的能动性实践。将手持行动装置用于通信与信息的的成长, 或许会增进来自于较高社经地位的青年人口的独立能动性与可及性。我们在佛蒙特邻近伯灵顿的两所呈现出些许不同土地使用脉络的高中, 与年轻人进行的一项计画中, 检视青少年使用信息与通信技术 (ICTs) 来安排运输的频率与方式, 他们什麽样的移动需求受到满足, 使用何种运输模式, 以及家户境况如何概念化将 ICTs 运用至能动性的使用。我们探讨使用手机和车辆的方式, 如何影响郊区与乡村脉络中, 高中青少年安排并执行其日常生活的方式。我们根据当前青少年在建成、数码与社会地景中形成新交汇之见解, 运用连结能动性、通信与地方的概念架构。

En el contexto de la descontrolada expansión urbana y dependencia en el carro en las áreas metropolitanas de los EE.UU., la gente joven—especialmente los adolescentes de los suburbios de clase media—crean nuevas prácticas de movilidad con la adopción casi universal de los teléfonos celulares y los altos niveles de acceso a los automóviles. El aumento del uso de aparatos móviles de mano para las comunicaciones y la información podría fortalecer la movilidad y accesibilidad independiente para los segmentos socioeconómicos más altos de la población joven. En un proyecto con adolescentes de dos escuelas de secundaria, cerca de Burlington, Vermont, que representan contextos de usos de la tierra algo diferentes, examinamos qué tan a menudo y cómo usan los jóvenes las tecnologías de la información y las comunicaciones (TICs) para acordar transporte, qué necesidades de transporte se están abocando y cuáles medios de transporte se usan, y de qué modos las situaciones del hogar contextualizan el uso de las TICs para movilidad. Exploramos las maneras como el acceso a los teléfonos celulares y a los carros afecta el modo como los jóvenes de la escuela secundaria organizan y viven sus vidas cotidianas, en contextos suburbanos y rurales. Empleamos un marco conceptual que conecta movilidad, comunicación y lugar a partir de la noción de que los adolescentes contemporáneos generan nuevas intersecciones entre los paisajes construidos, digitales y sociales.

Drawing from research completed in 2011 and 2012 with teens (ages fourteen to eighteen) in two suburban school districts in the Burlington, Vermont, metropolitan area, we find evidence that contemporary teens in suburban settings develop their own "mobility practices" (Kesselring 2006), in which they negotiate and traverse the intersections, boundaries, and sometimes the contradictions of daily life. We propose, based on this work, that the most relevant landscapes for our teen respondents are the built environment (characterized by land use, density, and transportation infrastructure), the digital networks of electronic information and communication (characterized by smartphone access, cellular coverage, and communication practices), and the social environment (composed of family, friends, school, sports, and work). Teens in our study balanced opportunities enabled by these against a context of various constraints to organize their time, mobility, and obligations.

In this project we wondered: How is the proliferation of instant, mobile, digital technologies seen in the ways that high school teens organize and enact their daily lives? How do they do this in different geographic contexts with varying access to transportation? Kesselring's (2006) notion of mobility practice fits our findings well in terms of negotiations between broader structures of opportunity and constraint, along with individual goals and decisions. Kesselring (2006) explained:

> [M]obility practice is structured by contextual situations, economic and social conditions, and power relations in general. ... Mobility is often conceived of as a form of freedom, but in fact mobility results from the dichotomies of autonomy and hetero-nomy, production and adaptation. This is the very reason why mobility must be conceptualized in relation to flexibility as the ability of actors to adapt to the direction of flows. (270, italics added)

Although space constraints here allow only a brief hint at how a small sample of teens solved the dilemma of managing time, mobility, and activities, we hope that they provoke further exploration.

Critical Youth Mobilities

Critical youth studies identify young people as active and knowledgeable social agents (Holloway and Valentine 2000); critical youth geographies further account for the ways in which space and place affect young people's lives while identifying how young people themselves construct spaces of opportunity, fun, work, and even resistance. Existing literature on the intersection of teens, transportation, and technology is disparate and interdisciplinary, ranging from transportation planning to critical youth studies, from the recent mobilities literature to cultural critiques of information and communication technologies (ICTs). We draw from these bodies of scholarship by exploring how teen mobility is influenced by changing technologies but also rooted in place-based infrastructure and family rules, practices, and expectations. Our contribution to these fields lies, in part, in blending critical youth geographies (seeing age as socially constructed and young people as active, mobile, social agents) with transportation planning perspectives (integrating the effects of land-use patterns, transportation infrastructure, and travel behavior).

Many studies examine the intersections of the built and social environments for teens, finding that teens' use of walking and bicycling and their engagement with public transportation are highly context-specific, varying by race, gender, class, etc. (Bungum et al. 2009; Blumenburg et al. 2012; Emond and Handy 2012). Looking at a national level, Clifton (2003) was quite clear about the disempowering places experienced by U.S. teens: "The limited transportation choices and the low-density, segregated land use patterns of American cities may hold teens and their chauffeuring parents captive to the private automobile" (1). Arguably, U.S. policies and practices have generated landscapes that prioritize the car-driving, commuting population to the detriment of independent mobility of teens and other social groups who cannot or do not drive, ultimately producing a spatially unjust hierarchy of access. We have previously argued that this constitutes an "adultist urbanism" (Lee and Cope 2012).

Since 2000 there has been a growing literature on the diverse ways in which young people blend physical mobility with use of ICTs, thus adding the digital landscape that includes cellphones and home-based Internet. In their exploration of the social networks, ICT use, and travel among "youngish" adults, Larsen, Urry, and Axhausen (2006) examined "how social networks are spatially distributed and how they are produced through networking practices of travel, communications and meetings" (5). Focusing on preteens and teens, Pain et al. (2005) found that some young people in the United Kingdom experienced an expansion of their geographic realm if they had their cellphones with them; being contactable by parents allowed greater freedom, within a context of other space–time restrictions and curfews. In work with eleven- to sixteen-year-olds, Holloway and Valentine (2003) found that, rather than fostering social isolation, ICT use allowed participants to complement and facilitate offline, in-person interactions.

Another significant study of young people, ICTs, and mobility is the longitudinal work of Thulin and Vilhelmson (2005, 2007, 2009, 2012), which began in 2000 with a sample of Swedish eighteen-year-old students and followed them over the years. The authors' key findings are diverse but can be summarized by the understanding that individuals' negotiations and navigation through daily life are influenced by geographic context, social–cultural power relations, personal agency, and material factors (e.g., availability and affordability of technologies). Through their case study approach, Thulin and Vilhelmson provided a corrective to early speculations that ICTs would either completely substitute for physical travel (functioning as a sort of "immobilizing" factor in work and social lives) or would so vastly improve our efficiency that we would use leisure time for nonelectronic activities. In fact, their participants demonstrated a wide range of balances among cellphone use, home-based ICT use, physical travel, and face-to-face social interactions that were often intensified or complemented by ease of communication. For example, Thulin and Vilhelmson (2009) noted a trend we saw repeatedly in our research, in which cellphones produced a sense of "permanent access and instant reach" (139):

> The mobile [phone] now makes it possible to constantly negotiate and renegotiate agreements for meetings and joint activities in real time as circumstances change. Plans for the day become more flexible and schedules become less fixed in time and space, allowing for more spontaneous or impulsive decision making. (Thulin and Vilhelmson 2009, 143)

In our own analyses we identified this phenomenon as the cellphone providing both certainty and flexibility for teens, their parents, and their friends.

The work of these authors and others writing in the first decade of the 2000s (e.g., Grinter and Palen 2002; Oksman and Turtiainen 2004; Hjorthol 2008) is valuable for this analysis, for their ground-breaking examination of technological impacts on daily life. Indeed, even with high levels of access to ICTs and cars, we found that geography still matters (distance, land use, residential density, cell-signal gaps) and that even teens with good access to resources negotiate multiple obligations, social groups, and family rules.

Method

We collected data in 2011 and 2012 through a two-phase, mixed-methods approach to examine the mobility practices of teens. The first phase (October 2011) consisted of linked online surveys, one for parents and guardians and another for their teenaged children, distributed to the entire student populations of two high schools outside Burlington, Vermont. The parent survey asked about household composition; the availability and use of different travel modes, particularly vehicle access and sharing; family rules about teen mobility, curfews, and driving privileges and responsibilities; communication and information resources (e.g., smartphones, Internet access); and sociodemographics. The teen survey was concerned with driving

Table 1. Travel and communications indicators for student respondents (based on surveys, October 2011)

	School A	School B
Number of teen respondents[a]	60	86
Teens with driver's license	25%	37%
Average age teens received license	16.13	16.09
Walk or bike to/from school	14%	3%
Ride bus to/from school	35%	34%
Drive or ride in car to/from school	51%	63%
Teen cellphone ownership	95.0%	81.4%
Teens' phones that are smartphones (with data plan)	40.4%	42.9%

[a]Respondents were evenly split between male and female and all four grades were equally represented.

licensure and access to household vehicles; rules about where they can go, with whom, when, and by what modes; the places they go and places they cannot access and why; their friends and where they get together; and the availability and use of cellphones, particularly for activity and travel planning.

The second phase (June 2012) involved a half-day meeting to gather in-depth mobility information from a small group. We invited all 146 teen respondents (see Table 1) from the first phase who had agreed to being contacted for follow-up activities. Of these, five teens were interested and able to attend: two girls from one high school and a girl and twin boys from the other. First, an individual web-mapping exercise asked participants to locate their home and school and routes between them, other places teens went in the past week, and local places they would like to go but could not ("denied places"). Teens annotated their maps with notes about travel modes, companions, things to do on the way, and transportation obstacles. Second, the participants collaboratively identified important places in their communities and common transportation routes on large paper orthographic maps (Figure 1). This group setting encouraged the students to share their personal perspectives while revealing communal experiences and diverging views. Third, to gain insights on the interaction between communication and mobility, we created a "text review" methodology, based on long-standing oral history techniques, where each teen was interviewed and asked to review the text messages on her or his phone to identify discussions with others about going to a place or doing an activity in the past week.[1] For those discussions, we asked the students with whom and when they were messaging, the destinations and activities being planned, other people involved, and the

Figure 1. Ramani and Eryn mapping common routes and destinations in their town. Photo by Meghan Cope. (Color figure available online.)

outcomes of their plans. Not only did the text review exercise reveal how teens use various forms of messaging to coordinate activity and travel plans, but it also provided insights on an emerging hierarchy of communication tools used (talking in person, e-mailing, texting, phoning) to execute these plans. Finally, we facilitated a focus group discussion with the students on mobility in their lives and constraints they face in getting where they want or need to go.

To summarize and analyze this wide variety of data and contextualize the quantitative data from the closed-ended survey questions, we wove information gathered in both phases of the study into short narratives we call *vignettes*[2] for the five students who joined us in the second phase; illustrative selections from three of those vignettes are presented here. To create the vignettes, we pulled household and contextual data from the surveys, responses to open-ended questions from the parents and guardians and the teens themselves, and data gathered from the second-phase meeting. Stitching these together also gave us a brief longitudinal perspective (seven months between the first and second phase), a period during which several participants experienced significant mobility changes by attaining a driver's license, getting a car, or getting a smartphone, thus greatly influencing their mobility and communications. Constructing the vignettes constitutes analysis because it allowed us to identify themes and generate concepts

from this rich assortment of data that were relevant to the teens with respect to their mobility practices.

Regional Context

The two high school districts we engaged with are both part of the Burlington metropolitan area with similar student enrollments of about 1,000 students and demographics of mostly white, fairly well-off, educated populations. The districts differ in housing and commercial densities, land uses and public transit options, and catchment areas. School A serves a single, suburban town of 18,500 people in thirty square miles and is located on a mixed-use, arterial corridor with regularly scheduled public bus service connecting it to other parts of the town and Burlington. Although the school district does provide busing service, many students also use public transit because it is free for them. The immediate surrounding area of the school has good nonmotorized infrastructure connecting to common teen destinations: part-time employment, recreation, civic activities, entertainment, food, and shopping. Housing density is mixed, with some areas of car-oriented subdivisions within the town but also denser clusters of residential neighborhoods with good bus and sidewalk and path connectivity.

School B serves four suburban and rural towns on the region's outer edge, with a slightly larger combined population of 22,500 in 166 square miles. It is located half a mile from a small village, off a two-lane state highway with no public transit service and limited access by travel modes other than automobile. Short segments of multiuse path in front of the school and road shoulders provide walking access to a gas station and convenience store and a sandwich shop nearby. Most students rely on school buses or private automobiles.

Table 1 shows licensure rates and mode share for getting to and from school (there was no appreciable difference between modes to school and getting home, so the rates are combined here). The data provide some coarse indications of the mobility practices and communication options of teens related to the context of the place: School A students are more likely to walk or bike to school and less likely to ride in a car, and they are somewhat less likely to have their licenses. Students at School B are more car-dependent and have lower rates of phone ownership, in keeping with the rural location of the school and poor cell coverage in some contributing towns. Although some of these findings suggest differences in land use, density, and accessibility, they provide little insight into the trade-offs and decisions that teens and their families make about daily travel or the role that ICTs play in mobility; for a better understanding of these, we turn to the vignettes.

Results: Vignettes

Three vignettes demonstrate how a small sample of teens construct and maintain mobility practices and how these are affected by digital technologies. We generated these narrative accounts of the teens' situations by combining information from both the parent and teen survey responses, the mapping and text review exercises, and the focus group (the sources of data for quotes and specific data are included in parentheses). By combining multiple sources of data we found more and deeper connections between teens' opportunities and constraints in the built environment, their navigations of the technological landscape, and their interlocking social relations. We chose to highlight two students from School A, which represents a highly favorable and resource-rich local context for teens, and one student from School B, with a relatively more challenging set of physical and digital landscapes for teens to navigate.[3]

The teens represented here have some common and contrasting characteristics. The two from School A are both females with one younger sibling, and both rely heavily on their cellphones to coordinate their busy schedules. Eryn,[4] however, has her own car with an associated set of family obligations, and most of her activities are social; Ramani shares her mother's car and her

activities are mostly college-oriented pursuits. The student from School B is male, also with a sibling, but does not have his license or depend on a cellphone for activity and travel coordination because of poor cellular coverage in his rural home area. His nonschool times are filled with sports and other extracurricular activities and he completely depends on his mother to drive him.

Eryn

Eryn was sixteen, white, in Grade 11 at School A, and lived with her single mother, an "executive" earning more than $100,000 annually, and brother, age fifteen (parent survey). The household had broadband access and all three had smartphones. Eryn got her license as soon as she turned sixteen; her mother bought a new car and gave Eryn her older SUV because "I'm a single parent and needed help transporting my children. Eryn having a car made this possible" (parent survey). Her mother said the biggest household change since Eryn got her license was "I'm not spending hours in my car each day. Our gas usage has actually gone down as [Eryn] drives the less economical vehicle much less than when I was driving it everywhere" (parent survey).

Using her smartphone extensively, Eryn texted and used social media a great deal. She used ICTs daily to plan travel (teen survey), which was reflected in her network of friends and activities plotted out in the map exercises. In her survey she said that getting her license allowed her to "be involved in more activities and be with my friends more." The teen survey asked for information about the participants' five closest friends and where and how often they get together; for the friends she listed, Eryn visited three girlfriends daily outside of school and she saw two boys three or four times a week. She drove herself alone or with others to friends' houses. At the time of the focus group (June 2012) Eryn did not participate in afterschool sports, but in her survey (October 2011) she mentioned field hockey practice. Other places she listed going include another school to watch a soccer game, a food store, and two other friends' houses (teen survey). Eryn mostly drove herself to these destinations and all trips except for one were with others.

According to both her and her mother's surveys, Eryn had no restrictions on where she could go, whether by her driving or using other travel modes, and there were no agreements on what activities Eryn could or could not do. Her mother said that Eryn was required to tell her what time she would be home before leaving and Eryn had to follow state laws regarding who else can be in the car. When asked about other driving-based rules, her mother wrote "obviously, no use of phone" while driving (illegal in Vermont). The understood deal between Eryn and her mother was that she would transport her brother where he needs to go (text review and focus group discussion). Eryn did not have to pay for operating or insuring the car but her mother provided only one tank of gas per month—beyond that Eryn had to pay for fuel. As Eryn said in summary in her survey, "I use my car all the time."

Ramani

Ramani was seventeen, South Asian, in Grade 11 at School A, and lived with her parents, grandmother, and sister, age fourteen (surveys). Her activities included cross-country running and golf, violin, and a university math class; these took her to four different towns (teen survey). At the time of the focus group, her texting communications were about prepping for college entrance exams in study groups and doing fundraising at school for the lacrosse team. Ramani claimed that she had no curfew (teen survey). In the parent survey, her father said that they had agreements about going places with parents or "valid" drivers.

When she only had her driving permit (survey, October 2011), Ramani recorded a movie theater, shopping, a friend's house, and food store (all in neighboring towns) as "denied places" due to having "no way to get there." She also indicated that she hung out with her five listed friends primarily at her house or their houses. In the final comments of the survey, she wrote:

> It is difficult to get around when you have to rely on your friends and family for rides. In the summer, I am able to bike to more places because the weather is permitting. But, many places are too far away and I need to drive. But, with a permit, I can't drive by myself so I rely heavily on my parents for transportation. A lot of time is wasted waiting for rides as well. With my own car and a license, I could be much more efficient and get everywhere I need to go.

The following spring, having attained her license, Ramani was able to access these places, although she had to negotiate for use of her mother's car and noted she could get rides from friends who were also new drivers (paper map exercise). She stated that despite having her license, she still had to wait for rides with some frequency (focus group).

Ramani commented on infrastructure in several areas, both positively and negatively. In the digital

mapping exercise she mentioned a bike path she used in the summer. There is a short sidewalk near her neighborhood but it ends abruptly before any useful destinations can be reached. She repeatedly lamented (teen survey, digital mapping) that the public buses do not come near her neighborhood and she was prevented from going places.

Ramani mostly texted or used Facebook messaging. At the time of the survey, she had a cellphone but no data plan, but by the time of the focus group, she had a smartphone. For transportation she usually texted and sometimes called, but when her cellphone was broken she relied on Facebook on her computer (focus group). She texted for urgent things but used Facebook for events that were farther out in time or with friends from other schools; as events became closer she switched to text. Ramani used e-mail primarily with adults. Two of the six threads of text messages she reviewed with us (text review) were group texts to organize gatherings. Four threads were with peers and, of those, three were follow-ups from conversations in school hallways. Four threads involved Ramani looking for a ride somewhere, and another involved organizing an event at her home.

Jacob

Jacob is seventeen, white, in Grade 11 at School B, and lives with his parents and twin brother. His father earns more than $200,000 per year; his mother is home. There is little transportation infrastructure near their rural home, with no sidewalks, paths, or bus stops (parents' survey).

Jacob's life was filled with sports; his travel was all for practices and competitions, his friends were mostly teammates, and his communications were primarily about coordinating these activities (teen survey). He had no agreements or rules about where he could or could not go or how he got around, likely because he was totally dependent on his mother for transportation to school, sports practices, and games. At the time of the survey (October 2011), Jacob did not have his license or a cellphone, but by the focus group (June 2012) he had a learner's permit and a smartphone (although he mentioned poor cellular service at home).

In the survey, Jacob listed five important places that he went in the sample week: doctor's office, a friend's house, two parks, and a movie theater; all of these travels were as a car passenger. Looking at the friend network, Jacob listed five friends: four boys and his brother. Jacob saw his friends either at school, home (his brother), or out running. Jacob said his hometown "is out in the middle of nowhere. Lack of public transportation is a given, and everything is far away, so walking, or biking is impractical. You have to drive with either a friend or a family member to get around" (teen survey).

The mapping exercises reinforce the sense that Jacob mostly traveled to sports-related locations. In the text review, Jacob mentioned that he and his friends e-mailed well in advance for making arrangements, but texting is shorter and you can do it when you need to be quiet. His sample text was one to his brother inquiring how a tennis match was going at another court while Jacob himself was at the track and they were trying to figure out where to meet up afterward. He used the phone to call home for rides but not much with friends because most of his friends were on the same teams and he talked with them in person at and before and after competitions (text review, focus group).

Analysis

The vignette accounts enable us to generate some new insights on teens' combinations of mobility, activity, and communication as well as confirm previous work by others. First, as Dal Fiore et al. (2014) found for adult "digital nomads," rather than substituting for trips, ICTs are both substituting and complementing teen travel behavior. Mobile devices allow teens to stay connected but digital interactions do not replace in-person activities, particularly because teens often have little discretion over where they are supposed to be at a given moment (school, home, sports practice, etc.). Teens do make discretionary arrangements in the social realm: Ramani made arrangements first in the hallways at school, which were then followed up and refined through messaging, to ultimately meet together in person. Eryn used her cellphone and car to make social gatherings possible and frequent, even in the context of available public transportation. For Jacob, texts and phone calls allowed him to stay in touch with his family and friends while he participated in sports and coordinated rides.

Second, teens engage in advance planning and they organize events at the last minute, seeming to use different communication methods (e-mail vs. text) depending on the event's time frame; this is similar to Ling and Haddon's (2008) finding that mobile devices impact teens' "spontaneous, fluid nature of planning of *if, where* and *when* to meet" (145, italics added). Although ICTs allow teens to stay connected with family, friends, teachers, and coaches, tensions remain between the need to plan in advance and constraints

due to decisions and resources that are out of teens' control. For instance, even with Ramani's license and smartphone, she was limited in her ability to finalize plans because she did not always know whether or not she would have access to her mother's vehicle. In addition to the influence of timing on the type of communication, we found evidence of teens using e-mail and telephone calls with adults but texting and messaging with peers, leading to an even more complicated kaleidoscope of time–space organization.

Third, older teens (sixteen and older) who have access to cellphones and cars operate in complex sets of social obligations with family, friends, extracurricular activities, work, and school. With these communication and transportation privileges also come responsibilities, such as Eryn's driving her brother around and paying for fuel beyond the first tank. Ramani's negotiations for her mother's car involved getting up very early to drive her mother to work so that Ramani could then use the vehicle or, on the other end, having to wait long periods after sports practice for a ride. Jacob's friend network overlapped entirely with his sports activities, therefore achieving a sort of efficiency in the context of a great deal of time spent in practices and long journeys to competitions. Thus, although access to cellphones and cars could conceivably ease the challenges of middle-class teens, evidence here shows that these tools actually foster heightened expectations and intensity of activities, responsibilities, and time–space management.

In examining the mobility practices of teens and how they organize and enact their daily lives, we noticed important intersections among the built environment, digital networks, and social groups. Although the built environment and digital networks are primarily designed and constructed by adults, and even teens' social groups are influenced by adults, our findings hint at ways that teens blend whatever resources they have to manage busy, mobile lives.

First, the built environments of the two school districts present opportunities and constraints to high school teens. In the context of School A, the relatively compact housing, mixed-use commercial development, strong transit, and bicycle and pedestrian facilities meant that teens were able to get around to some degree without a car, as reflected in Table 1. The low density and long distances of School B's district made teen destinations harder to reach, lengthened trips, and required more advanced planning. Second, the digital landscape is similarly varied in its combinations of opportunities and constraints. Cellphone signals at the time of the study (late 2011 and early 2012) were fairly stable for teens in School A but very spotty in School B's catchment area. Broadband availability in the relevant areas also varied based on population density, terrain, remoteness, and the cable/ digital subscriber line coverage (Broadband VT 2010), resulting in near-universal access for School A residents but differential availability for School B teens. Uneven cellphone signals were a significant issue reported by students from the more rural towns served by School B and, indeed, seem to have influenced the rates of teens having cellphones: 95 percent of students at School A had cellphones, whereas 81 percent of students at School B did. Both rates were well above the U.S. national average among teens at the time of 77 percent (Madden et al. 2013), perhaps reflecting the wealthy demographics, but the difference between schools was, according to students themselves, due to poor rural cell coverage.

Further, the social landscape of teens is characterized by diverse relations between them and their parents, siblings, friends and peers, teachers, coaches, and employers. We found that individual teens occupied simultaneous positions of responsibility and dependence, particularly among sixteen- to eighteen-year-olds who had greater spatial mobility and, in most cases, driver's licenses. Perhaps in keeping with the upper-middle-class demographics, most respondents in both schools demonstrated complicated schedules with extracurricular sports and other commitments; jobs and internships; homework and academic pursuits; socializing with friends; and, of course, school and family filling their days. In the teen survey, many respondents referenced the challenges they face in managing their activities, mobilities, and time–space commitments. Some activities were more rigid (e.g., sports games or performances) and required planning ahead, whereas other activities lent themselves to last-minute or flexible plans. Among the older cohort (sixteen and older), the diversity and intensity of activities increased, which in many cases was both made possible by and, in turn, necessitated teens' access to cellphones and cars.

Finally, the built, digital, and social landscapes are conditioned by individual contexts, and teen mobility practices depend on access to both public and private resources. For Eryn, this meant a trade-off of high levels of spatial freedom (her own car, largely paid for by her mother) for responsibility (driving her brother around). For Ramani, this manifested in daily negotiations (often by texting and phoning) for rides or access to her mother's car; despite the fact that her school is

in a relatively transit-rich location, her spatially extensive and time-sensitive activities meant that she constantly employed digital and social resources to overcome challenges in the built environment. For Jacob, living in rural isolation, having his mother act as a driver and being friends with his teammates afforded him the ability to negotiate the challenges of distance and poor cellphone coverage. Thus, we found that teens navigate different contexts using multiple resources and strategies based on their situations and household and personal decision making.

Conclusions

Although the development and adoption of new digital technologies vastly outpaces changes in the social and built environments, the fact that these dimensions overlap and interact with each other means that teen mobility practices will also continue to evolve. Building on research from the 2000s, cited earlier, we hope to provide a snapshot in time of a complex set of factors that are continually shifting, which in turn prompt and necessitate adaptation and innovation by future teens and ongoing research to understand societal shifts.

Research on teens' mobility practices that focuses on the intersections of the built environment, digital technologies, and social groupings and that highlights teens' agency in navigating, negotiating, and resisting constraints and opportunities will help us understand new combinations of mobility, communication, and place. Future investigations will also help destabilize, or at least contextualize, stereotypes about young people's mobility, such as the popular notion that driving is less important for millennials than for previous generations (U.S. Public Interest Research Group 2013), and instead pay attention to making all places more accessible to all groups. Similarly, as unevenness in the digital landscape continues to be relevant, there is a corresponding need for the public to negotiate important equity and social justice questions around the digital divide. Although the teens we worked with occupied positions of relative privilege, particularly in socioeconomic status, they experienced marginalization through the cumulative effects of adultist urbanism. Jacob, Ramani, and Eryn were able to make their busy lives work in part through the negotiations they made daily using access to cellphones and cars, but many other teens in our survey, especially those without licenses and in lower income households, were far more anchored to the key sites of teen life (home, school, etc.) and had little access to ICTs. As Gilbert et al. (2008) pointed out, "By learning what strategies are being employed successfully by marginalized populations, both at the individual and collective scales, we can gain a better understanding of how ICTs can be a part of improving quality of life" (923) for all populations.

Funding

This research was funded in part by a grant from the U.S. Department of Transportation through the UTC Program at the New England University Transportation Center (at the Massachusetts Institute of Technology) and the University of Vermont Transportation Research Center.

Notes

1. The phones remained in the hands of the students at all times and each teen was asked to reveal only information he or she was comfortable sharing. Leaving the phone in the teens' hands allowed them to curate their own summary of messaging and was important for respecting the teens' privacy.
2. This use of the term is distinguished from the practice of introducing hypothetical stories ("vignettes") to research subjects for the purposes of data collection, which is common in psychology and health sciences.
3. Due to space considerations we did not include all five students' vignettes. We feel that the three chosen represent sufficient richness and variety to illustrate our points. Jacob's twin brother had virtually the same daily patterns and friend groups, so we only chose to include one of the boys. The other female participant (School B) was fascinating and vocal in her resistance to using a car or a cellphone (although she owned both) but admitted that she was very much in a minority at her school. Although her perspectives were useful in putting her peers' views in more stark relief, her self-proclaimed "outlier" status made it difficult to incorporate her story succinctly and within the page limit. We hope to revisit her position in a future paper.
4. Pseudonyms are used throughout.

References

Blumenburg, E., B. Taylor, M. Smart, K. Ralph, M. Wander, and S. Brumbaugh. 2012. *What's youth got to do with it? Exploring the travel behavior of teens and young adults.* Los Angeles: Institute of Transportation Studies, UCLA. http://www.uctc.net/research/papers/UCTC-FR-2012-14.pdf (last accessed 20 October 2014).

Broadband VT. 2010. Coverage statistics and maps. http://publicservice.vermont.gov/topics/connectivity/broadbandvt/12312010 (last accessed 29 January 2016).

Bungum, T., M. Lounsbery, S. Moonie, and J. Gast. 2009. Prevalence and correlates of walking and biking to school among adolescents. *Journal of Community Health* 39:129–34.

Clifton, K. 2003. Independent mobility among teenagers: An exploration of travel to after-school activities. *Transportation Research Record* 1854. http://www.ltrc.lsu.edu/TRB_82/TRB2003-001412.pdf (last accessed 7 July 2014).

Dal Fiore, F., P. L. Mokhtarian, I. Salomon, and M. E. Singer. 2014. "Nomads at last"? A set of perspectives on how mobile technology may affect travel. *Journal of Transport Geography* 41:97–106.

Emond, C., and S. Handy. 2012. Factors associated with bicycling to high school: Insights from Davis, CA. *Journal of Transport Geography* 20 (1): 71–79.

Gilbert, M., M. Masucci, C. Homko, and A. Bove. 2008. Theorizing the digital divide: Information and communication technology use frameworks among poor women using a telemedicine system. *Geoforum* 39:912–25.

Grinter, R., and L. Palen. 2002. Instant messaging in teen life. Paper presented at the Association of Computing Machinery 2002 Conference on Computer Supported Cooperative Work, New Orleans, LA.

Hjorthol, R. 2008. The mobile phone as a tool in family life: Impact on planning of everyday activities and car use. *Transport Reviews* 28 (3): 303–20.

Holloway, S., and G. Valentine, eds. 2000. *Children's geographies: Playing, living, learning.* London and New York: Routledge.

———. 2003. *Cyberkids: Children in the information age.* London and New York: Routledge-Falmer.

Kesselring, S. 2006. Pioneering mobilities: New patterns of movement and motility in a mobile world. *Environment and Planning A* 38:269–79.

Larsen, J., J. Urry, and K. Axhausen. 2006. *Mobilities, networks, geographies.* Aldershot, UK: Ashgate.

Lee, B. H. Y., and M. Cope. 2012. Teens on the move: Implications of land-use and transportation practices for youth mobility in northern Vermont, USA. Paper presented at the annual meeting of the Association of American Geographers, New York.

Ling, R., and L. Haddon. 2008. Children, youth and the mobile phone. In *International handbook of children, media and culture*, ed. K. Drotner and S. Livingstone, 137–51. London: Sage.

Madden, M., A. Lenhart, M. Duggan, S. Cortesi, and U. Gasser. 2013. *Teens and technology 2013.* Washington, DC: Pew Research Internet Project. http://www.pewinternet.org/Reports/2013/Teens-and-Tech/Main-Findings/Teens-and-Technology.aspx (last accessed 25 September 2013).

Oksman, V., and J. Turtiainen. 2004. Mobile communication as a social stage: Meanings of mobile communication in everyday life among teenagers in Finland. *New Media Society* 6:319–39.

Pain, R., S. Grundy, and S. Gill, with E. Towner, G. Sparks, and K. Hughes. 2005. "So long as I take my mobile": Mobile phones, urban life and geographies of young people's safety. *International Journal of Urban and Regional Research* 29 (4): 814–30.

Thulin, E., and B. Vilhelmson. 2005. Virtual mobility of urban youth: ICT-based communication in Sweden. *Tijdschrift voor Economische en Sociale Geografie* 96 (5): 477–87.

———. 2007. Mobiles everywhere: Youth, the mobile phone, and changes in everyday practice. *Young* 15:235–53.

———. 2009. Mobile phones: Transforming the everyday social communication practice of urban youth. In *The reconstruction of space and time: Mobile communication practices,* ed. R. Ling and S. Campbell, 137–58. London: Transaction.

———. 2012. The virtualization of urban young people's mobility practices: A time-geographic typology. *Geografiska Annaler B* 94 (4): 391–403.

U.S. Public Interest Research Group. 2013. A new direction: Our changing relationship with driving and the implications for America's future. http://uspirg.org/reports/usp/new-direction (last accessed 5 October 2015).

Latin@ Immobilities and Altermobilities Within the U.S. Deportability Regime

Marta Maria Maldonado,* Adela C. Licona,[†] and Sarah Hendricks[‡]

*School of Language, Culture & Society, Oregon State University
[†]Department of English, University of Arizona
[‡]Department of Behavioral Sciences, University of Tennessee–Martin

In this article, we explore how racialized constructions of a "Latin@ threat" serve as ideological underpinning for the practices of the U.S. deportability regime and also fuel broader practices of policeability, with consequences for Latin@ mobilities and immobilities. Drawing from ethnographic observation and in-depth interviews with Latin@s in Perry, Iowa, we discuss "the border within" as an extension of border politics and borderlands rhetorics to the U.S. heartland, explore imposed mobilities and immobilities, and also recognize tactical immobilities and altermobilities undertaken by Latin@s.

我们于本文中探讨 "拉丁裔威胁" 的种族化建构, 如何作为美国将移民驱逐出境的体制实践的意识形态基础, 同时催化了更为广泛的警备维安实践, 并导致了拉丁裔的能动性与不动性。我们透过对爱荷华州佩里的拉丁裔进行民族志观察与深度访谈, 探讨 "内部边界" 作为对美国心脏地带而言的边界政治和边境修辞的延伸, 探究被强加的能动性和不动性, 并认定拉丁裔所采取的策略性不动性与改变的能动性。

En este artículo exploramos el modo como las construcciones racializadas de una "amenaza Latin@" sirven de apuntalamiento ideológico para las prácticas del régimen de deportabilidad estadounidense, al tiempo que alimentan prácticas de policibilidad de mayor alcance, con consecuencias para movilidades e inmovilidades Latin@s. Con base en observación etnográfica y entrevistas a profundidad con los Latin@ en Perry, Iowa, presentamos una discusión acerca de "la frontera de adentro" como una extensión de políticas fronterizas y de retórica de fronteras para el núcleo continental estadounidense, exploramos las movilidades e inmovilidades impuestas, y también registramos las inmovilidades y alternomovilidades tácticas emprendidas por los Latin@s.

Nation-states routinely design and deploy techniques and practices of deportation and deportability. Each nation has its own history of relations with other nation-states and its unique history of ethnoracialized and classed labor relations. In large part, such histories provide the ideological content, the storylines that justify who is authorized entry into national space and under what conditions, who is more or less free to move, and who must be removed. Understanding the everyday production of immigrant mobilities and immobilities thus necessitates elucidating the histories and ideologies that anchor and animate particular national deportation and deportability regimes. Similarly, mobility and immobility are key dimensions of immigrants' lives, shaping the ability to make connections to host societies and receiving communities (see Coleman and Kocher 2011; Bose 2013; Stuesse and Coleman 2014).

Latin@s[1] are one of the fastest growing ethnoracialized populations in the United States. In this article, we draw from a case study in Perry, Iowa, a rural "new gateway" in the U.S. heartland, to explore mobilities and immobilities as everyday aspects of the simultaneously racialized and spatialized politics of Latin@ incorporation. We begin by discussing the historical and ideological contexts that give meaning to Latin@ experiences on U.S. soil. We then offer an interpretation of how historical and ongoing racialized constructions of Latin@s shape the focus and practices of the U.S. deportability regime. Subsequently, we introduce the context of Perry, Iowa, and go on to consider mobilities and immobilities as experienced and narrated by Latin@s who live and work there.

The Threat of Latin@ Strangers

Although Latin@s have been integral to U.S. development since the mid-1800s, and although 62 percent are U.S. citizens by birth as of the 2010 U.S. Census, dominant rhetorical representations frequently portray them as foreigners, recent arrivals, and outsiders to U.S. society. Ahmed's (2000) discussion of the sociocultural construction of "strangers" in relation to notions of home is useful for contextualizing constructions of Latin@s as unbelonging in the United States and elucidating consequences for their mobility. Ahmed argued that a stranger is not someone we do not recognize but someone we recognize precisely as a stranger. She suggested that there are techniques that allow for reading and recognizing some people as "not belonging, as being out of place" (Ahmed 2000, 21). The politics of mobility are about such notions of unbelonging and "who gets to move with ease across the lines that divide spaces [which] can be re-described as the politics of who gets to be at home" (Ahmed 2007, 162). Neighbors belong. Strangers don't. Strangers are, Ahmed explained, those "whose behavior seems unpredictable and beyond control" (quoting Merry 1981, 125). Such unpredictability produces anxiety. Expressed spatially, this anxiety looks like and effects boundary making; expressed rhetorically, it is pathologizing, criminalizing, and illegalizing.

Stranger anxiety contributes to what Shamir (2005) called a *paradigm of suspicion*, which provides ideological justification for a mobility regime:

> [T]he primary principle for determining the "license to move," both across borders and in public spaces within borders, has to do with the degree to which the agents of mobility are suspected of representing the threats of crime, undesired immigration, and terrorism, either independently or, increasingly, interchangeably. ... This mistrust has been an important engine in the increasing formal criminalization of mobility itself. (202)

We argue that the Latin@ threat is the paradigm of suspicion foregrounded in the U.S. mobility regime, with roots in the historical and ongoing incorporation of Latin Americans into the United States as colonized and racialized subjects. Although immigrants from all over the world make their home in the United

States, not all immigrants are equally visible or central in official and popular representations of immigrants and therefore targeted by underlying mistrust. Typically, immigration in the United States is treated as a Latin@ issue. U.S. concerns with a Latin@ threat are long-standing (Chavez 2013). Recently, however, a particularly virulent anti-Latin@ and related anti-immigrant climate has become entrenched (see Beirich 2011; Hartman, Newman, and Bell 2014). After 11 September 2001 (hereinafter 9/11), we have also witnessed an alignment of racialized discourses about immigration and immigrants (with a focus on Latin@s) and discourses about terror and terrorists (see Puar 2007; Bloodsworth-Lugo and Lugo-Lugo 2014; Lugo-Lugo and Bloodsworth-Lugo 2014).

A culture and politics of fear associated with the "war on terror" and related concerns about a Latin@ threat provide the racialized and gendered storylines that animate the paradigm of suspicion in the United States, with significant implications for movement and mobility, broadly framing Latin@s as dangerous strangers and therefore as perpetually deportable. In such a context, deportation emerges as a way for the state to manage concerns about social proximity of those perceived as threatening and dangerous (Shamir 2005) and to create an illusion of secure space for those for whom detention and deportation are ideologically constructed as safety measures (e.g., U.S. citizens, white women, and children).[2]

Latin@s and the U.S. Deportability Regime

In the United States, deportation practices are increasingly deployed and managed as part of a national security strategy. In his 18 July 2008 congressional testimony, then–Department of Homeland Security (DHS) Secretary Michael Chertoff alluded to the dramatically increased number of immigrant apprehensions and "removals" as evidence that DHS was succeeding in its stated goal to "secure the homeland and protect the American people." Concomitant with "stepped up" immigration enforcement after 9/11 has been the accelerated expansion of an immigration industrial complex (Huling 2002; Golash-Boza 2009; Doty and Wheatley 2013). On average, DHS detains 34,000 individuals daily, the vast majority of whom are Latin@, mostly Mexican.

These trends have clear significance for Latin@ mobilities and immobilities. In a regime of deportability, everyday space interacts with spaces of (threatened) detention, deportation, and deportability to produce a population of seemingly suspicious persons that is mobile and immobile in particular ways (for a discussion of spatiality and subject formation, see Shabazz 2009). In effect, physical containment of suspicious persons constitutes one form of the emerging global mobility regime, as it functions to draw boundaries around populations in the forms of prisons, urban ghettoes, and quarantines (Shamir 2005; Shabazz 2009). For populations criminalized through the racialized and gendered rhetorics of illegality, the regime of deportability creates a material as well as a psychic threat of detention and deportation: It works to make their threatening bodies visible, even hypervisible (Licona and Maldonado 2014), and therefore more readily surveilled and contained. The regime of deportability also produces "coercive spatial arrangements" that are "laden with architecture, techniques, and rules that circumscribe workers' mobility" (Shabazz 2009, 281). At various points throughout the past decade, the bodies of Mexican and Central American immigrant workers have been brought out from the shadows in which they are forced to live (by virtue of being illegalized and deportable), in a series of very visible raids (the largest was undertaken in Iowa), as part of the reproduction of the narrative of a looming menace and threat of terrorism and terrorists. Surveillance tactics, biometric monitoring, policeability, the prisonization of space and labor, and the constant and expanding power of deportation form specific racialized containment strategies (Shamir 2005; Shabazz 2009), which rely on spotlighting immigrant persons. Immigrant bodies are rendered visible to demonstrate not only the ongoing need for the protection of nonimmigrant, noncriminal, "deserving" and normative populations from perceived dangerous strangers but also the power of the state to offer this protection.

Managed in large part by deportability regimes, the production of racialized, gendered, and class-based immigrant mobilities and immobilities is both a translocal and transnational process, linked to global political economic forces (Nelson and Nelson 2010). This becomes apparent when one considers the seeming incongruence between increasingly restrictive immigration policies, continued labor demand, and heavy reliance on immigrant labor from Latin America in the United States. Despite the nationally visible "spectacle" of raids as tough immigration enforcement and the sweep of regressive legislative measures across the nation, Latin@ immigrant workers continue to be heavily recruited and effectively forced to pursue a surreptitious mobility that enables the continuation of business as usual—the continued global production, accumulation, circulation, and consumption of capital (Sassen 1988; Nelson and Nelson 2010).

De Genova (2010) argued that the freedom of movement is inseparable from labor, the capacity to creatively transform our objective circumstances. Labor is also always necessarily relational, so individuals enact varying degrees of freedom of movement within particular forms of familial and social organization, embedded in and always constituted by multiscalar relations of power. Large-scale deportation practices have profound impacts on a wide range of individuals. Deportability and practices of removal and deportation affect not just unauthorized Latin@ immigrants. Many Latin@ families include both documented and undocumented members, so the threat of deportability entails potential family separation, with its emotional and material consequences. Likewise, U.S.-born Latin@s are affected by the rhetorics and practices of the regime of deportability because, given the racialization of Latin@s as foreigners and immigrants, being "read" as Latin@ immediately renders one "suspect" of illegality.

Within the context of the deportability regime, a subcontext of *policeability* (Rosas 2006) is produced, whereby both practices of surveillance (Bajc 2013) and tactics of resistance can be identified. Policeability can be described as a state of constant surveillance predicated on the hyperregulation of routine activity, evident in displays of state power, vigilantism, and the informal management of everyday life. A focus on policeability helps discern the everyday production of Latin@ mobilities and immobilities in three major ways. First, it calls attention to how the disciplining practices of the state extend beyond (federal) government agencies directly charged with enforcement of immigration controls, highlighting or allowing for consideration of the role of local authorities in surveilling, containing, and policing immigrants. Second, it allows for examination of the informal and everyday ways in which borders are enacted and enforced. Third, it allows for consideration of questions of legality and illegality in relation to the discourses and actions of the state and, more broadly, of those in power. In what follows, we turn to a discussion of Latin@s' reported experiences of policeability, mobility, and immobility in Perry, Iowa.

Methods

This article is based on data obtained through ethnographic observation and in-depth, semistructured interviews with fifty-five Latin@s who reported living, working, or both in Perry, Iowa. Initial contact points for the interviews were Latin@s active in community organizations and local service agencies. Other contacts were accessed through the local Catholic church. The sample was expanded in a snowball fashion. In an effort to attend to Latin@ heterogeneity, the sample included twenty-two men and thirty-three women, different age groups and national origins (including both U.S.- and foreign-born), different lengths of residence in the United States and in Perry, and varied English-language proficiencies. Although participants were never asked about their immigration status, in the course of the interviews it became apparent that the sample included people with different immigration and citizenship statuses.

Participants were asked a range of questions about their day-to-day lives in Perry. Interviews were transcribed verbatim, and participants were assigned pseudonyms.[3] Data analysis entailed an initial phase of coding to identify data pertinent to the initial research questions (the original goal of the ethnographic research project was to explore how Latin@s understood, and the extent to which they experienced, integration) and also emergent themes. A second phase of coding involved condensing themes into categories. One of the major categories that emerged in participants' discussions was a range of everyday practices that Latin@s employ to manage conditions linked to policies and practices aimed at surveilling and controlling their presence, circulation, and behavior.

Perry, Iowa

Perry is a small town of 8,108 residents (U.S. Census Bureau 2013) located along the North Raccoon River, northwest of Des Moines, in central Iowa. Meat processing has been central to Perry's economy since the 1920s. Historically, whites of European ancestry constituted the vast majority of the town's residents and its meat processing workers. Beginning in the 1990s, changes associated with industrial restructuring made meat-packing jobs unattractive to this local population, and employers began aggressive recruitment of Latin@ workers from areas of established Latin@ settlement (Kandel and Parrado 2005). Such recruitment paved the way for the demographic transformation of Perry.[4] The local meat-packing plant continues to be the primary local employer. Census data show that there were forty-seven Latino residents in Perry

in 1990. By 2010, Latin@s had become 35 percent of the town's total population.

The Border Within

Latin@ incorporation into Perry and other new gateways across the U.S. heartland is an embodied and spatialized manifestation of *the border within*, by which the perception of some people as strangers serves to justify differential inclusion—and exclusion—within a nation whose ideals espouse equal rights (Espiritu 2003). Latin@ accounts of their day-to-day lives in Perry reveal how policeability and other practices of surveillance, not only by immigration authorities but also by local law enforcement and non-Latin@ residents, shape Latin@ subjectivities and everyday movements. This is true for immigrants and for entire Latin@ families. For example, José noted the relationship between Latin@s in Perry and local law enforcement:

> The thing is that this town is so small. . . . The police has us (Latin@s) hyperwatched. They're always observing us . . . we're always under custody.

Minerva, a U.S. citizen who has lived in Perry for more than a decade, spoke about her perception that sometimes police detain and subject Latin@s to procedures that exceed official authority:

> The majority of the Hispanic population in this town is undocumented, so they fear. There have been moments when the police have acted as immigration agents, and there have been some police officers who discriminate. . . . They were abusing Hispanics, going beyond what their job required them to do. For example, I have seen that, if I get stopped for an infraction, they don't bring the dogs out on me. But, if someone looks to them like they might be undocumented, and they commit an infraction, instead of addressing the infraction, they bring in the canine unit to see if they have drugs, or this, or that, or the other, and they take away their car.

Young Latin@s, too, reported feeling under heavy and constant surveillance by police and other local authorities, such as teachers and city facilities personnel, who presume that they are "always up to no good" and that they might be involved in criminal or gang activity. Common among young Latino men interviewed was a sense that they cannot be out in public places, especially with other Latin@s, without being detained by police. During our ethnographic observations, we witnessed several instances of police holding, intervening with, and dispersing groups of young Latin@s gathered in public spaces.

Beyond interactions with law enforcement and local authorities, a feeling shared by most Latin@ immigrants interviewed was that their whereabouts, movements, and actions are constantly and carefully monitored by their non-Latin@ (white) neighbors and residents. Many reported being the target of disapproving or "evil looks" at restaurants or stores, being confronted or scolded by non-Latin@s for speaking Spanish, for painting their house a particular color, or parking cars on their own lawn. Several immigrant women spoke about going grocery shopping at night to avoid the quizzical, disapproving stares of non-Latin@s, even though they know they miss out on the best, freshest produce. These experiences illustrate the ways in which policeability extends beyond formal contexts and official authorities to reach, discipline, and conscript the movements of immigrants in and through a wider variety of practices and community interactions.

Transportation, Automobility, and Policeability

The border within disproportionately affects the mobility of Latin@s in Perry, manifesting in the impoundment of private automobiles, the dispersion of youth from public spaces, and hostility that discourages Latin@s from frequenting commercial institutions. In addition to these manifestations of policeability, heavy reliance on personal automobiles in the United States and regulatory practices of space and transportation serve to conscript Latin@ movement.

Driving a personal automobile is vital to physical survival and social interaction in the United States. Latin@s in Perry travel less in private vehicles than do non-Latin@s, however. For instance, the foreign-born in Perry (nearly all from Latin America) are 36 percent less likely to drive alone to work than are the U.S.-born (U.S. Census Bureau 2013). Additionally, Latin@-headed households in Perry are 66 percent more likely to live in a household without a vehicle (U.S. Census Bureau 2000).

That Latin@s travel less by personal vehicle is due in part to low income and the high cost of owning, maintaining, and operating a vehicle. Latin@s in Perry make $11,797 per capita, compared to $20,802 per capita earned by non-Latin@ whites (U.S. Census Bureau 2013). Most recently arrived Latin@s in our study reported not owning a car and not being able to afford one, especially not a newer, reliable one.

State-level regulations that exclude undocumented immigrants from getting driver's licenses exacerbate the financial constraints on mobility and even create conditions of *forced immobilizations* (Stuesse and Coleman 2014) of Latin@ individuals and households. In Iowa, only those immigrants granted Deferred Action for Childhood Arrival (DACA) status are able to obtain a driver's license. Many Latin@ immigrants, whose labor is needed, actively sought, and welcomed in Perry, are not allowed to drive legally. As a result, Latin@s might not drive even when a serious need arises, and many experience considerable stress and fear when they do drive.

For example, several Latin@s who work at the local meat processing plant mentioned having experienced injuries at work and explained that a lack of transportation compounded by a fear of detection and detention by law enforcement discouraged them from traveling to obtain needed treatment. Overall, it is clear from Latin@ accounts that an ever-present fear of detection and detention by police or U.S. Immigration and Customs Enforcement (ICE) routinely shapes their movements and nonmovements, influencing decisions about when to leave home and when not to, how to circulate in town, and routes to obtain health care and medical assistance, shopping, and entertainment options. Asked about challenging aspects of his transition to Iowa, Vicente noted:

> Transportation. I didn't know how to drive, and here to hold a job one has to have a car. And one is always in fear . . . (The state) should allow one to get some kind of documentation that allows one to drive . . . one is constantly worried that the police might stop one, and take away one's car. . . . I've been stopped by the police, and because I don't have a driver's license they make me step out of the car . . . and then one is not allowed to drive, and has to call someone to come pick one up. It has happened to me where I haven't been able to get to work, and I have to call in to say that I won't make it.

Vicente neither knew how to drive nor had a car when he arrived in Iowa. Yet he drives out of sheer necessity, although he would prefer to abide by laws and get a driver's license. Because he cannot, he is caught in the contradiction of a local economy's need for his labor and laws restricting his ability to travel legally from his residence to work. This contradiction thus exerts a coercive force, rendering Vicente (and others like him) visible and vulnerable to policeability.

Vicente's comments highlight how in Perry (and arguably elsewhere in the United States) transportation space is increasingly linked to the regime of deportation and deportability, as public roads are one space where immigrants are particularly subject to surveillance. Automobiles create private spaces of individual containment in public space (Henderson 2006). The profiling and regulation of immigrant movement invades these private spaces and functions as a screen, sifting through those who are moving and identifying and removing those who are undesirable or perceived as threatening. The experiences of many of those interviewed in Perry suggest that moving in cars makes them more vulnerable to detention, because, as they describe, police can and often do stop them with almost any pretext.

Many who do not drive in the context of an automobile-dependent society turn to carpooling as an alternative transportation strategy. Carpooling could be touted as a form of social capital because it enables access to places otherwise inaccessible to those without automobiles (Charles and Kline 2006). Carpooling provides only a partial and limited solution, however. Several problematic aspects of carpooling are illustrated by comments from Carlos, a Guatemalan immigrant:

I: So how do you get to work?

R: To get to work, there are several ladies who work where we work, and they give us rides. They charge us five dollars daily, to help pay for gas. ... This week ... the lady was free on a day I was supposed to work, so I didn't have a ride. Another guy who lives down the street has told me that when I need it I can use his car.

Carpooling provides travel to a limited number of destinations, such as between home and work, whereas living entails additional travel purposes. Carlos has addressed the dilemma of how to commute to work without a car but not other needs, such as purchasing groceries, getting to the doctor's office, attending parent–teacher meetings, going to church, or attending other social and recreational functions. Carpooling also makes riders dependent on the availability and generosity of drivers, who might be relatives, neighbors, coworkers, or members of religious organizations. Carpooling arrangements also failed to meet Carlos's transportation needs when the driver did not need to work on a day Carlos did; many who depend on carpooling do not have an open offer to borrow a car. In addition, carpooling creates a power relationship in which the driver is granting a favor to the rider, and social reciprocity tends to be expected. The reciprocity involved in paying the driver often ends up costing much more than comparable public transit costs in places with public transportation (Bohon, Stamps, and Atiles 2008), and those who do not have the money are not able to reciprocate.

Many Latin@s respond to transportation barriers by avoiding travel when possible (Lovejoy and Handy 2011). Maria explained her predicament:

M: I don't know or mingle with a lot of people, only the families of my kids' friends.

I: How come you don't know or hang out with others?

M: Because I'm always here [home]. I never go out. I don't have a car to move around. If I have to go somewhere, it's difficult; I mostly have to stay nearby, here in Perry.

Latin@s who do not have access to personal automobility, such as Maria, simply do not make trips that they otherwise might and therefore exist in a geographically constricted area. As a result, lack of transportation severely restricts mobility and social life. Among those interviewed in Perry, Latina immigrant women tended to be the most isolated, rarely circulating outside their homes. The limitation on driver's licenses, policeability practices in public road spaces, and the U.S. culture of automobile dependency together shape the everyday mobility patterns of Latin@s in Perry.

Transgressive Latin@ Altermobilities

Despite the significant constraints on mobility imposed by the deportability regime and its everyday racializations, most Latin@s find ways to move about, even if only on a limited basis. Highlighting the relationship between deportability and mobility, Latin@ movements and their timing are necessarily strategic, aiming to avoid detection, discomfort, and trouble. For example, when needing to travel to the nearby city of Des Moines for health care or entertainment, immigrants spoke about taking back roads instead of traveling on highways. Also, many of the destinations Latin@s travel to within and outside of Perry tend to differ from the destinations frequented by non-Latin@ residents. For example, Latin@s and non-Latin@s frequent different local pubs. Some immigrant Latin@s reported shopping at local Latin@-owned establishments, even when shopping at the chain grocery store might be less expensive, mostly to avoid uncomfortable encounters with non-Latin@s. Latin@ immigrant families and some mixed-status families with automobiles pursue outdoor leisure options (e.g., fishing) together, in remote locations, where they are less likely to encounter others. Sometimes several families carpool to pursue such options.

One important tool that facilitates Latin@ mobilities in Perry are cell phones, often in conjunction with Spanish radio broadcasts. These technologies enable Latin@s to monitor the presence and movements of police and immigration authorities and so the possibilities for detainment and deportability. For example, Margarita recounted a recent event:

> Last weekend, they said on the radio that there were going to be police all along Highway 141, but they didn't know exactly at which points or for how long. What they knew was that it was going to be that way until three in the morning. So then we communicated, nobody go out—if we have to buy food, do so right here, even though it's more expensive. But that way we avoid the police.

The threat of deportability, and sometimes knowledge about the movements of police and immigration authorities, the quick dissemination of which is enabled by mobile technologies and radio broadcasts, is directly related to immigrants undertaking tactical immobilities and crafting altermobilities, often pursuing isolated routes for automobile travel and for daily movements.[5]

Conclusions

The need to bridge the micro and macro levels theoretically, empirically, and methodologically has been identified as a key challenge facing mobility scholars (e.g., D'Andrea, Ciolfi, and Gray 2011). This is, indeed, a formidable challenge, to which we have sought to respond by attending to the relationship between racialized historical representations of Latin@s, practices in and of the U.S. deportability regime, and everyday life, with a focus on mobility and immobility at the local level. Our case study helps trace the particular ways in which anti-immigrant rhetorics are related to the rhetorics of anti-terrorism and to related practices of surveillance and policeability in a regime of deportability and how these rhetorics justify the production of compromising, potentially, and actually exploitative conditions for the mobility and immobility of immigrants and of racialized Latin@ populations in the United States.

The conditions and kinds of mobilities and immobilities we have highlighted here (related to the conditions and kinds of racially coded visibilities and invisibilities; see Licona and Maldonado 2014) are produced not just by detention and deportation practices but also, in large part, by the threat of deportability. Imposed mobilities and immobilities render immigrants—and some nonimmigrants, given both racialization and familial and other social ties—isolated and vulnerable, by limiting or altogether curtailing access to business transactions, social interactions, entertainment, and health care. In so doing, they effectively conscript the expression of Latin@ subjectivities spatially, simultaneously reproducing power structures and exploitative social arrangements across geographic scales. The case of Perry shows also the "unofficial arm" of the U.S. deportability regime, the ways in which the rhetorics and practices of the regime shape cultural scripts in communities extending to the informal management of everyday life.

Our research also illustrates the tensions and the relationalities among policing, deportability, and mobilities and immobilities, revealing the everyday form and political content of the movements and nonmovements of Latin@ immigrant and nonimmigrant populations vis-à-vis those of normative, nonimmigrant, white populations in the racialized U.S. context. Conscripted mobility and forced immobilities are routinely imposed on Latin@ immigrants as well as on U.S.-born Latin@s by the U.S. deportability regime. In turn, Latin@s craft and pursue a range of alternate, transgressive mobilities and also tactical immobilities, as strategies of survival and tactics of resistance. The experiences of Latin@s in Perry highlight how race or ethnicity and class are embodied and become manifest rhetorically and spatially through movements and nonmovements and routings and reroutings that result from the relation and tension between power and resistance. Perry, Iowa, provides but one local example of the border within at a time of intensified immigrant policing in the United States by federal, state, and local authorities with impacts on the everyday movements, material conditions, and subjectivities of Latin@ populations.

Acknowledgments

The authors wish to thank the editors of the special issue and anonymous *Annals of the American Association of Geographers* reviewers for their helpful comments and suggestions that improved and strengthened this article.

Notes

1. We use the term Latin@ to acknowledge gender as fluid and to interrupt "masculine generics", the practice of using the male form of names to represent all experiences.
2. See, for example, documentary films *Farmingville* (produced by Camino Bluff, 2003) and *Los Trabajadores/The Workers* (produced by Independent Television Service, 2001), which document residents' fears in relation to the growing presence of Latin@ immigrants in Farmingville, New York, and Austin, Texas, respectively.
3. Excerpts from interviews in Spanish were translated by one of the researchers, who is a native speaker.
4. Other contributing factors channeling Latin@s to Perry and rural towns throughout the Midwest include the militarization of the U.S.–Mexico border, the passage of restrictive policies and growing anti-immigrant sentiment in established gateways, social networks, lower cost of living, availability of affordable housing, and the desire to move away from areas of high crime and gang activity (Fennelly and Leitner 2002; Zúñiga and Hernández-León 2005; Massey 2008).
5. This finding is consistent with findings from studies in other new gateways (see, e.g., Stuesse and Coleman 2014).

References

Ahmed, S. 2000. *Strange encounters: Embodied others in postcoloniality*. London and New York: Routledge.
———. 2007. A phenomenology of whiteness. *Feminist Theory* 8:149–68.
Bajc, V. 2013. Sociological reflections on security through surveillance. *Sociological Forum* 28 (3): 615–23.
Beirich, H. 2011. The anti-immigrant movement. Southern Poverty Law Center, Intelligence Files. http://www.splcenter.org/get-informed/intelligence-files/ideology/anti-immigrant/the-anti-immigrant-movement (last accessed 5 October 2011).
Bloodsworth-Lugo, M., and C. Lugo-Lugo. 2014. *Project(ing) 9/11: Productions of race, gender, and citizenship in recent Hollywood films*. Lanham, MD: Rowman & Littlefield.
———. 2010. *Containing (un)American bodies: Race, sexuality, and post-9/11 constructions of citizenship*. New York: Rodopi.
Bohon, S., K. Stamps, and J. Atiles. 2008. Transportation and migrant adjustment in Georgia. *Population Research and Policy Review* 27 (3): 273–91.
Bose, P. 2013. Building sustainable communities: Immigrants and mobility in Vermont. *Research in Transportation Business and Management* 7:81–90.

Charles, K. K., and P. Kline. 2006. Relational costs and the production of social capital: Evidence from carpooling. *Economic Journal* 116 (511): 581–604.
Chavez, L. R. 2013. *"The Latino threat": Constructing immigrants, citizens, and the nation*. 2nd ed. Stanford, CA: Stanford University Press.
Coleman, M., and A. Kocher. 2011. Detention, deportation, devolution and immigrant incapacitation in the U.S., post 9/11. *Geographical Journal* 177 (3): 228–37.
D'Andrea, A., L. Ciolfi, and B. Gray. 2011. Methodological challenges and innovations in mobilities research. *Mobilities* 6 (2): 149–60.
De Genova, N. 2010. Theoretical overview. In *The deportation regime: Sovereignty, space, and the freedom of movement*, ed. N. De Nova and N. Peutz. Durham, NC: Duke University Press.
Doty, R. L., and E. S. Wheatley. 2013. Private detention and the immigration industrial complex. *International Political Sociology* 7:426–43.
Espiritu, Y. L. 2003. *Home bound: Filipino American lives across cultures, communities, and countries*. Berkeley: University of California Press
Fennelly, K., and H. Leitner. 2002. How the food processing industry is diversifying rural Minnesota. Working Paper No. 59, Julian Samora Research Institute at Michigan State University, East Lansing, MI.
Golash-Boza, T. 2009. The immigration industrial complex: Why we enforce immigration policies destined to fail. *Sociology Compass* 3 (2): 295–309.
Hartman, T., B. Newman, and C. S. Bell. 2014. Decoding prejudice toward Hispanics: Group cues and public reactions to threatening immigrant behavior. *Political Behavior* 36 (1): 143–63.
Henderson, J. 2006. Secessionist automobility: Racism, anti-urbanism, and the politics of automobility in Atlanta, Georgia. *International Journal of Urban and Regional Research* 30 (2): 293–307.
Huling T. 2002. Building a prison economy in rural America. In *Invisible punishment: The collateral consequences of mass imprisonment*, ed. M. Mauer and M. Chesney-Lind, 197–213. New York: The New Press.
Kandel, W., and E. Parrado. 2005. Restructuring of the U.S. meat processing industry and new Hispanic migrant destinations. *Population and Development Review* 31:447–71.
Licona, A. C., and M. M. Maldonado. 2014. The social production of Latino/a visibilities and invisibilities: Geographies of power in small town America. *Antipode* 46 (2): 517–36.
Lovejoy, K., and S. Handy. 2011. Social networks as a source of private-vehicle transportation: The practice of getting rides and borrowing vehicles among Mexican immigrants in California. *Transportation Research Part A: Policy and Practice* 45 (4): 248–57.
Lugo-Lugo, C. R., and M. K. Bloodsworth-Lugo. 2014. "Anchor/terror babies" and Latina bodies: Immigration rhetoric in the 21st century and the feminization of terrorism. *Journal of Interdisciplinary Feminist Thought* 8 (1): 1–21.
Massey, D. 2008. *New faces in new places: The changing geography of American immigration*. New York: Russell Sage Foundation.

Merry, S. E. 1981. *Urban danger: Life in a neighborhood of strangers.* Philadelphia: Temple University Press.

Nelson, L., and P. B. Nelson. 2010. The global rural: Gentrification and linked migration in the rural USA. *Progress in Human Geography* 35 (4): 441–59.

Puar, J. 2007. *Terrorist assemblages: Homonationalism in queer times.* Durham, NC: Duke University Press.

Rosas, G. 2006. The managed violences of the borderlands: Treacherous geographies, policeability, and the politics of race. *Latino Studies* 4 (4): 401–18.

Sassen, S. 1988. *The mobility of labor and capital: A study in international investment and labor flow.* New York: Cambridge University Press.

Shabazz, R. 2009. "So high you can't get over it, so low you can't get under it": Carceral spatiality and black masculinities in the United States and South Africa. *Souls* 11 (3): 276–94.

Shamir, R. 2005. Without borders? Notes on globalization as a mobility regime. *Sociological Theory* 23 (2): 197–217.

Stuesse, A., and M. Coleman. 2014. Automobility, immobility, altermobility: Surviving and resisting the intensification of immigrant policing. *City & Society* 26 (1): 51–72.

U.S. Census Bureau. 2000. Census 2000, Summary File 3. http://factfinder2.census.gov (last accessed 30 June 2015).

———. 2013. 2009–2013 American Community Survey 5 year estimates. http://factfinder2.census.gov (last accessed 30 June 2015).

Zúñiga, V., and R. Hernández-León, eds. 2005. *New destinations: Mexican immigration in the United States.* New York: Russell Sage Foundation.

Connected Mobility in a Disconnected World: Contested Infrastructure in Postdisaster Contexts

Mimi Sheller

Department of Sociology, Drexel University

Drawing on research in postearthquake Haiti, with reference to other postdisaster situations, this article examines how uneven mobility and communication systems often reinforce unequal distributions of network capital and thereby reproduce uneven physical and informational space. The reflexive mobile methodology highlights how postdisaster humanitarian mobilizations, including interventions by diaspora members and researchers, could inadvertently intensify uneven access to blended physical and digital infrastructures. Focusing on the intersection of disaster logistics with systems for mobile communication, remote data collection, aerial vision technologies, and data visualizations assisted by satellites and aerial photography, the article draws on two specific local examples of contested water and energy infrastructure to explore how recipients of international aid contest unequal network capital and struggle against the reproduction of uneven spatialities and mobilities. In conclusion, it suggests that critical awareness of uneven network capital and more reflexive efforts to build connectivity across differentiated mobility systems, communication platforms, and scales might help lessen the negative retrenching of differential mobilities during postdisaster recovery.

本文运用在海地地震后所进行的研究，以及有关其他的灾后境况，检视不均的能动性和沟通系统，如何经常加深了不均等的网络资本分佈，因而再生产了不均等的实体与信息空间。反身性的能动性方法，凸显出灾后的人道主义动员，包括离散社群成员和研究者的介入，如何可能不经意地加剧获取各种实质与数码建设的不均管道。本文聚焦灾害运筹和移动通信系统、遥测信息搜集、航拍技术以及由卫星和航拍图所促成的信息可视化之间的相互交织，运用两个水资源与能源建设争夺的特定地方案例，探讨国际救援接收者，如何争夺不均等的网络资本，并对抗不均空间性与能动性的再生产。本文于结论中主张，对于不均网络资本的批判意识，以及更具反身性地建构差异化的能动性系统、沟通平台和尺度之间的连结之努力，或许有助于减轻灾后重建中的差异化能动性的负面紧缩。

A partir de investigación efectuada en Haití después del terremoto con referencia a otras situaciones posteriores a aquella catástrofe, este artículo examina la manera como a menudo los desiguales sistemas de movilidad y comunicaciones fortalecen distribuciones inequitativas del capital social de red, reproduciendo así el espacio físico e informativo desigual. La metodología móvil reflexiva destaca cómo las movilizaciones humanitarias posdesastre, incluyendo las intervenciones de miembros de la diáspora e investigadores, sin proponérselo podrían intensificar la desigualdad de acceso a las infraestructuras físicas y digitales combinadas. Con enfoque en la intersección de la logística de desastre con los sistemas de comunicación móvil, la recolección de datos remotos, tecnologías de visión aérea y visualizaciones de datos apoyadas con fotografía satelital y aérea, el artículo se apuntala en dos ejemplos locales específicos de infraestructura disputada para agua y energía con el fin de explorar el modo como disputan la desigualdad del capital social de red quienes reciben ayuda internacional, y cómo combaten contra la reproducción de espacialidades y movilidades inequitativas. Como conclusión, el artículo sugiere que la conciencia crítica sobre desigualdad del capital social de red y los esfuerzos más reflexivos para construir conectividad a través de sistemas móviles diferenciados, plataformas de comunicación y escalas podrían ayudar a suavizar el recorte negativo de movilidades diferenciales durante la recuperación posdesastre.

At all phases, up to and including reconstruction, disasters don't simply flatten landscapes, washing them smooth. Rather they deepen and erode the ruts of social difference they encounter.

—N. Smith (2006)

Natural disasters strike at mobility systems, cutting off roadways, electricity, and communication networks. More than that, though, they also engender unique mobilities and immobilities that often deepen uneven spatialities. Recent approaches to the geography of disasters and postdisaster recovery emphasize how prior distributions of advantage and disadvantage lead to uneven reconstruction and redevelopment, often exacerbating preexisting inequalities and reinforcing their spatial forms through the recovery process itself (Schipper and Pelling 2006; N. Smith 2006; Johnson 2011; Gotham and Greenberg 2014). This is now known as the Katrina Effect after the hurricane, levee failure, and uneven recovery in New Orleans in 2005 (Taylor et al. 2015). Not only do economically disadvantaged communities bear a disproportionate burden of destruction, but they also "bear the concomitant repercussions of who gets what, when, and how much in the emergency relief and reconstruction process" (Ruwanpura 2011, 246). As emergency responders, armed forces, relief workers, and even researchers move into and out of affected areas, they might inadvertently "deepen and erode the ruts of social difference," as N. Smith (2006) put it, reproducing and exacerbating uneven access to shelter, water, food, transport, and communication. This produces new social conflicts over infrastructure.

A vignette from postearthquake Haiti exemplifies this process. After the January 2010 earthquake, the international charitable organization Samaritan's Purse installed a water filtration system at a privately owned community center. The center had long hosted a school, botanical garden, and community activities but now housed several hundred internally displaced people (IDP) in a tent camp on an agricultural field outside of the city of Leogane. An adjacent guesthouse housed international visitors (including my own research team) who paid for rooms, food, electricity, WiFi, and use of the same water. A diesel generator powered the high-tech filtration system, which made potable water via reverse osmosis (see Figure 1). After five months, however, the landowner (who had lived in New York for thirty years as a member of the Haitian diaspora) decided that it was time to evict the camp because she was seeking international funding to build a women's center on the same land. After a tense confrontation, the displaced people were forced to leave the site. A "mob" returned, however, and seized the water equipment, dismantling it and carrying it down the road to a new location. It was unlikely that they would be able to get it working again, yet its seizure served symbolic purposes. So, too, did the delivery to the field, shortly thereafter, of two empty steel shipping containers that were planned to serve as the basis for the imminent women's center.[1]

The eviction conflict entailed a number of different mobilities and immobilities, both physical and digital. Serial displacement from temporary shelters was

Figure 1. Samaritan's Purse water filtration system installed at an internally displaced persons' camp outside of Leogane, Haiti, 2010. Credit: Photo by author. (Color figure available online.)

typical throughout postearthquake Haiti (Schuller 2010). Access to potable water also depended on either free distribution at temporary camps or purchase by the bucket or the sachet, entailing a whole logistics for the transportation of water, with much of it carried by women and children (Sheller et al. 2013). The arrival of international responders, including humanitarians, international researchers, and the United Nations peacekeeping forces stationed just down the road, mobilized airplanes, armored trucks, rented sport utility vehicles, shipping containers, and an improvised logistics system for the delivery of food, water, satellite communications, and electricity. Foreigners were also dependent on access to passports, plane tickets, and all of the data-readiness that air travel demands, plus vaccinations, malaria pills, insect repellents, special clothing, and first aid kits to secure the traveler's body; access to rental cars and places to stay with clean water and toilets; and usually an international cell phone that also functions as a camera and wireless Internet device, loaded with apps to ease their mobility. Finally, there is the circulation of the Haitian diaspora, who sought to help their country in distress, including via social media, Web sites, and mobile phone services to exchange information, news, and money, while also seeking access to international funding and resources.

The landowner's eviction of the IDP camp was a reminder that she had access to financial capital and legal power but also to extensive "network capital" that made the land valuable in ways beyond its agricultural use value or as a simple field to place tents. Elliott and Urry (2010) defined "network capital" as consisting of access to instruments for mobility such as legal documents (driver's licenses, passports, residency rights), private vehicles and premium infrastructure, location-free information and connection to networks at a distance, and the time and means of complex coordination. Building on literature in critical mobilities theory on the relation among mobilities, immobilities, and moorings (Hannam, Sheller, and Urry 2006; Sheller and Urry 2006; Sheller 2014), I return to this and other examples later to think through how natural hazards that become "disasters" both demobilize and remobilize, reinforcing and spatially etching elements of network capital that produce blended physical and digital space at multiple scales. Disasters generate complex "processes and connections which demand that transnational, national, urban and bodily scales be kept in view at the same time" (Graham 2011, xxvi; M. P. Smith 2001). My own research practice

spanned these scales and locations and demands critical reflection on its materialities, mobilities, subjects, spaces, systems, and events (Adey et al. 2014).

Reflexive Mobile Methods

This article draws on the author's experience of working on two research projects funded by the National Science Foundation (NSF) in Haiti from 2010 through 2013, collaborating as a social scientist with engineers and hydrologists. The first related to local participation in planning water and sanitation infrastructure around Leogane after the January 2010 earthquake, and the second concerned the impacts of climate change on two lakes on the Haitian–Dominican border, Lake Azuei in Haiti and Lake Enriquillo in the Dominican Republic, in 2012.[2] This is not a report on the empirical findings of those studies, however, but a meta-analysis of the geographies of mobility they entailed and revealed. I consider this to be a reflexive mobile method (Büscher and Urry 2009; Büscher, Urry, and Witchger 2011) in which reflection on my own research teams' (im)mobilities and (dis)connectivity at multiple scales and those of the subjects and locations we encountered can instigate critical analysis.

Various analyses have suggested that philanthropic, nongovernmental, humanitarian aid organizations are "least of all accountable to the locale within which they operate" and their proliferation after disasters solidifies a neoliberal "shift towards the non state sector" (Ruwanpura 2011, 260; see also Kennedy 2004; Agier 2011; Narkunas 2014; Taylor et al. 2015). The role of nongovernmental organizations (NGOs) in postearthquake Haiti has been particularly problematic, often described as compounding the disaster (Schuller 2010; Katz 2013; Polyné 2013). The role of researchers is also pertinent. I also draw on work as a co-organizer of the workshop reviewing all NSF Rapid funding in postearthquake Haiti and as an expert advisor from January to May 2012 to the World Bank's Global Facility for Disaster Risk Reduction, meeting at the World Bank in Tokyo, Japan (World Bank 2012), and at the Earthquake Engineering Research Institute in Oakland, California, to prepare Knowledge Notes on the Japanese earthquake and tsunami of 2011. What can we learn from the critical analysis of disaster responses if we take into account not just their effectiveness but also their uneven spatial effects?

This article builds on existing literature on the violence and countergeographies of uneven postdisaster redevelopment in two ways. First, I examine the role of postdisaster humanitarian and nongovernmental activity in responding to disaster as a form of uneven mobility. One crucial element of "uneven redevelopment" concerns the role of im/mobilities in reshaping space and delimiting forms of access. Insofar as "urban crises lay bare the underlying power structures, long-neglected injustices, and unacknowledged inequalities of contemporary cities" (Gotham and Greenberg 2014, 223; see also Graham 2009), then uneven mobilities are a prime example of the social mechanisms by which such inequalities are reproduced in the wake (and in the name) of such crises (Cook and Butz forthcoming). The seizure of the Samaritan's Purse water system alerts us to the contestation over these uneven infrastructures that pitted immediate physical needs at the scale of the body (for water and shelter) and of the locality against a landowner's desire to mobilize financial opportunities at a larger scale, drawing on her international network capital, which was reinforced by the presence of a foreign research team at her guest house, who were also drawing on and building their own "global" network capital.

Second, focusing on the use of mobile communication, remote data collection, aerial vision technologies, and high-tech visualizations assisted by satellites and aerial photography by disaster responders, I also explore how disaster logistics contribute to uneven mobility via highly skewed communication systems that might contribute to the exclusion of local participants (including governmental and civil society actors) in key recovery and rebuilding activities. Finally, I consider how we might prepare for and bridge these uneven topologies of postdisaster network capital through building open connectivity and more democratic participatory processes across diverse communication platforms. By bringing into view the contested material grounding and spatial frictions of infrastructures (including those supporting academic research), I seek to show how information flows are not simply smooth and seamless but through their texture actually shape space, states, and subjects in highly uneven and unequal ways.

Mobilizing for Disaster Response

With the rise of materialist approaches to media and mobility there has been an effort to investigate the significance of communication not simply as image, message, or content to be relayed from one point to another but rather as an embodied spatial practice that produces space–time and is itself constitutive of social orders (Packer and Wiley 2012). Such an approach builds on the early work of Carey (1989) on the materiality of communication, especially in its relation to transportation networks such as the train and the telegraph (Packer and Robertson 2007), while also incorporating more recent mobilities theory that pays attention to the material infrastructures and "moorings" of mobility and communication systems, including their spatiotemporal dimensions (Hannam, Sheller, and Urry 2006; Sheller and Urry 2006). This suggests a geographical approach to mobilities that combines aspects of physical movement and informational movement into a hybrid understanding of space (de Souza e Silva and Sheller 2015). Although the materiality of infrastructure is often "backgrounded," the relation between informational and physical mobilities moves to the foreground during disasters, when it breaks down. The material turn in media studies and mobilities research helps to highlight political questions surrounding contested infrastructuring processes (Star 1999), the inequalities of humanitarian logistics, and their contribution to uneven redevelopment in postdisaster situations.

The implementation of disaster response in Haiti involved a huge mobilization of information technologies to track and map the distribution of aid, the actors involved, and the actions taken (Sheller 2013). First, humanitarian responders, researchers, engineers, and armed forces arrived equipped with various kinds of mobile informational technologies for communicating as well as gathering, geotagging, and mapping information. In addition, the United Nations peacekeeping force, known as MINUSTAH, had bases with very strong communications infrastructure, including satellite dishes and radio or cell towers. The humanitarian responders were themselves organized into clusters such as Shelter and Water, Sanitation and Hygiene (WASH), which in Leogane held meetings on the United Nations base and used mobile phones and laptop computers with Internet connections to communicate and share information. This linked them to an entire field known as "crisis informatics" involving "digital humanitarian organizations," crowd-sourced information, open-source mapping, and the emergence of "collective intelligence" (Büscher, Liegl, and Thomas 2014), as discussed further later.

Cowen (2014) argued that "humanitarian affect is a powerful feature of contemporary military missions" and that both share a deep concern with "supply chain logistics" (134–35). The intermixing of U.S. military troops, MINUSTAH peacekeeping forces, and foreign NGO responders in Haiti produced a particular kind of security apparatus that divided the humanitarian operation from the local civilian population with physical walls, fences, and gates, as well as more symbolic divides of language, culture, and mutual incomprehension. Cowen argued that the "new models of security prioritize flow" but "are organized through new forms of containment—new kinds of borders and security zones" (56). The same can be said for humanitarian flows, which also entail new kinds of borders, security, and containment. With the rise of "supply chain security" that reshapes "politics, space, and citizenship" (Cowen 2014, 55), we can reflect on how the global humanitarian supply chain mobilizes for disaster in ways that demobilize and contain local recipients of aid, both spatially and politically.

Humanitarian aid appears to be about the arrival of a flow of assistance into a disaster zone but is equally about immobility and fixity: the staging of a series of camps, fenced warehouses, containment areas, secured ports, and secure servers. The logistical flow of aid during humanitarian relief operations simultaneously concerns a logistics of immobility, securitization, and privileged mobility within insecure zones of action. When armed forces (and accompanying NGOs) mobilize to secure roads, airports, ports, and warehouses, they also have access to different infrastructures of Hertzian space (cell phones, radio waves, satellite communications, mobile geographic information system [GIS] platforms, Google Earth maps, etc.). Hertzian space refers to the interface for the physical interactivity between electronic devices and people (Dunne 2001). Postdisaster processes of im/mobility and in/security leverage and recombine these uneven material and digital spaces into a hybrid reproduction of differential landscapes.

Second, aerial surveillance and information gathering were rapidly deployed immediately after the earthquake by external institutions such as the World Bank (as well as the U.S. military), including visioning technologies such as GeoEye satellite imaging coupled with Google Earth. This allowed the technologically empowered a virtual mobility to zoom in and out of topographical satellite maps of Haiti geotagged with information, photographs, and other GIS data, as humble as the placement of latrines. But the capacity to "zero in" and access communication networks or aerial vision is unevenly distributed. The combined use of aerial views, mobile GIS, and data visualization technologies reproduces uneven spaces and differential network capital, which becomes especially problematic in the process of postdisaster decision making, planning, and rebuilding. For example, whereas I was able to navigate the back roads of Leogane by downloading highly detailed local maps using a free iPhone app called Gaia Earth (coupled with a relatively costly AT&T international data plan), very few Haitians had broadband access, smartphones, or easy access to disrupted electricity.

Third, after the earthquake a community known as the Global Earth Observation Catastrophe Assessment Network (GEO-CAN) formed to use crowd-sourcing techniques to have engineers and scientists around the world compare before and after satellite images and later aerial photographs of building damage. Sources of imagery included the World Bank WB-IC-RIT aerial missions, aerial missions flown by Google, Pictometry, and the National Oceanic and Atmospheric Administration (NOAA), "as well as large volumes of high-resolution satellite imagery being transferred into the public domain by DigitalGlobe and GeoEye." This variety of high-resolution and very-high-resolution imagery was complemented by "LiDAR data (WB-IC-RIT aerial mission) and later by oblique aerial imagery (Pictometry data) and in-field survey photos" (Ghosh et al. 2011, S190). Although extremely valuable, these externally controlled modes of surveillance, aerial visioning technologies, and GIS integration into real-time mapping nevertheless reinforce the very technologies for mobility control and infrastructure management by outside experts that work together to reinforce social inequalities, remake uneven spatialities, and re-create subjects with differential network capital.

All of these applications of virtual mobility and informational mobility are directly related to the operationalization of mobility regimes that enable foreign travel into Haiti and foreign control of logistics, while largely preventing poorer Haitians from leaving the country, marginalizing their own self-representations, and interfering with their self-determination of rebuilding processes. As Adey points out, the aerial survey and its associated aerial gaze arose out of colonial authority and the administration of territory, while aerial photography led to modes of revealing and scrutinizing that have been intensified via biometric capita (Adey 2010, 88–89). Only some groups have access to surveillance technology and its powerful aerial views, informational databases, and related mobility and communication technologies. This enabled outsiders to channel information controlled via disaster management organizations, technologies, and infrastructures based outside of Haiti, whether connected to foreign NGOs or to the Haitian diaspora. Engineers and social scientists working in Haiti have called for much stronger local participation in post-disaster decision-making and planning processes (Earthquake Engineering Research Institute 2010), yet such participation was not taking place at any level (Sheller et al. 2014).

Some digital humanitarians attempted to connect their data collection with communities on the ground, using open-source maps, crowd-sourced information, and shared verification processes to comb through localized reports from various sources. Groups like Ushahidi worked to aggregate, verify, and curate the data into open-source GIS mapping platforms (haiti.ushahidi.com). This kind of open mapping project potentially makes microlevel disaster news and information accessible and searchable by location, so that interested parties can zero in on specific sites or types of information. The crowd-sourcing of information has become a crucial element of current recommendations for disaster management, which assists in enhancing mobility capabilities. Such tools are now considered crucial for crisis mapping and indeed have been described as a kind of bottom-up and emergent "social collective intelligence" that can complement more top-down orchestration (Büscher, Liegl, and Thomas 2014). In the next section I turn to local struggles over the appropriation of such informational resources.

Local Appropriations of Hybrid Spaces and Communication Geographies

Constellations of people, devices, networks, laws, regulations, and everyday practices together enable any communication to take place but in ways that are unevenly produced, distributed, and consumed. All communication infrastructures, including the mobile interfaces that we rely on every day, are "a dynamic process that is simultaneously made and unmade" (Horst 2013, 151). Theorists of mobility refer to "constellations of mobility" (Cresswell 2010), including not just flows and connections but also turbulence and disruptions, slowness and waiting (differential speeds), and friction (Adey et al. 2014). There is always an ongoing struggle over not just who has access to infrastructure but also over what shapes and modes it will take.

Whereas states might try to exercise control "from above," people might try to appropriate, hack, or game the system "from below" (Horst 2013), through various appropriations of technology that redirect infrastructure into everyday social practices (de Souza e Silva et al. 2011; McFarlane, Desai, and Graham 2014). Through these struggles over "infrastructuring" (Star 1999), users might create fissures and new possibilities for connection, which could have important effects on national space, on scalar relations, and on governance and control (Horst and Miller 2006; Baptiste, Horst, and Taylor 2011). The evicted community who tried to take control of "their" water filtration system enacted a struggle over water infrastructure that brought into view the wider transnational and informational networks that imbued foreign and diaspora actors (who generally also have far greater access to financial capital) with network capital that serves to extend their mobility even as it serially and coercively demobilizes and remobilizes those they are purportedly helping. Local people criticized and acted against their exclusion from such actor networks and enacted different senses of the timing of urgency, action, and needs.

Consider another on-the-ground appropriation of a solar-powered water kiosk in a rural village near Leogane. In 2012 OxfamItalia built a sturdy cement-block kiosk for solar-powered ultraviolet filtration of water. When our research team visited, though, the taps were dry. Why had this international humanitarian intervention failed to result in a working potable water system? Could our expert engineer fix it? When we looked inside the locked kiosk, however, instead of a technical problem what we found was a social reorganization of infrastructure: Inside were a dozen cellphones and a few laptop computers charging off the solar panel (Figure 2).

It became apparent that this community did not want a water filtration kiosk as was assumed by the humanitarian organization that built it; what they wanted was a phone-charging station. They appropriated the solar power and repurposed it for uses that exceeded the plans of the benefactors. Perhaps there was someone running a business selling treated water, or there might have been more powerful social sectors in the community who appropriated the electricity for themselves or maybe made money from charging phones. Whatever the reason, the water kiosk had been decommissioned and repurposed.

This story alerts us to several things: (1) if local people do not make decisions about infrastructure, it might well fail, at least in its stated aims; (2)

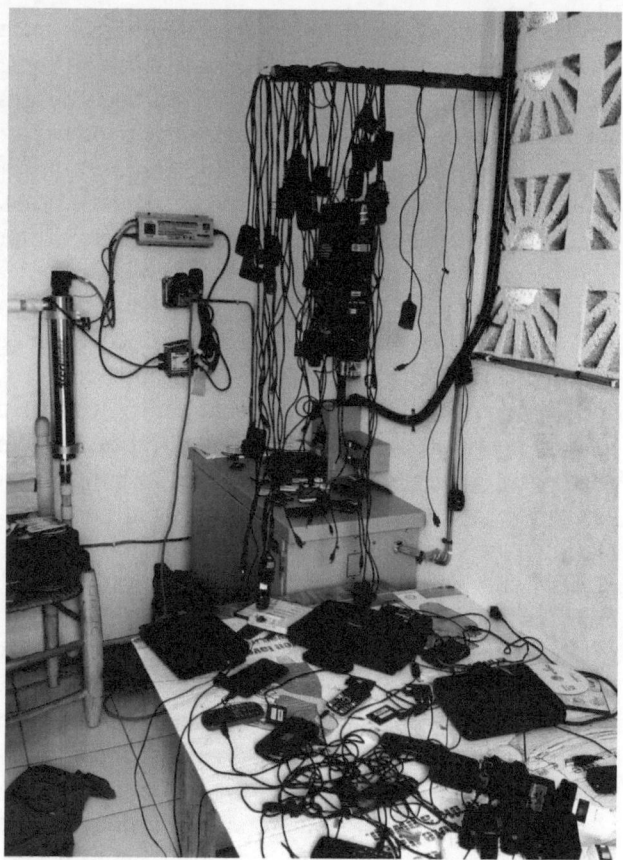

Figure 2. Inside the solar-powered water kiosk: An ad hoc phone and laptop charging station. Credit: Photo by author. (Color figure available online.)

communications infrastructure might be more essential in the medium term after a disaster than even water; and (3) even small rural villages might have pretty sophisticated mobile communications technology at hand, which could serve many important purposes, including, perhaps most important, organizing financial remittances from abroad, which can be done in Haiti through the innovation of "mobile money" (Baptiste, Horst, and Taylor 2011). Mobile money is itself an important appropriation of cell phone networks for remote banking and person-to-person wire transfer purposes. Access to cell phones suggests conduits for pouring information, resources, and money back into where it is needed most—among the disaster-affected population—which ultimately might be more effective than depending on foreigners' network capital. We need to ask how local network capital can be built on and integrated into disaster response as more than just a source of data for others outside Haiti to use. We also need to recognize that collective struggles over infrastructure in postdisaster situations can help reveal the fault

lines of uneven mobilities and unequal network capital.

Conclusion

In concluding, my argument is not that we should stop all satellite and aerial mapping or humanitarian data collection but that researchers need to more carefully consider the ethics of data collection, data sharing, visualization technologies, and the uses to which they are put in restructuring spatial relations and mobility systems that are already grounded in uneven relations of power. Only if such technologies are reflexively put into practice with the democratizing aim of maximizing mobility capabilities and network capital for all will they contribute to greater mobility justice, which ultimately can reduce future vulnerability to disaster. Researchers and humanitarians should reflect on their role in the uneven production of blended physical and Hertzian space. Electronic infrastructures are instrumental in reshaping physical spatial relations and especially uneven mobilities during processes of postdisaster recovery and rebuilding.

Even highly developed countries are not immune to these issues. A government review of the experience in Japan after the earthquake, tsunami, and nuclear meltdown of March 2011 indicates that even a well-prepared country needs mobile electric charging and satellite phone stations, "information rangers" to help people submit and access data from remote areas, and better equipped emergency evacuation centers. Most users of social media were concentrated in urban areas, and were generally young, so there was less news coming from rural areas (where the vast majority of fatalities were among people sixty-five and older). This suggests an age-based and urban–rural digital divide in which network capital is also concentrated in urban areas.[3]

Similar concerns arise in relation to 3D printing for development, known as 3D4D. Birtchnell and Hoyle (2014) argued that 3D4D "offers a solution to the disconnected supply chains, collapsed economic markets, and vulnerability of the citizenry characteristic of disaster settings and clusters of poverty in the Global South" (7). They offer inspiring examples of how it is being used as "a bottom-up development option in the Global South," including the case of iLab/Haiti (http://www.ilabhaiti.org), which 3D prints recycled plastic into goods like prosthetic hands and umbilical cord clamps. The gaps in network capital and the uneven spatial processes they underwrite still present serious obstacles to the success of such grassroots technology projects, however. As funding priorities shift, and the flow of aid ebbs, initiators of such projects might move on to other areas and local appropriation might fail to take hold.

This presents new challenges for theorizing the kinds of production of space and scale that N. Smith (2008) theorized in his influential *Uneven Development*. The struggles over humanitarian infrastructure discussed in this article were not simply spatial struggles for access to water. Instead, they blend multiple kinds of physical access (to shelter, water, land, and energy) and digital access (to smartphones, digital maps, satellites, and WiFi). This is more than a simple digital divide between the Global North and Global South (or between urban and rural regions), because many parts of the Global South (and nonurban places) are highly but unevenly connected, although there are also disconnected "digital deserts" in the Global North. It implies instead that there are complex forms of splintered infrastructure (Graham and Marvin 2001) that extend both globally and locally, in which some groups remain highly connected even in the midst of general disconnection—moving through the same physical topographies but connected to different Hertzian topologies.

Bridging such hybrid spaces of electronic interactivity requires paying closer attention to the capabilities that people already have and how these might be built on in ways that might strengthen their network capital and extend their existing modes of action. Using communication technology to build grassroots development networks only works, though, if there are communities organized to appropriate technology and adapt it to their needs, rather than the imposition of high-tech solutions from outside. Building wider access to communication networks should be as much of a priority as delivering water, because this is one of the means through which community-based organizations most effectively reshape uneven spatial relations. Democratizing digital access is not simply about creating open maps and crowd-sourced data. It requires meeting people halfway, at the connected locations where they join mobile phones to energy, WiFi, and satellites; where they transport bodies, goods, and information to physical and virtual places; and where they turn mobile money into access to land, water, and shelter.

Notes

1. The shipping containers were donated by USAID, and the organization is currently raising funds to build a Women's Center whose mission statement emphasizes self-organization, microcredit and savings enterprises, and English, computer, sewing, and cooking classes.

2. Although I gratefully acknowledge the NSF support and the work of my colleagues that enabled me to be part of these two projects, this article is not directly related to either of them but will reflexively use my experience to elucidate critical observations of the humanitarian response in post-earthquake Haiti. NSF-RAPID: Supporting Haitian Infrastructure Reconstruction Decisions with Local Knowledge, PI F. Montalto, Award No. 1032184, and NSF-RAPID: Understanding Sudden Hydro-Climatic Changes and Exploring Sustainable Solutions in the Enriquillo Closed Water Basin SW Hispaniola, PI J. Gonzalez, Award No. 1264466. For published outcomes see Sheller et al. (2013), Sheller et al. (2014), and Galada et al. (2013, 2014).

3. Based on participation in World Bank Seminar "Disaster Risk Management and Social Media," 17 January 2012, Tokyo, including presentations by Junya Ishikawa, CEO Dreamdesign Co., Ltd., Toru Takanarita of Sendai University, and Hiroyasu Ichikawa, SocialCompany, Inc.

References

Adey, P. 2010. *Aerial life: Spaces, mobilities, affects.* Chichester, UK: Wiley-Blackwell.

Adey, P., D. Bissell, K. Hannam, P. Merriman, and M. Sheller, Eds. 2014. *The Routledge handbook of mobilities.* London: Routledge.

Agier, M. 2011. *Managing the undesirables: Refugee camps and humanitarian government,* trans. D. Fernbach. London: Polity.

Baptiste, E., H. Horst, and E. Taylor. 2011. Earthquake aftermath in Haiti: The rise of mobile money adoption and adaptation. *Lydian Journal* 7, n.p.

Birtchnell, T., and W. Hoyle. 2014. *3D printing for development in the global south: The 3D4D challenge.* Basingstoke, UK: Palgrave Macmillan.

Büscher, M., M. Liegl, and V. Thomas. 2014. Collective intelligence in crises. In *Social collective intelligence computational social sciences,* ed. D. Miorandi, V.

Maltese, M. Rovatsos, A. Nijholt, and J. Stewart, 243–65. Zurich, Switzerland: Springer.

Büscher, M., and J. Urry. 2009. Mobile methods and the empirical. *European Journal of Social Theory* 12 (1): 99–116.

Büscher, M., J. Urry, and K. Witchger. 2011. *Mobile methods*. London and New York: Routledge.

Carey, J. W. 1989. Technology and ideology: The case of the telegraph. In *Communication as culture: Essays on media and society*, J. Carey, 201–30. London and New York: Routledge.

Cook, N., and D. Butz. Forthcoming. Mobility justice in the context of disaster. *Mobilities*. Advance online publication. doi:10.1080/17450101.2015.1047613.

Cowen, D. 2014. *The deadly life of logistics: Mapping violence in global trade*. Minneapolis: University of Minnesota Press.

Cresswell, T. 2010. Towards a politics of mobility. *Environment and Planning D: Society and Space* 28 (1): 17–31.

de Souza e Silva, A., and M. Sheller, eds. 2015. *Mobility and locative media: Mobile communication in hybrid spaces*. London and New York: Routledge.

de Souza e Silva, A., D. M. Sutko, F. Salis, and C. de Souza e Silva. 2011. Mobile phone appropriation in the favelas of Rio de Janeiro, Brazil. *New Media & Society* 13 (3): 363–74.

Dunne, A. 2001. *Hertzian tales: Electronic products, aesthetic experience, and critical design*. Cambridge, MA: MIT Press.

Earthquake Engineering Research Institute (for the National Science Foundation). 2010. *The 12 January 2010 Haiti earthquake: Emerging research needs and opportunities*. Oakland, CA: EERI.

Elliott, A., and J. Urry. 2010. *Mobile lives*. London and New York: Routledge.

Galada, H. C., F. A. Montalto, P. L. Gurian, M. Piasecki, M. Sheller, T. Ayalew, and S. O'Connor. 2013. Attitudes toward post-earthquake water and sanitation management and payment options in Leogane, Haiti. *Water International* 38 (6): 744–57.

———. 2014. Assessing preferences regarding centralized and decentralized water infrastructure in post-earthquake Leogane, Haiti. *Earth Perspectives: Transdisciplinarity Enabled* 1 (5). http://www.earth-perspectives.com/content/1/1/5 (last accessed 14 December 2015).

Ghosh, S., C. Huyck, M. Greene, S. Gill, J. Bevington, W. Svekla, R. DesRoches, and R. Eguchi. 2011. Crowdsourcing for rapid damage assessment: The Global Earth Observation Catastrophe Assessment Network (GEO-CAN). *Earthquake Spectra* 27 (S1): S179–98.

Gotham, K., and M. Greenberg. 2014. *Crisis cities: Disaster and redevelopment in New York and New Orleans*. Oxford, UK: Oxford University Press.

Graham, S., ed. 2009. *Disrupted cities: When infrastructure fails*. London and New York: Routledge.

———. 2011. *Cities under siege: The new military urbanism*. London: Verso.

Graham, S., and S. Marvin. 2001. *Splintering urbanism: Networked infrastructures, technological mobilities and the urban condition*. London and New York: Routledge.

Hannam, K., M. Sheller, and J. Urry. 2006. Mobilities, immobilities and moorings. *Mobilities* 1 (1): 1–22.

Horst, H. 2013. The infrastructures of mobile media: Towards a future research agenda. *Mobile Media and Communication* 1 (1): 147–52.

Horst, H., and D. Miller. 2006. *The cell phone: An anthropology of communication*. Oxford, UK: Berg.

Johnson, C., ed. 2011. *The neoliberal deluge: Hurricane Katrina, late capitalism, and the remaking of New Orleans*. Minneapolis: University of Minnesota Press.

Katz, J. 2013. *The big truck that went by: How the world came to save Haiti and left behind a disaster*. New York: Palgrave Macmillan.

Kennedy, D. 2004. *The dark sides of virtue: Reassessing international humanitarianism*. Princeton, NJ: Princeton University Press.

McFarlane, C., R. Desai, and S. Graham. 2014. Informal urban sanitation: Everyday life, poverty, and comparison. *Annals of the Association of American Geographers* 104 (5): 989–1011.

Narkunas, J. P. 2014. Human rights and states of emergency: Humanitarians and governmentality. *Culture, Theory and Critique* 56 (2): 208–27.

Packer, J., and C. Robertson. 2007. *Thinking with James Carey: Essays on communications, transportation, history*. New York: Peter Lang.

Packer, J., and S. C. Wiley, eds. 2012. *Communication matters: Materialist approaches to media, mobility and networks*. London and New York: Routledge.

Polyné, M., ed. 2013. *The idea of Haiti: Rethinking crisis and development*. Minneapolis: University of Minnesota Press.

Ruwanpura, K. N. 2011. Squandered resources? Grounded realities of recovery in post-tsunami Sri Lanka. In *The neoliberal deluge*, ed. C. Johnson, 245–65. Minneapolis: University of Minnesota Press.

Schipper, L., and M. Pelling. 2006. Disaster risk, climate change, and international development: Scope for, and challenges to, integration. *Disasters* 30 (1): 19–38.

Schuller, M. 2010. *Unstable foundations: Impact of NGOs on human rights for Port-au-Prince's internally displaced people*. Boston: Institute for Justice and Democracy in Haiti.

Sheller, M. 2013. The islanding effect: Post-disaster mobility systems and humanitarian logistics in Haiti. *Cultural Geographies* 20 (2): 185–204.

———. 2014. The new mobilities paradigm for a live sociology. *Current Sociology* 62 (6): 789–811.

Sheller, M., H. C. Galada, F. A. Montalto, P. L. Gurian, M. Piasecki, S. O'Connor, and T. Ayalew. 2013. Gender, disaster and resilience: Assessing women's water and sanitation needs in Leogane, Haiti, before and after the 2010 earthquake. *wH2O: The Journal of Gender and Water* 2 (1): n.p.

Sheller, M., S. O'Connor, H. C. Galada, F. A. Montalto, P. L. Gurian, and M. Piasecki. 2014. Participatory engineering for recovery in post-earthquake Haiti. *Engineering Studies* 6 (3): 159–90.

Sheller, M., and J. Urry. 2006. The new mobilities paradigm. *Environment and Planning A* 38:207–26.

Smith, M. P. 2001. *Transnational urbanism: Locating globalization*. New York: Blackwell.

Smith, N. 2006. There is no such thing as a natural disaster. http://understandingkatrina.ssrc.org/Smith/ (last accessed 14 December 2014).

———. 2008. *Uneven development: Nature, capital and the production of space.* 3rd ed. Athens: University of Georgia Press.

Star, S. L. 1999. The ethnography of infrastructure. *American Behavioral Scientist* 43 (3): 377–91.

Taylor, W. M., M. P. Levine, O. Rooksby, and J. Sobott. 2015. *The 'Katrina Effect': On the nature of catastrophe.* London: Bloomsbury.

World Bank. 2012. *Lessons from the Japanese earthquake and tsunami for developing countries.* Washington, DC: World Bank's Global Facility for Disaster Risk Reduction, with the Government of Japan.

Contesting Street Spaces in a Socialist City: Itinerant Vending-Scapes and the Everyday Politics of Mobility in Hanoi, Vietnam

Noelani Eidse, Sarah Turner, and Natalie Oswin

McGill University

In 2008, Hanoi's municipal government banned street vending from numerous sites, significantly delineating and redefining access to urban space. The ban privileges certain forms of movement by designating streets and sidewalks for the fluid movements of "modern" transportation, rather than the staccato "traditional" mobilities of street vendors who stop frequently to ply their trade. In this article, we explore the everyday mobilities of Hanoi's vendors in light of this ban, focusing on the careful negotiations vendors undertake to secure rights to the city's streets and highlighting how vendor mobilities are socially, politically, and culturally produced and reworked. We combine Cresswell's six facets of mobility with Kerkvliet's everyday politics to form a hybrid everyday politics of mobility. In doing so we highlight vendors' daily experiences of mobility and the politics affecting itinerant vendors compared to their stationary counterparts. Based on eight months of fieldwork in Hanoi, incorporating interviews, mobile ethnographic methods, and vendor journaling, this article contributes an in-depth examination into the politics of (im)mobility in the Global South, considering how mobility is framed and produced in a distinctly socialist context. By focusing on the everyday politics of vending in Hanoi and the tactics undertaken to carve out mobilities in the urban landscape, we illustrate these vendors' daily lived realities as well as their connections with and contestations of local, regional, and global political–economic systems. We find mobility is a mechanism of resistance, as vendors strive to maintain mobile livelihoods despite threats of state sanctions and exclusion.

2008 年, 河内市政府禁止了多处的街头贩卖, 显着地勾勒并重新定义了获得城市空间的途径。该禁令透过将街道与人行道指定作为流畅的 "现代" 交通移动之用, 而非经常停驻与顾客进行交易的街头小贩所拥有的断断续续的 "传统" 能动性, 以此偏好特定的移动形式。我们于本文中, 探讨河内小贩面对此一禁令时的每日生活能动性, 聚焦这些小贩为了确保城市街道权所进行的谨慎协商, 并强调小贩的能动性如何在社会上、政治上与文化上进行生产与再製。我们结合克瑞斯威尔 (Cresswell) 的能动性六大面向与克弗列特 (Kerkvliet) 的每日生活政治, 形构混杂的每日能动性政治。我们藉由这麼做, 凸显出相较于其定着的对照者而言, 小贩每日生活中的能动性之经验, 以及影响游走小贩的政治。本文根据在河内为期八个月的田野工作, 包含访谈、移动式的民族志方法, 以及小贩日志, 对于全球南方的 (不) 能动性政治, 做出深度检视之贡献。我们透过聚焦河内进行贩卖的每日生活政治, 以及在城市地景中开拓能动性的策略, 描绘这些小贩的每日真实生活, 以及他们与在地、区域和全球政治经济系统的连结及竞逐。我们发现, 当小贩面对国家禁令和排除的威胁, 仍力图维系其动态生计之时, 能动性便是一种反抗的机制。

El gobierno municipal de Hanoi proscribió en 2008 las ventas callejeras en numerosos sitios, delineando y redefiniendo significativamente el acceso al espacio urbano. La prohibición privilegia ciertas formas de movimiento designando calles y calzadas para los movimientos fluidos del transporte "moderno", contra las caprichosas movilidades "tradicionales" de los vendedores callejeros que paran aquí y allá para ejercer su oficio. En este artículo exploramos las movilidades cotidianas de los vendedores de Hanoi a la luz de esta prohibición, concentrándonos en las cuidadosas negociaciones que ellos emprenden para asegurar derechos sobre las calles de la ciudad, y destacar cómo las movilidades de los vendedores son producidas y reelaboradas social, política y culturalmente. Combinamos las seis facetas de movilidades de Cresswell con las políticas cotidianas de Kerkvliet para formar una política de movilidad cotidiana híbrida. Al hacerlo resaltamos las experiencias cotidianas de movilidad de los vendedores y las políticas que afectan a los vendedores itinerantes en comparación con las de sus contrapartes estacionarios. Con base en ocho meses de trabajo de campo en Hanoi, que incorporó entrevistas, métodos etnográficos móviles y apuntes de los vendedores, este artículo contribuye un examen a profundidad de las políticas de (in)movilidad en el Sur Global, considerando el modo como la movilidad es enmarcada y producida en un contexto distintivamente socialista. Al enfocarnos en las políticas cotidianas de ventas en Hanoi y en las tácticas emprendidas para forjar las movilidades en el paisaje urbano, ilustramos las realidades vividas a diario por los vendedores, lo mismo que sus conexiones y retos de los sistemas político-económicos locales, regionales y globales. Hallamos que la movilidad es un mecanismo de resistencia, en tanto los vendedores se esfuerzan por mantener fuentes de subsistencia móviles a pesar de las amenazas de sanciones y exclusión del estado.

Within municipalities in the Global South, modernist and revanchist policies often pit the state's vision of urban development against those of the informal economy, restricting informal vendors' livelihood security and options for a safe trading space (Bromley 2000; Brown 2006; Hansen, Little, and Milgram 2013). In 2008, in the capital city of the Socialist Republic of Vietnam, Hanoi's municipal government enacted a ban on street vending in sixty-two streets and forty-eight public spaces.[1] This ban effectively defined sidewalks and pavements as spaces to move through rather than move in—spaces of flow rather than spaces of place—positioning vendors as obstructions to, instead of part of, the flow (Castells 1996; Blomley 2011; Graham 2014). Along with previous legislation limiting sidewalk commerce, Hanoi's 2008 ban has resulted in vendors' visions of fair and reasonable governance and control colliding with those of the municipal government. In response, vendors often develop ingenious tactics, digging in their heels in opposition to state and developers' plans, as part of an everyday politics of livelihood survival (Cross 2000).

Building on previous investigations of vendor livelihoods in Hanoi that have produced meaningful findings regarding trade practices (cf. Drummond 1993; DiGregorio 1994; Mitchell 1995; Tana 1996; Higgs 2003; Koh 2008; Jensen, Peppard, and Thang 2013), our case study focuses on street vending as an important, but until now conceptually neglected, form of mobility in a socialist context. First, we briefly review the mobilities and everyday politics literature, demonstrating how a combination of these concepts can help us to better understand vendor mobilities and actions in the Global South. Then we detail the rapidly transforming socioeconomic context in which Hanoi's vendors are operating, especially since the 2008 vending ban. We introduce the vendors at the heart of this piece and then frame our analysis according to Cresswell's six facets of mobility, highlighting the differential mobility politics between fixed and mobile traders. We reveal how itinerant vendors have complied with, negotiated, or resisted the ban and examine the everyday politics of making do in this restrictive trade environment.

To tease out the points of collision between the lived mobilities of street vendors and the imagined, ideal mobilities of Hanoi's state officials, this article draws on eight months of fieldwork involving conversational and semistructured interviews, solicited journals, mobile ethnographic methods, and participant observation. The first author carried out 265 conversational street vendor interviews (during 2010, 2012, and 2015), as well as fifteen semistructured interviews with law enforcement officers and policymakers. We also draw on semistructured interviews completed with forty additional street traders in 2009 (Turner and Schoenberger 2012). Participant journals with ten vendors contribute in-depth accounts of vendors' day-to-day mobilities, and walking-while-talking interviews with two itinerant vendors add further nuance regarding mobile patterns of trade. Of our 305 vending participants, 265 are women and 40 are men. Stationary, local vendors account for 83 of our participants, whereas 222 are itinerant, migrant vendors. All itinerant participants come from farming households and describe vending as their primary source of income, whereas the majority of fixed vendors (90 percent) work this way for pleasure or to supplement their income. Vendors sell a range of products including produce, prepared food, beverages, convenience items, and services, usually earning around US$5.00 per day, compared to a national average income of approximately US$5.50 per day (World Bank 2015).

Conceptualizing Mobilities and Everyday Politics for the Vietnam Case

Mobility is by no means a new phenomenon. Nonetheless, the ways in which people, ideas, and materials move have undergone increasingly intensive investigation in recent years. Because "all the world seems to be on the move," scholars across the social sciences and humanities have advanced a "new mobilities paradigm" (Sheller and Urry 2006, 207; also see Adey et al. 2014) to train analytic attention on the character and quality of movements and flows. The resulting body of work offers a particular take on the process of movement, one that unravels the entanglements of movement with power and meaning and interrogates its social, cultural, and political production. In other words, the concept of mobility facilitates critical discussions of the politics and power dynamics that animate processes of movement, raising questions about who is or is not able to move, what forms of movement are privileged and desired over others, and how the same movement can take on drastically different meaning depending on the positionality of the mobile subject and the motive force behind their movement (Cresswell 2006, 2010; Uteng 2009; Tanzarn 2012; Oswin 2014).

McCann (2011, 121) argued that "mobility is stratified and conditioned by access to resources and by one's identity (classed, racialized, gendered, etc.)." Elaborating on the gender dynamics at play, Hanson (2010) noted the interdependence of mobility and gender, examining the ways in which they both shape and are shaped by one another. Hanson argued that mobilities are reflective of the positionality of the mobile subject and that as unequal power relations are often drawn along lines of gender, mobilities are differentially accessed by male and female subjects. This is clear in the case of street vending in Vietnam, predominantly undertaken by female laborers. In this context, Leshkowich (2014) argued that attempts to render vending livelihoods as chaotic, disorderly, and problematic are closely linked to attempts at regulating and controlling female mobility. Leshkowich added that street vendors are targeted by state reprisals because of who vendors are, rather than how they make a living.

Within mobilities scholarship, increasing attention is being paid to the street as a space of flows and movements, barriers and moorings (Blomley 2011). Yet, as is true of the mobilities turn more broadly, there is a geographical bias within work on the politics of the street. That is, with some notable exceptions (see Gough and Franch 2005; Porter et al. 2010; van Blerk 2013), it tends to focus on the Global North. This is surprising because, as Cresswell (2006, 20) noted, "mobility is central to what it means to be modern." Indeed, developmental imperatives put issues of modernization high on the agenda in the Global South. Narratives of progress and modernization are particularly evident in state attempts to regulate and reorder street spaces as a means of increasing global connectedness (Hutabarat 2010). As increased automobility is imagined as the foundation for the modern city, urban streets in the Global South become representational spaces linked to the assertion of national identity (Short and Pinet-Peralta 2010). What results is a contested urban landscape, in which multiple stakeholders compete for use of the streets, often resulting in an encroachment on spaces of place that are central to the social, economic, and political livability of the streetscape (Khayesi, Monheim, and Nebe 2010). The dissonance between the imagined and lived mobilities of street spaces is imbued with inequity, as everyday users—such as vendors—are neglected by vehicle-centered planning initiatives and presented as obstructions to the flow and, by extension, inhibitors of progress toward modernity (Short and Pinet-Peralta 2010).

As we explore later, Hanoi's municipal government exercises social control in part through the deployment of a particular and narrow notion of mobility as instrumental and productive. This mobile imaginary poses significant challenges to residents who rely on street vending as an economic survival strategy. Nonetheless, vendors do not receive these regulations passively. Although overt resistance to livelihood restrictions is fairly futile in Vietnam's semi-authoritarian context, vendors advance under-the-radar approaches that either bring them into closer compliance with the law in a manner that suits them or allow them to work around problematic regulations and enforcement. To explore these negotiations of mobile proscriptions, in addition to the critical mobilities framework already detailed, we draw on Kerkvliet's (2009) notion of everyday politics, which he described as "people embracing, complying with, adjusting, and contesting norms and rules regarding authority over, production of, or allocation of resources and doing so in quiet, mundane, and subtle expressions and acts that are rarely organised or direct" (232). Kerkvliet divided everyday politics into four categories; namely "support, compliance, modifications and evasions, and resistance" (233). By focusing on the everyday politics of street vending in Hanoi and the tactics undertaken to carve out mobilities in the urban landscape, we can thus illustrate these vendors' daily lived realities as well as their connections with and contestations of local, regional, and global political–economic systems.

Hanoi: A Socialist Planned Context

In Hanoi, informal livelihoods such as street vending provide a much needed means for residents to earn a living amidst growing disparity associated with urban development, rising cost of living, and stagnant agricultural profits (Jensen, Peppard, and Thang 2013). The informal sector provides a means for survival and entrepreneurship, as informal laborers respond to market demands, doing so in a way that is both beneficial to the provider and convenient for the consumer (Sassen 1994). Regardless of livelihoods it provides, however, state initiatives seek to dismantle Hanoi's informal sector without adequately addressing the complex factors at play in the persistence of informal livelihood activities (Jensen and Peppard 2003; Turner and Schoenberger 2012).

Hanoi's vendors have been subject to numerous regulations over the past thirty years, the enforcement of which has varied substantially. In 1984, city authorities announced that pavements were only for walking, charging a fee for other activities (Koh 2008). In 1991, a national traffic and pavement order campaign (Decree 135/CT) was applied in Hanoi's inner-city areas (Order 57/UB), with police fining street vendors who were unable to flee (Drummond 1993). This, and other decrees that followed, continued to be unevenly implemented throughout the 1990s, with vendors usually finding ways to outmaneuver the ward (neighborhood) officials in charge of implementing such laws at the local level (Koh 2008). Crackdowns were also increasingly linked to large-scale public events—including the 2003 Southeast Asia Games and 2006 Asia-Pacific Economic Cooperation summit—resulting in the implementation of tough measures aimed at clearing vendors from the streets and encouraging a "civilized lifestyle." Yet enforcement always remained uneven as ward officials were accustomed to a steady flow of payoffs (Jensen, Peppard, and Thang 2013).

More recently, in 2008, Hanoi banned street vending along sixty-two streets in the city center and in forty-eight public spaces around hospitals, schools, and bus and train stations (People's Committee of Hanoi 2008). The ban forbids vendors from "blocking transportation" on "national highways, roads, pavement in the city, road for transportation in community areas" (People's Committee of Hanoi 2008). This policy reflects the state's approach to urban development, positioning vendors as inhibitors of traffic flow, unproductive, and obstacles in the state's modernization discourse. Moreover, the state's fixation with automobility, made evident by the ban, mirrors a long-standing urban planning rationale that equates progress with fluid movement (Castells 1996). Policies like the 2008 ban that reimagine streets as channels for the efficient flow of traffic have been introduced at the cost of everyday users of street spaces such as pedestrians, cyclists, and vendors (Khayesi, Monheim, and Nebe 2010). One urban planner working in Hanoi noted, "Street vendors who take up space in the streets or sidewalks disrupt other people from being able to use that space for transportation—to walk or drive their motorbikes. [Vending] is not the intended use, so it's not allowed." Similarly Phi Thai Binh, Vice-Chairman of Hanoi's People's Committee, described the ban as an effort to "re-establish urban order in a civilized way" ("Hanoi's Street Vendors" 2008).

In Hanoi, the *enforcement* of the vending ban, aiming to reduce "slow mobility," has mainly targeted itinerant traders, who overwhelmingly originate from Hanoi's periurban zones (cf. Agergaard and Thao 2011; Jensen, Peppard, and Thang 2013). These periurban zones have been subject to drastic modernization policies and plans since August 2008 when the official land area of Hanoi was expanded from 920 to 3,345 km^2, increasing the city's population overnight from approximately 3.5 million to 6.23 million (Prime Minister of Vietnam 2008). The merger of Hanoi with its periurban environs is part of the government's ongoing bid to create an economic superhub—more populous than Singapore or Kuala Lumpur and rivaling Ho Chi Minh City—through rapid modernization ("Supersized Hanoi" 2008). In the process, Hanoi is engulfing periurban villages and market gardens, refashioning them as private high-rise office and apartment complexes. New waves of migrants emerge from these periurban areas as residents now see street trading as one of their few livelihood means, despite having to compete with long-time Hanoi residents already using public areas for fixed vending (van den Berg, van Wijk, and Hoi 2003).[2] Simultaneously, authorities have tagged certain central corridors for demolition to create new transportation links. Such links—including highways, expressways, and a metropolitan railway system—privilege modern mobilities and gesture toward patterns of disparity emerging alongside increased urbanization (Smart and Smart 2003; Cresswell 2006).

Six Facets of an Everyday Politics of Vendor Mobility

Cresswell (2006) described mobilities as a conceptual triad formed from movement, representation, and practice. In Hanoi, vendor *movement* refers to their day-to-day motions through the streets, in turn encoded and *represented* by the state as an obstacle to modernity and development, resulting in an experience, or *practice*, imbued with everyday politics. Cresswell's theorization explains mobility as a highly regulated and contested resource and offers an analytical vantage point for examining the power dynamics and inequity characterizing mobile hierarchies on Hanoi's streets. Additionally, by drawing on Kerkvliet's notion of everyday politics, we can better understand how those relegated to the bottom of the mobility hierarchy enact everyday politics in a bid to push back against the state structures and imperatives that seek to immobilize them. In this section we combine Cresswell's six facets of mobility (motive force, route, speed, rhythm, experience, and friction) with a discussion of everyday politics in order to examine vendors' daily experiences of mobility. In doing so, we highlight the mobility politics affecting itinerant vendors compared to their stationary counterparts.

Motive Force

The motives for vending differ between Hanoi's fixed and itinerant traders. Stationary vendors tend to be longtime residents of Hanoi, supplementing their household or pension income by selling from fixed stalls near their homes. Alternatively, the majority of itinerant vendors are migrants, frequently sharing rented rooms in Hanoi with other vendors from the same village and returning home to visit their families monthly. Itinerant traders lack the social and financial capital to secure a fixed trading spot. One young shoe vendor explained, "Locals have more rights to the sidewalk than we [migrant vendors] do—they wouldn't sell

itinerantly, because they are able set up stalls, and we wouldn't dare sell in one place." Itinerant vendors trade to support their households in response to changing conditions in their home villages, including decreasing land access and a growing inability to survive on farming alone. Vending also offers flexibility in contrast to the strict schedules of factory work that can conflict with familial and farming responsibilities. Nonetheless, itinerant vendors primarily undertake this trade due to lack of alternative livelihood options. Although itinerant vendor Kiều[3] faces harsh regulations because of her conical hats being so conspicuously for sale, she does not know what else to do: "Everyone from my home sells these, all these vendors you see with hats and bracelets—we are like family. This is all I know."

Route

The 2008 ban attempts to foster fluid traffic flow, channeling vendor routes away from sixty-three streets on which vending is prohibited. Signs on the sidewalks and daily announcements over loudspeakers remind vendors to stay clear of these streets. Yet these major thoroughfares are lucrative sites for trade and hubs for foot and vehicle traffic. Itinerant traders thus carefully adjust their daily routes, taking note of which streets and spaces are more frequently targeted by ward officials—including larger, artery roads and those that permit ease of access for police vehicles. In strategic acts of everyday resistance, some vendors continue to trade on

Figure 1. Balloon vendors trading in the middle of a busy and highly regulated intersection on the edge of Hanoi's Old Quarter (a location included in the 2008 ban). *Source:* Photograph by Noelani Eidse, 2015. (Color figure available online.)

highly regulated streets by noting restrictions on police mobilities; vendors identify nearby side streets too narrow for police vehicles to pass through and use them as escape routes to flee police. One young vendor, Năm, described how she and her friends are able to sell on a bustling banned street (Figure 1), regardless of frequent raids: "We go into small streets or alleys where they cannot go in with their trucks, or we run to one-way streets where trucks cannot come in either." Alternatively, fixed local vendors draw on social capital with ward police, who they have often known for years, and on perceived rights to the pavement as long-term residents to trade on regulated streets.

Speed

Cresswell (2010) argued that a discussion of the politics of speed must address choice, cautioning against the simplistic equation of high-speed movement with the "kinetic elite." This is certainly the case in Hanoi, where stationary traders are able to choose fixity, and itinerant vendors must adjust their speed strategically, moment by moment, to avoid reprisals for their trade. One itinerant pomello vendor, Hạnh, explained:

> Locals have more rights to the sidewalk than us migrants because they are from here. They often sell in front of their homes. They are able to use the pavement without being troubled by the police—not like us.

Itinerant vendors are frequently told to keep moving by shop owners, residents, and fixed vendors, in an act of exclusion from remaining stationary (Figure 2). At the same time, their movement is considered too slow and out of place in the city. Fixed stall owner Hoai noted, "Street vendors using a bamboo carrying pole are dangerous for the traffic and they should not clog the street." Vendor equipment and goods are often cumbersome, making sustained or rapid movement physically strenuous or impossible. For instance, Cúc, an elderly vendor selling clothing with a carrying pole, noted that her stock could weigh up to 45 kg, explaining why she is unable to walk more than a few meters at a time. Moving slowly or stopping for long periods increases itinerant vendors' vulnerability, and during police raids they must run quickly to avoid being caught and fined. Itinerant vendors' ability to choose their speed of movement is thus filtered through a plethora of physical, social, and political constraints.

Rhythm

Itinerant vendors' daily rhythm is staccato, divided into intervals of rest and motion. By remaining on the move almost constantly, itinerant vendors are able to cover more ground and access a greater number of customers, but brief periods of rest are necessary to manage the physical stress of their trade. Hùynh, a DVD

Figure 2. A stationary vendor displays her cleaning products on the sidewalk in front of her house, while an itinerant guava vendor pauses momentarily on the road in front of a closed shop. *Source:* Photograph by Noelani Eidse, 2015. (Color figure available online.)

vendor, explained, "I get so tired from walking around, but the longer I sit in one place, the more sales I miss." Because the state imagines streets as spaces of fluid movement and flow, as introduced earlier, the improvised and unpredictable rhythms of itinerant vendors are branded as antimodern, uncivilized, and a hindrance to development. Some vendors we interviewed have internalized this state narrative, reiterating that they are an obstruction to the flow of goods and people in the city, rather than being part of the flow itself. Siu, a vendor using a bicycle to ply her wares, noted:

> I think I'm not allowed to sell because I take up too much space, I block traffic. ... The cars might take up more space, but I stop more than they do. But what can I do? I do this because I don't have any other way to get by.

Siu's statement highlights the tensions between these public spaces as imagined spaces of flow and as lived spaces entwined with livelihood practice (Brown 2006; Cresswell 2010).

Harnessing the benefits of their itinerancy, vendors strategically adjust their daily rhythms to trade under the radar. They are savvy to the daily and weekly rhythms of ward officials and police and have created micromobility patterns to avoid fines and retributions. Police repeatedly patrol at fixed times and are not particularly inventive in their routines. Vendors build on these repeated customs to create their own trade mobilities. Siu noted, "When the police are out, even if I'm just walking I'll get a fine—so I don't go out then." By resting when the police are less active and moving when the police patrol, vendors minimize the physical strain of their mobile livelihoods while reducing their interaction with police. Itinerant vendors likewise adjust their rhythms according to power dynamics between themselves and locals. One mango vendor, Linh, explained, "We sometimes rest in front of that pharmacy during lunch, but if we don't move quickly enough when they open for business again, they will curse, push our carts into the street, throw tea in our faces and kick us." As such, vendors are compelled to stay on the move and pause cautiously, always ready to pick up their goods and become mobile again.

Experience

Itinerant vendors' experiences of mobility are characterized by a sense of being out of place. They are considered outsiders by fixed vendors, who often disparage itinerant vendors as migrants, with lower socioeconomic status. One itinerant migrant clothing vendor explained:

> We face many hardships here, just because we're from outside Hanoi and don't know the rhythm of the city. The vendors who live in Hanoi do not accept us.

Tensions rise over the favorable treatment Hanoi residents can receive from local officials. Hạnh, an itinerant hawker, put it bluntly: "Street vendors with stalls are always from Hanoi; they think they're above us vendors who walk around." Despite many itinerant interviewees having worked in Hanoi for fifteen years or more, they still feel poorly treated and excluded by native Hanoians who sometimes threaten to call the police, damage their products, or take goods on credit without repaying their debt. Even when not actively trading, itinerant vendors experience restricted mobility; their positionality as vendors affects what shops they can enter, where they can sit, and which streets they can pass through. Kiều, an itinerant vendor in her sixties introduced earlier, noted that by simply walking through certain streets with her goods she is a possible target for police action, irrespective of whether or not she is pursuing a sale. The positionality and mobility of an itinerant vendor thus becomes paralleled with a fixed social position in the minds of local residents and officials (Cresswell 2006).

Friction

Clearly, the greatest source of friction for Hanoi's vendors is the enforcement of the 2008 ban. Due to the relationships fixed vendors forge with local city officials, however, they rarely face the same friction as migrant vendors. Stationary vendors often have an established informal relationship with local law enforcement officers—such as regular unofficial payments—and a degree of social capital from connections that enable them to maintain their trade. One stationary cigarette vendor, Thạnh, noted, "It is a privilege to be able to pay the police," explaining that engaging in bribe networks enables her to occupy street spaces with reduced fear of retribution. Without the social connections needed to negotiate the illegality of their livelihoods, migrant vendors' mobilities are disproportionately restricted by the 2008 ban. Indeed, by defining street vendors as those who use streets and pavements to undertake "buying/selling activities *without a fixed* space" (People's Committee of Hanoi 2008, italics added) police can manipulate their take on the ban, focusing on migrant vendors who trade itinerantly and consequently experience harsher fines and restrictions.

The friction that some itinerant vendors experience in relation to the ban has led them to an everyday politics response of compliance (Kerkvliet 2009), discontinuing their trade permanently, whereas others temporarily modify their trade to minimize risks. Danh, an itinerant vendor selling lighters, noted that when the police increase their presence in the streets—for instance, during Vietnam's Independence Day celebrations—he stops trading for a week or two, returning home to Hu'ng Yên. Like countless others, though, once the police ease their presence again, Danh resumes his trade. For Cresswell (2010), friction is about asking when and how mobilities stop, yet in the context of our study, the question becomes: To what *extent* does friction immobilize vendors? Itinerant vendors respond to friction with various acts of everyday politics, doing so to negotiate external sources of pressure and continue to trade in spite of friction. By adapting their route, rhythm, and speed, itinerant vendors push back against sources of friction that threaten to immobilize them.

Concluding Thoughts

We have demonstrated that in Hanoi, when delineating between acceptable and unacceptable mobilities, those in positions of power demonstrate a rationale of functionality, favoring modern, fluid mobilities over those that are both traditional and staccato. Inhabitants who do not conform to acceptable mobilities become obstructions to, rather than part of, the flow in the eyes of the state. Some vendors have internalized this state discourse, although they are also quick to note that they have few alternative livelihood options. The imagined mobilities of state planning clash with vendors' everyday experiences of mobility, emphasizing the tensions between mobile subjects who compete for the same space (Cresswell 2010). Informal mobilities thus become entwined with processes of negotiation and resistance as vendors undertake forms of movement outside the state's view. Indeed, it is the very fact that their mobilities are informal and exist beyond regulation that enables Hanoi's street vendors to carve a trading space.

In sum, we have highlighted the informal power plays and processes undertaken by long-term Hanoi residents, selling from fixed street stalls, compared to those carried out by migrant itinerant traders. To understand the underlying factors contributing to the everyday politics of disparate mobilities, one must consider positionality, addressing how mobilities are embodied and gendered. As our case demonstrates, in the Socialist Republic of Vietnam's capital, certain forms of mobility are strongly privileged over others, and mobility hierarchies are not only drawn according to movement but fixity, boundaries, and moorings as well (Cresswell 2010). Mobility is socially produced and reworked, inherently political and differential—both produced by and reproductive of hierarchies of power and social exclusion (Uteng 2009; Tanzarn 2012). Echoing McCann's (2011) suggestion, we argue that by applying the concept of mobility in conjunction with other concepts, we are able to more fully understand the spatial politics at play. By incorporating Kerkvliet's everyday politics with Cresswell's six mobility facets we have teased out the ways in which street vendors contest their (im)mobilities, enacting everyday politics to exert claims to the city's streets. By integrating everyday politics into an examination of mobility, we contribute new understandings of how compliance, contestation, negotiation, and evasion and resistance can underscore mobility and in turn prove essential for informal economy livelihood options. We have highlighted the power dynamics that underpin hierarchies of itinerant trade in the Global South, while also uncovering the ways in which street vendors—members of the "kinetic underclass" (Cresswell 2010)—push back against the very structures that seek to immobilize them.

Acknowledgments

We are grateful to the participants in Hanoi who shared their stories with us. We also thank the anonymous referees, the *Annals of the American Association of Geographers* special issue editors, and Jana Eidse for their valuable suggestions.

Funding

Funding for this research was provided by the Social Science and Humanities Research Council of Canada and the International Development Research Centre, Canada.

Notes

1. In 2009, the vending ban was revised to include an additional street, with sixty-three streets banned in total. The updated ban can be viewed online at http://m.thu vienphapluat.vn/archive/detail/84303.
2. This migration raises a range of complex issues, not least that having registration papers with official residence status in the city provides access to government-funded health care, schooling, and other amenities. Few street vendor migrants receive official urban residential status (see Agergaard and Thao 2011).
3. All participants have been assigned pseudonyms.

References

Adey, P., D. Bissell, K. Hannam, P. Merriman, and M. Sheller. 2014. Introduction. In *The Routledge handbook of mobilities*, ed. P. Adey, D. Bissel, K. Hannam, P. Merriman, and M. Sheller, 1–20. London and New York: Routledge.

Agergaard, J., and V. T. Thao. 2011. Mobile, flexible, and adaptable: Female migrants in Hanoi's informal sector. *Population, Space and Place* 17:407–20.

Blomley, N. 2011. *Rights of passage: Sidewalks and the regulation of public flow*. London and New York: Routledge.

Bromley, R. 2000. Street vending and public policy: A global review. *International Journal of Sociology and Social Policy* 20 (1–2): 1–28.

Brown, A. 2006. Challenging street livelihoods. In *Contested space: Street trading, public space, and livelihoods in developing cities*, ed. A. Brown, 3–16. Warwickshire, UK: ITDG.

Castells, M. 1996. *The rise of the network society*. Oxford, UK: Blackwell.

Cresswell, T. 2006. *On the move: Mobility in the modern western world*. London and New York: Routledge.

———. 2010. Towards a politics of mobility. *Environment and Planning D: Society and Space* 28 (1): 17–31.

Cross, J. C. 2000. Street vendors, modernity and postmodernity: Conflict and compromise in the global economy. *International Journal of Sociology and Social Policy* 1 (2): 30–52.

DiGregorio, M. R. 1994. Urban harvest: Recycling as a peasant industry in northern Vietnam. East-West Center Occasional Papers, No. 17, East-West Center, Honolulu, HI.

Drummond, L. 1993. Women, the household economy, and the informal sector in Hanoi. Unpublished master's thesis, University of British Columbia, Vancouver, Canada.

Gough, K. V., and M. Franch. 2005. Spaces of the street: Socio-spatial mobility and exclusion of youth in Recife. *Children's Geographies* 3 (3): 149–66.

Graham, S. 2014. Disruptions. In *The Routledge handbook of mobilities*, ed. P. Adey, D. Bissel, K. Hannam, P. Merriman, and M. Sheller, 468–71. London and New York: Routledge.

Hanoi's street vendors given temporary reprieve. 2008. *Vietnam News* 24 January 2008. http://vietnamnews.vn/society/173216/ha-nois-street-vendors-given-temporary-reprieve.html (last accessed 31 May 2015).

Hansen, K. T., W. E. Little, and B. L. Milgram, eds. 2013. *Street economies in the urban Global South*. Santa Fe, NM: School for Advanced Research Press.

Hanson, S. 2010. Gender and mobility: New approaches for informing sustainability. *Gender, Place & Culture: A Journal of Feminist Geography* 17 (1): 5–23.

Higgs, P. 2003. Footpath traders in a Hanoi neighbourhood. In *Consuming urban culture in contemporary Vietnam*, ed. L. Drummond and M. Thomas, 75–88. London and New York: Routledge Curzon.

Hutabarat Lo, R. 2010. The city as a mirror: Transport, land use and social change in Jakarta. *Urban Studies* 1 (27): 1–27.

Jensen, R., and D. M. Peppard. 2003. Hanoi's informal sector and the Vietnamese economy: A case study of roving street vendors. *Journal of Asian and African Studies* 38 (1): 71–84.

Jensen, R., D. Peppard, and V. T. M. Thang. 2013. *Women on the move: Hanoi's migrant roving street vendors*. Hanoi, Vietnam: Women's Publishing House.

Kerkvliet, B. J. T. 2009. Everyday politics in peasant societies (and ours). *Journal of Peasant Studies* 36 (1): 227–43.

Khayesi, M., H. Monheim, and J. M. Nebe. 2010. Negotiating streets for all in urban transport planning: The case for pedestrians, cyclists and street vendors in Nairobi, Kenya. *Antipode* 42 (1): 103–26.

Koh, D. 2008. The pavement as civic space: History and dynamics in the city of Hanoi. In *Globalization, the city and civil society in Pacific Asia: The social production of civic spaces*, ed. M. Douglass, K. C. Ho, and G. L. Ooi, 145–74. London and New York: Routledge.

Leshkowich, A. M. 2014. *Essential trade: Vietnamese women in a changing marketplace*. Honolulu: University of Hawai'i Press.

McCann, E. 2011. Urban policy mobilities and global circuits of knowledge: Toward a research agenda. *Annals of the Association of American Geographers* 101 (1): 107–30.

Mitchell, D. 1995. The end of public space? People's park, definitions of the public, and democracy. *Annals of the Association of American Geographers* 85 (1): 108–33.

Oswin, N. 2014. Queer theory. In *The Routledge handbook of mobilities*, ed. P. Adey, D. Bissel, K. Hannam, P. Merriman, and M. Sheller, 85–93. London and New York: Routledge.

People's Committee of Hanoi. 2008. 02/2008/QD-UBND Quyết định. Ban hành Quy định về quản lý hoạt động bán hàng rong trên địa bàn Thành phố Hà Nội [Decision: Promulgating regulation on management of street-selling activities in Hanoi]. Revised policy No. 46/2009/QD-UBND, 15 January 2009, Hanoi, Vietnam.

Porter, G., K. Hampshire, A. Abane, E. Robson, A. Munthali, M. Mashiri, and A. Tanle. 2010. Moving young lives: Mobility, immobility and inter-generational tensions in urban Africa. *Geoforum* 41:796–804.

Prime Minister of Vietnam. 2008. *Quyết định 490/QD-TTg phê duyệt quy hoạch xây dựng vùng Thủ đô Hà Nội* [Decision of the Prime Minister on the approval of construction planning for Hanoi capital zone]. Hanoi, Vietnam: Prime Minister of Vietnam.

Sassen, S. 1994. The informal economy: Between new developments and old regulations. *Yale Law Journal* 103:2289–304.

Sheller, M., and J. Urry. 2006. The new mobilities paradigm. *Environment and Planning* 38 (2): 207–26.

Short, J. R., and L. M. Pinet-Peralta. 2010. No accident: Traffic and pedestrians in the modern city. *Mobilities* 5 (1): 41–59.

Smart, A., and J. Smart. 2003. Urbanization and the global perspective. *Annual Review of Anthropology* 32:263–85.

Supersized Hanoi. 2008. *The Straits Times.* 7 June 2008. http://www.streetnet.org.za/index.html (last accessed 5 June 2009).

Tana, L. 1996. *Peasants on the move: Rural–urban migration in the Hanoi region.* Singapore: Institute of Southeast Asian Studies.

Tanzarn, N. 2012. Gendered mobilities in developing countries: The case of (urban) Uganda. In *Gendered mobilities,* ed. T. P. Uteng and T. Cresswell, 159–72. Burlington, VT: Ashgate.

Turner, S., and L. Schoenberger. 2012. Street vendor livelihoods and everyday politics in Hanoi, Vietnam: The seeds of a diverse economy? *Urban Studies* 49 (5): 1027–44.

Uteng, T. P. 2009. Gender, ethnicity, and constrained mobility: Insights into the resultant social exclusion. *Environment and Planning A* 41 (5): 1055–71.

van Blerk, K. 2013. New street geographies: The impact of urban governance on the mobilities of Cape Town's street youth. *Urban Studies* 50:556–73.

van den Berg, L. M., M. S. van Wijk, and P. V. Hoi. 2003. The transformations of agriculture and rural life downstream of Hanoi. *Environment and Urbanization* 15:35–52.

World Bank. 2015. Vietnam overview. 15 April. http://www.worldbank.org/en/country/vietnam/overview#3 (last acc-essed 31 May 2015).

Mobilizing a Spatial Politics of Street Skating: Thinking About the Geographies of Generosity

Elaine Stratford

Discipline of Geography and Spatial Sciences, University of Tasmania

Estimates suggest that tens of millions of people skateboard for transport and pleasure—it is a mobility practice both instrumental and playful. That play is important for creativity, connection, and positive affect is known. Yet skating is often typified as mere vandalism, despite the fact that, intrinsic worth aside, its hybridity is instructive: It invites consideration of the spatial politics of the street and the possibility of accommodating this and, indeed, other forms of "alternative" movement. Arguably, the prospect of such generous geographies is fundamental to ideas about the right to the city, an entitlement embracing responsibilities to one another. Nevertheless, given the ongoing dominance of automobility and widespread anxieties about skating, the tendency has been to try and contain it in parks and regulate its presence on streets, not least by creating design solutions to render it difficult to engage in. A corollary of these strategies, in combination with skaters' own resolve to claim rights to the city, is that skaters move on to roadways. These armatures have not been designed generously to accommodate forms of mobility apart from motor vehicles—and sometimes pedestrians and cyclists. Consequently, skaters are among the millions who die on the roads annually. In relative terms, the number is minute; nevertheless, each death invokes this question: How can we mobilize a spatial politics of street skating by thinking about the geographies of generosity in ways that might avoid such events? Reflecting on that question is the purpose of this article.

据估计，有数以千万计的人为了交通运输和乐趣而玩滑板——这是一个同时具有工具性和游乐性的能动性实践。该活动对创造力、连结与正向情感的重要性已众所週知。滑板经常被认为仅只是破坏公物的行为，儘管除了内在价值之外，事实上滑板的混杂性是具有啓发性的：它邀请思考街道的空间政治和容纳此般活动的可能性，当然还有其他"另类"运动形式之考量。此般丰富的地理展望，可以说是城市权概念的基础，亦即拥抱对他人责任的权利。然而，有鉴于汽车能动性的持续性支配地位，以及对于玩滑板的广泛疑虑，目前的趋势仍试图将滑板侷限于公园内，并规范它在街道的出现，特别是藉由创造使其难以使用的设计方案。这些策略的必然结果，加上滑板使用者自身争取城市权的决心，则是滑板使用者移动至道路上。这些防御设计并未能宽厚地容纳除了机动车辆——有时则是行人和自行车骑乘者之外的能动性形式，最终导致滑板族成为每年在路上死于非命的千万人之一。相对而言，该数字是微小的；但每件死亡却仍引发了以下问题：我们如何能够透过思考宽厚的地理来避免此般事故，以此动员街道滑板的空间政治？反思上述问题，则是本文的目的。

Hay estimativos que sugieren que decenas de millones de personas se desplazan en patineta, por transporte y por placer—es una práctica de movilidad instrumental y de distracción. Se sabe que este juego es importante en términos de creatividad, conexión y afecto positivo. Sin embargo, a menudo el patinaje se tipifica como mero vandalismo, a pesar del hecho de que, dejando de lado su valor intrínseco, su hibridad es instructiva: El juego invita a la consideración de la política espacial de la calle y la posibilidad de admitir esta y, en verdad, otras formas de movimiento "alternativo". Es discutible, pero el prospecto de tales geografías generosas es fundamental en ideas acerca del derecho a la ciudad, un privilegio que implica responsabilidades para con los demás. No obstante, dado el corriente dominio de la automovilidad y las ansiedades generalizadas acerca del patinaje, la tendencia es por ensayarlo y ubicarlo en parques, y regular su presencia en las calles, no menos importante que crear soluciones de diseño que convierten al patinaje en algo muy difícil de practicar. Un corolario de estas estrategias, combinadas con la propia resolución de los patinadores de reclamar derechos sobre la ciudad, es que éstos se desplacen a las calzadas. Estas armaduras no han sido diseñadas generosamente para acomodar formas de movilidad diferentes de los vehículos de motor—y a veces peatones y ciclistas. En consecuencia, los patinadores figuran entre los millones que mueren en las carreteras todos los años. En términos relativos, el número es diminuto; sin embargo, cada muerte invoca esta pregunta: ¿Cómo podemos movilizar una política espacial de patinaje en las calles pensando sobre las geografías de la generosidad de modo que se logre evitar tales eventos? Reflexionar sobre esta pregunta es el propósito de este artículo.

As part of my ongoing studies on the question of how to flourish in the world, this article responds to a call "for a more finely developed politics of mobility [that attends] motive force, speed, rhythm, route, experience, and friction" (Cresswell 2010, 17). It also pays heed to an observation that "studies of the rhythms and other temporalities of mobility are in their infancy . . . [and that future] study could explore in greater detail these multiple and contested mobile rhythms and the values and practice surrounding them" (Edensor 2014, 169).

In such light, this work is simultaneously provocation, reflection, and invitation and is based on several suppositions. First, rights to the city should extend to all citizens—broadly meant, acknowledging that rights are also entitlements implying responsibilities to one another. In the city, as elsewhere, such rights extend to mobility and, perhaps because of limited observance of responsibilities to one another, we often experience movement as stressful (Bissell 2014). Yet, and second, the city is a place of enchantment, a term Bennett (2001) used to mean enacted and embodied commitments to change and be generous. Arguably, the city could be more enchanting if experienced playfully and by means of generous spatial politics of shared existence—not least when we are mobile. Third, however, automobility is a predominant feature of city life and a chief way in which exchange values are valorized (Urry 2006). In the process what gets circumscribed, ignored, or indifferently viewed are varied use values such as might arise from playfully alternative mobilities. Simultaneously, the multitude of roadway design and engineering solutions that do provide for green space and nonmotorized traffic are often underfunded or piecemeal; tend to privilege walking and cycling; and accept as given automobility's dominance and constraining effects on other mobilities and rights to the city. Doubtless, at some points in time and in some situations, these constraints make sense at several levels; some, for example, are intended to protect us from harm, even while shoring up the status quo and allied vested interests. Fourth, these dynamics make it difficult to think into existence alternative ontologies of mobility, for they seem to require a great unraveling of the present fabric and logic of the city.

On the basis of the foregoing, the claim made here is that stasis and political immobilization are precisely what is to be avoided: There is a need to consider how to create generous geographies that allow for more, and more playful, mobilities in the city, no matter how modest. Here, the focus is on skateboarding on the grounds that at its best, and alongside other alternative ways of moving, it provides ludic and enchanting ways of engaging with the city. Intrinsically interesting, skateboarding also serves as an exemplar by which to push back against discourses and practices that deny to those who skate (or ride, scoot, trace . . .) rights to the city and, of central interest in this article, that result in road deaths. Such matters are, in fact, at the crux of what follows: How can we mobilize a spatial politics of street skating by thinking about the geographies of generosity in ways that might avoid such events?

Mobilizing the Right to the City

According to Lefebvre (1996), the idea of the right to the city gathers the interests of members of society who are conscious that they dwell and move among one another. Gazing down from his balcony to the streets of Paris, Lefebvre (2004) later observed that pedestrians seemed only to engage in muted murmurings—one does "not chatter while crossing a dangerous junction" (28). These apparently unrelated ideas—rights to the city and dangers of movement—are connected by Lefebvre's desire for a transformative approach to understanding everyday life. Consider the commonplace act of crossing the road: Lefebvre (2004) was sure such acts had "ethical, which is to say practical, implications" (18) and could be subject to hopeful, generous experiments to live both differently and with difference (Kofman and Lebas 1996).

Dwelling and moving are powerfully coconstitutive spatialized ontologies: All sorts of "meaning, identity, and cultural signification" get mobilized between "fixed and bounded sites" that are cities' enclaves and "infrastructure channels and transit spaces" that are their armatures—tunnels, roadways, and sidewalks, for example (Jensen 2009, 139, 141). Skateboarding is among the most dynamic mobility practices in the city, and although Lefebvre might not have thought about the right to the city with it in mind, his habit of gazing from his apartment window might mean that he witnessed skateboarding on the streets below. Perhaps he considered the different speeds and spatial needs of skaters and walkers or the sonorous strike of polyurethane wheels or well-shod heels on pavements while writing about the city or rhythmanalysis (Lefebvre 2004). Perhaps he could be persuaded that skateboarding is an assemblage produced by movement, positive affect for the skater, and a spatial politics invoking an emancipatory agenda close to his own.

In fact, claims about the right to the city made by or on behalf of skateboarders are regularly dismissed as impractical, rebellious, or marginal (Stratford 2002). Although skateboarding's visibility and popularity mean that those with commercial interests will not call for its complete removal from the city, strenuous attempts are made to contain it to official skate parks, not least because it is deemed risky for consumers. Yet this urban governance strategy is ineffectual: Assuredly, skaters use provided parks, but they also create ephemeral sites to skate and use streets and sidewalks. For them, skateboarding is recreation and transportation, the urban playful and utilitarian; "loose space" in which use values trump exchange values and enable deep play and tactical urbanism (Lefebvre 1991; Franck and Stevens 2007; Hou 2010).

This idea of play is central to the reflections advanced here because, as Huizinga ([1949] 2014) reminded us, people are *homo ludens*—the playful species. We use play to create forms of order that might be temporary but over which we exact some control; we use it to give expression to a desire to enliven the banal, breathe life into the fabric of existence, and enjoy moments of freedom and reflection. For Woodyer (2012), play is "a vehicle for becoming conscious of practices and relationships we enact or engage without thinking" (317). She pointed out two possibilities for play when it is first understood in such terms and then unsettled: It becomes a form of "coming to consciousness" and an invitation to cultivate transformative "ethical generosity" (322).

Barnett and Land (2007) posited that generosity is best understood not as a kind of weak universalism, but as deeply partial, on the basis that "partiality and finitude might be the conditions for any ethical–political project that is at once both geographically expansive and geographically sensitive" (1066) and thus generous and caring. Partiality is necessary in caring practices that attentively involve being responsible, competent, and responsive. Such capacities mean acknowledging that generosity is a "political concept . . . a power akin to forgiving or promising" (1070) and is also "a mundane, ordinary, and everyday practice always undertaken in the company of others" (1072): It needs to be formalized and normalized. How then might those charged with urban governance and spatial planning consider being more partial to playful forms of active transport in the city and then regulating for such shifts so that they become commonplace and valued for their foundations in a kind of spatialized largesse?

That generosity is and should be quotidian is important here if one takes seriously the idea that enchantment is possible. Like Barnett and Land (2007), Bennett (2001) has interests in such matters, inquiring whether "the very characterization of the world as disenchanted ignores and then discourages affective attachment to that world. The question is important because the mood of enchantment may be valuable for ethical life" (3). Although the chief focus here is on skateboarding, other forms of mobility are implicated and collectively they might have the capacity to effect change, leading to more enchanting urban environments and convivial interactions—more generous geographies and mobilities. Imagine, for instance, significantly more elderly people using motorized wheelchairs on roadways, given estimates of a 5 percent per annum growth in their use in the United States between 1990 and 2005 (LaPlante and Kaye 2010). Imagine, too, significantly more people choosing to ride bicycles rather than drive vehicles; or greater numbers of pedestrians blading, skating, or practicing parkour, combining recreation and transportation to stay healthy or contributing to lowered carbon emissions. Such ideas of shared positive affect, shared commons, and individual and collective adaptations signify ethical aspirations that Bennett insisted will support enchantment. In this sense, generosity requires "rendering oneself more open to . . . other selves and bodies . . . [and being] willing and able to enter into productive assemblages with them" (Bennett 2001, 131). Ultimately, such impulses are possible because they "tap into a sub-intentional disposition to favor life" (158) and, I contend, liveliness and the ludic.

Bennett's work has broad parallels with Thrift's reflections on the spatial politics of affect and Thorpe and Rinehart's (2010) engagement with his ideas vis-à-vis skateboarding. Thrift (2004) asserted that cities "may be seen as roiling maelstroms of affect" (57); noted that "anger, fear, happiness and joy are continually on the boil, rising here, subsiding there" (57); and observed that these affects manifest in both grand and mundane political settings. At the time, Thrift argued that there was little or no work on affect in relation to the city. That gap has been partly addressed, yet more is warranted because—Thrift suggested—grounds exist to develop an "ethico-political perspective which attempts to instill generosity towards the world by using some of the infrasensible knowledges that we have already encountered" (72). As Thorpe and Rinehart (2010) later observed, the objects used in alternative sports—which is how they classified skateboarding—"define the very activity itself" and offer those who engage in it "a sensuous experience of being a mobile hybrid" (1273). But, they queried, how to convey the split-second joy that arises from "sensual, affected, and affecting" activities (1269) in ways that engender what Thrift called a "politics of affect" and a "politics of hope." These spatial politics, Thorpe and Rinehart posited, are precipitated by new ways of thinking and experiencing beyond habits and norms and, crucially, "may carry over to other realms" (1277).

Consider, then, what might arise were newly reconfigured enclaves and armatures to enable more of us to generously sense and respond to others. In general terms, it could be possible more often and readily to exhibit habitually generous acts such as giving way, stepping aside, and making space. It could be possible to refine a sensitivity to the differential velocities at which different people move—and thus to respect others' claims to enclaves and armatures, including at the slowest of paces or outside the churn of capital in ways reminiscent of Lefebvre's emancipatory agendas. In more specific terms, it might be feasible to engage younger skaters' empathy when asking them to understand in an embodied sense the serious implications for an elderly person of being knocked down and breaking a hip joint. Simultaneously, it might be possible to remind elders of their Billycart days and the ongoing value of play (Powerhouse Museum n.d.). Not just rhetorically but infrasensibly it could be possible to persuade pedestrians that once skaters feel confident to street skate they will navigate competently around us, if we keep moving and do not panic. Yet even were there a shift to heightened sensibilities about one another's mobilities and their spatial politics, other labors would be needed in terms of governmental and infrastructural terms for different forms of mobility. Given this need, it is not impossible to think that we could engage in the wholesale retrofit of the armatures of the city, especially if generosity is partial—as Barnett and Land (2007) asserted—and if partiality means that certain generosities be withheld (e.g., those that would further entrench automobility).

The hopeful prospect of marshaling a spatial politics that could encompass mobilities such as street skating by thinking about the geographies of generosity must be reiterated if any gains are to be made for inhabitation, play, and active transport. Such insistent optimism is important not least because of the profoundly negative effects of deaths on the roads. Short and Pinet-Peralta (2009) noted that about 10 million road injuries and 1.2 million road deaths occur annually; for the latter that is an estimated 3,250 deaths daily. World Health Organization (2015) data corroborate these findings, which represent in Australia 6.1 per 100,000 population ($n = 1,363$) and in the United States 11.4 per 100,000 population ($n = 35,490$). A small proportion of those deaths was among skateboarders, but given the powerful ways in which skateboarding focuses attention on cities' enclaves, armatures, flows, boundaries, and capacity for playful engagement, is it possible that they are not in vain and can inform us how to use scholarship to advocate for material change?

An Immobilizing Trend: Skateboarder Deaths on the Roads

Skaters' playful interactions with street life seem to signify a clash of different mobility practices and spatial politics, not least those resulting in and from close engagements with motor vehicles. Some such engagements result in collisions, and some in death. In those cases, one witnesses the initial and disastrous convergence of moving bodies and entities—skater, board, driver, vehicle. There is the subsequent heart-rending meeting of other moved and moving bodies and entities—paramedics and ambulances, police officers and cars, loved ones, funeral directors and hearses. Here, the temporal and spatial limits of embodiment and mobility are cruelly exceeded, even as they are slowed and stopped.

Both the U.S. and Australian governments collect data on pedestrian deaths—skaters being classified as such (Australian Bureau of Statistics 2014; U.S. Department of Health and Human Services, Centers for Disease Control and Prevention, and National Center for Health Statistics 2010). Although the numbers of skateboarding fatalities are small in relative terms, each death is sad and arguably unnecessary, and each invites consideration about how all of us move through the city in ways that could be more generous. Yet without questioning the status quo in relation to the design of the city's armatures, significant numbers of people continue to press for more skate park provision—a containment strategy to minimize risk. In the United States, Portland-based Skaters for Public Skateparks (SPS) is one such group, and its focus on skate park provision derives, in part, from a collective and understandable concern for safety. Allied to this view, SPS has published data on skateboarder fatalities for 2006 (forty-two deaths), 2011 (forty-two), 2012 (thirty), and 2013 (twenty-one; SPS n.d.; Waters 2012, 2013, 2014). Acknowledged by SPS as limited, the data are collected from media reports and online and other populist sources and show that most skaters killed were aged fourteen to seventeen but ranged in age from six to fifty. In ninety-two cases (68 percent), they were hit by vehicles while street skating: Such instances included being hit by drunk or hit-and-run drivers and also involved riding at night and skitching—skating while hanging on to a moving vehicle. For this article, Australia's National Coronial Information Service (NCIS) provided data on skateboarder fatalities from 1 July 2000 to 31 December 2012. Of twenty-four individuals killed over the period, nineteen (79 percent) were aged fifteen to forty-four and most were fifteen to twenty-four; all but one were males; and nineteen were on roadways when mortally hurt or killed (Daley and Saar 2014).

Granted, the numbers of fatalities are small: In both the United States and Australia they represent 0.0001 percent of the nations' total populations of around 313 million and 22 million people, respectively. Of the 36 percent of all deaths attributable to external causes in Australia in 2012, 15 percent were caused by traffic accidents (Australian Bureau of Statistics 2012). For Americans aged five to thirty-four, motor vehicle accidents are the leading cause of death and account for thousands of deaths annually (U.S. Centers for Disease Control and Prevention 2013). Yet, to focus on comparative statistics is to miss the point; rather, it is important to acknowledge the intrinsic worth that skateboarding has as a playful and transporting practice. Setting aside (but not trivializing) that skating could be harmful to property, offends certain sensibilities and "accepted" norms of public engagement, and can be risky, the point remains that skating invites reflection about what might be done to provide for more creative city spaces and spatial politics.

A Spatial Politics of Street Skating?

Generations of skaters have enjoyed playful mobility even though skateboarding has been shaped by the "expansion of disciplinary practices associated with the management of public spaces, as in the proliferation of regulatory signs such as 'No Skateboards' ... [which represents] the constant possibility of a moralization of 'risky' conduct: skateboards are 'risky' in a way tennis

is not deemed to be" (Hunt 2003, 178). As skateboarding has become widespread, urban authorities and property owners have shown concern about its effects, with risky conduct, noise, and damage to infrastructure often deemed most objectionable (Stratford and Harwood 2001; Chiu 2009; Kern et al. 2014). Many governments declare skating a misdemeanor, controlling it using regulatory systems with ordinances, fines, powers of confiscation, and restriction to parks—a strategy sensitive to fluctuating insurance and construction costs and one that waxes and wanes accordingly. Nevertheless, in the United States alone in 1997 there were some 165 such parks; by 2008 that number had grown tenfold to over 2,000 parks (Howell 2008) catering to as many as 20 million skaters (Bradley 2010). A key logic of these strategies is that skaters are "out of place" in the city, their practices characterized by a loose use of space in ways unanticipated by urban authorities—use that generates both hybridities of bodies, boards, and infrastructure and particular affects and spatialities that nonskaters find difficult to read, navigate, and negotiate. Not least, skaters move at speeds rendering their pedestrian status a paradox.

Clearly, then, skate parks are meant to isolate skating's impact. Parks have positive attributes for skaters and families and are useful training and meeting grounds, but those who skate beyond childhood usually move to street skating in public spaces with increased visibility, volume, affect, and consequence. Simultaneously, many skaters have reached voting age, and growing numbers work in urban governments, as built environment professionals, and as scholars. Many are now reflecting on how public spaces might be differently designed and materially experienced (Beal 1995; Borden 2001; Carr 2010; Dinces 2011; Woolley, Hazelwood, and Simkins 2011; Sharpe 2012). At least some of them conclude that skating has transformative potential in the city: It is both playful and transporting.

These prospects are important when thinking about the affective and spatial politics of skateboarding given its capacity to move across, in line with, and against the grain of different streetscapes, infrastructures, age groups, places, spaces, and systems of governance. In this vein, Jenson, Swords, and Jeffries (2012) described how skateboarding encapsulates a kind of Lefebvrian desire for the city mobilized by the characteristics of space: This desire reveals to skaters all manner of pathways and playful obstacles so they "transform mundane architecture into pleasurable and unique play zones ... [using] 'skaters' eye'" (371–72). In the process, exchange and use values of the built environment are unsettled; the specificity of public space as adult and commercial space is challenged; and the -scapes of the city are rewritten and reread in unanticipated ways. Wax and scuff marks promise playful engagements of body, board, street, curb, bench, banister, or ledge, offering spaces to be appropriated for varying lengths of time. Of course, those with "skater's eye" also apprehend other signifiers: metal ridges, studs, cleats, gravel aprons, and roughened surfaces: Known as "skate haters," they tell skaters to move on (Vivoni 2009; Carr 2010, 2012). Indeed, rejection is the salient message conveyed to skaters and in case they miss that, signs declare "No skateboarders" instead of alternatives such as "Carry your board." Extensive observation of them over more than two decades suggests that most skaters are amenable to shifting gear in shared precincts but do not (and should not) countenance wholesale bans that confuse skateboarding with skateboarders and render persons, rather than activities, unwelcome.

Alternatives certainly exist. As Rawlinson and Guaralda (2011) noted, it is absolutely feasible to build into the fabric of cities' enclaves and armatures more diversely playful spaces and "sacrificial zones"; again, such work—like that to retrofit the city's armatures—requires both formalizing and normalizing strategies and tactics. Importantly, Rawlinson and Guaralda's proposal is based on an explicitly generous understanding of an affective spatial politics. Indeed, if cities are to be used

> by citizens for play activities without conflict, then building elements need to be designed fit for (ludic) purpose; capable of wearing use and possible abuse with dignity without compromising their non-ludic purpose. By designing in such a manner, conflicts over damaged property as a result of play activities and some social stigmas that surround urban appropriation are removed. (Rawlinson and Guaralda 2011, 22)

Conclusions

Lefebvre's call for utopian experiments to herald new ways to live with difference and differently is one deeply concerned with habitat and habiter, dwelling and moving, and the city's enclaves and armatures. It is a call fundamentally informed by a strong desire for conditions in which flourishing is the norm and implies strong commitments to new spatial politics, new sensibilities, and generosity.

Skateboarding is immensely popular, highly visible, commercially important, playful, and a relatively sustainable form of transport. Yet its capacity to irrupt, interrupt, and disrupt have regularly rendered it unwelcome outside purpose-built skate parks, and streetscapes are now peppered with "skate-haters" and signs that, too often, are phrased to ban persons instead of guide the direction and intensity of activities. Among those skateboarders who continue the practice beyond childhood, there is an almost inevitable move from parks to streets. In principle, this move is not unreasonable. In practice, in certain jurisdictions such as Australia it is not illegal, either, although stringent conditions apply to where skating is permitted. In many parts of the United States, though, for example, skating remains illegal. Yet, skaters continue to claim the right to the city, despite the risk of infringement notices, board confiscations, and tragic and lethal entanglements with others, not least those driving automobiles.

Alternatives exist. For instance, it is possible to design building exteriors that invite playful engagement by skaters—and also by parkour practitioners, children, office workers, and citizens of all walks of life. It is also possible to design the armatures of the city in ways that would reduce the spatial dominance of automobiles and enable other forms of mobility to expand. Yet there is an overwhelming lack of political engagement in such an agenda among those with formal power to change laws and regulations and among most of the rest of us. Thus, it seems especially important to reflect on what it might mean to be more generous—specifically in relation to skateboarding but also with respect to other mobilities. Many of them are playful, health-giving, and supportive of people with limited body movements such as infants and elders. Many also involve lower levels of resource use and pollution than motor vehicles.

Beyond these instrumental justifications for thinking about the geographies of generosity and how they might contribute to reinscribing the spatial politics of skating and other mobilities, there is an equally important reason to engage in these reflections. It is this: Imagine how enchanting it might be to radically shift the ways in which we move through the armatures of the city, foster sensory civility toward one another, respect difference, and recognize that "giving up" is not always about loss but about opening spaces for new ways of being.

References

Australian Bureau of Statistics. 2012. 3303.0 Causes of death, Australia, 2012. External causes (V01-Y98). http://www.abs.gov.au/ausstats/abs@.nsf/Lookup/by%20Subject/3303.0~2012~Main%20Features~External%20Causes%20(V01-Y98)~10034 (last accessed 7 July 2014).

———. 2014. 3303.0 Causes of death, Australia, 2012. Table 1.2 Underlying cause of death, All causes, Australia, 2003–2012. http://www.abs.gov.au/AUSSTATS/abs@.nsf/DetailsPage/3303.02012?OpenDocument (last accessed 7 July 2014).

Barnett, C., and D. Land. 2007. Geographies of generosity: Beyond the "moral turn." *Geoforum* 38 (6): 1065–75.

Beal, B. 1995. Disqualifying the official: An exploration of social resistance through the subculture of skateboarding. *Sociology of Sport Journal* 12:252–67.

Bennett, J. 2001. *The enchantment of modernity: Crossings, energetics, and ethics.* Princeton, NJ: Princeton University Press.

Bissell, D. 2014. Encountering stressed bodies: Slow creep transformations and tipping points of commuting mobilities. *Geoforum* 51 (1): 191–201.

Borden, I. 2001. *Skateboarding, space and the city: Architecture and the body.* Oxford, UK: Berg.

Bradley, G. L. 2010. Skate parks as a context for adolescent development. *Journal of Adolescent Research* 25 (2): 288–323.

Carr, J. 2010. Legal geographies—Skating around the edges of the law: Urban skateboarding and the role of law in determining young peoples' place in the city. *Urban Geography* 31 (7): 988–1003.

———. 2012. Activist research and city politics: Ethical lessons from youth-based public scholarship. *Action Research* 10 (1): 61–78.

Chiu, C. 2009. Contestation and conformity: Street and park skateboarding in New York City public space. *Space and Culture* 12 (1): 25–42.

Cresswell, T. 2010. Towards a politics of mobility. *Environment and Planning D: Society and Space* 28 (1): 17–31.

Daley, C., and E. Saar. 2014. Fatalities involving skateboards: Time period 1 July 2001 to 31 December 2011. Report Reference DR14-08, National Coronial Information System, Melbourne, Australia.

Dinces, S. 2011. "Flexible opposition": Skateboarding subcultures under the rubric of late capitalism. *The International Journal of the History of Sport* 28 (11): 1512–35.

Edensor, T. 2014. Rhythm and arrhythmia. In *The Routledge handbook of mobilities*, ed. P. Adey, D. Bissell, K. Hannam, P. Merriman, and M. Sheller, 163–71. Abingdon and New York: Routledge.

Franck, K. A., and Q. Stevens. 2007. *Loose space: Possibility and diversity in urban life.* London and New York: Routledge.

Hou, J. 2010. *Insurgent public space: Guerrilla urbanism and the remaking of contemporary cities.* London and New York: Routledge.

Howell, O. 2008. Skatepark as neoliberal playground: Urban governance, recreation space, and the cultivation of personal responsibility. *Space and Culture* 11 (4): 475–96.

Huizinga, J. [1949] 2014. *Homo Ludens.* London and New York: Routledge.

Hunt, A. 2003. Risk and moralization in everyday life. In *Risk and morality*, ed. R. V. Ericson and A. Doyle, 165–92. Toronto: University of Toronto Press.

Jensen, O. B. 2009. Flows of meaning, cultures of movements—Urban mobility as meaningful everyday life practice. *Mobilities* 4 (1): 139–58.

Jenson, A., J. Swords, and M. Jeffries. 2012. The accidental youth club: Skateboarding in Newcastle-Gateshead. *Journal of Urban Design* 17 (3): 371–88.

Kern, L., A. Geneau, S. Laforest, A. Dumas, B. Tremblay, C. Goulet, S. Lepage, and T. A. Barnett. 2014. Risk perception and risk-taking among skateboarders. *Safety Science* 62:370–75.

Kofman, E., and E. Lebas. 1996. Part I: Introduction. Lost in transposition—Time, space and the city. In *Writings on cities*, H. Lefebvre, ed. E. Kofman and E. Lebdas, 3–62. Oxford, UK: Blackwell.

LaPlante, M. P., and H. S. Kaye. 2010. Demographics and trends in wheeled mobility equipment use and accessibility in the community. *Assistive Technology* 22 (1): 3–17.

Lefebvre, H. 1991. *The production of space.* Oxford, UK: Blackwell.

———. 1996. *Writings on cities*, ed. E. Kofman and E. Lebas. Oxford, UK: Blackwell.

———. 2004. *Rhythmanalysis: Space, time, and everyday life*, ed. S. Elden and G. Moore, trans. S. Elden. London: Continuum.

Powerhouse Museum. n.d. Object: Billycart, 1950–1970. http://www.powerhousemuseum.com/collection/database/?irn=167251 (last accessed 1 December 2014).

Rawlinson, C., and M. Guaralda. 2011. Play in the city: Parkour and architecture. Paper presented at the First International Postgraduate Conference on Engineering, Designing and Developing the Built Environment for Sustainable Wellbeing, Brisbane, Australia.

Sharpe, S. 2012. The aesthetics of urban movement: Habits, mobility, and resistance. *Geographical Research* 51 (2): 166–72.

Short, J. R., and L. M. Pinet–Peralta. 2009. No accident: Traffic and pedestrians in the modern city. *Mobilities* 5 (1): 41–59.

Skaters for Public Skateparks. n.d. About SPS. http://www.skatepark.org/about/ (last accessed 10 July 2014).

Stratford, E. 2002. On the edge: A tale of skaters and urban governance. *Social and Cultural Geography* 3 (2): 193–206.

Stratford, E., and A. Harwood. 2001. Feral travel and the transport field: Some observations on the politics of regulating skating in Tasmania. *Urban Policy and Research* 19 (1): 61–76.

Thorpe, H., and R. Rinehart. 2010. Alternative sport and affect: Non-representational theory examined. *Sport in Society* 13 (7–8): 1268–91.

Thrift, N. 2004. Intensities of feeling: Towards a spatial politics of affect. *Geografiska Annaler Series B: Human Geography* 86 (1): 57–78.

U.S. Centers for Disease Control and Prevention. 2013. Injuries and violence are leading causes of death: Key data and statistics. http://www.cdc.gov/injury/overview/data.html (last accessed 16 March 2014).

U.S. Department of Health and Human Services, Centers for Disease Control and Prevention, and National

Center for Health Statistics. 2010. Health, United States, 2010, with special feature on death and dying. http://www.ncbi.nlm.nih.gov/books/NBK54374/#specialfeature.s8 (last accessed 23 March 2014).

Urry, J. 2006. Inhabiting the car. *The Sociological Review* 54 (S1): 17–31.

Vivoni, F. 2009. Spots of spatial desire: Skateparks, skateplazas, and urban politics. *Journal of Sport and Social Issues* 33 (2): 130–49.

Waters, T. 2012. SPS 2011 skateboarding fatalities. http://www.skatepark.org/park-development/2012/01/2011-skateboarding-fatalities/ (last accessed 10 July 2014).

———. 2013. SPS skateboarding fatality report for the USA, 2012. http://www.skatepark.org/park-development/2013/03/2012-skateboarding-fatalities/ (last accessed 2 February 2014).

———. 2014. SPS 2013 skateboarding fatalities. http://www.skatepark.org/park-development/advocacy/2014/02/2013-skateboard-fatalities/ (last accessed 10 July 2014).

Woodyer, T. 2012. Ludic geographies: Not merely child's play. *Geography Compass* 6 (6): 313–26.

Woolley, H., T. Hazelwood, and I. Simkins. 2011. Don't skate here: Exclusion of skateboarders from urban civic spaces in three northern cities in England. *Journal of Urban Design* 16 (4): 471–87.

World Health Organization. 2015. Global health observatory data repository: Road traffic deaths—Data by country 2010. http://apps.who.int/gho/data/node.main.A997 (last accessed 6 June 2015).

Locked in Place: Young People's Immobilities and the Slovenian Erasure

Stuart C. Aitken

San Diego State University

The case of Slovenia's erased minority populations (*Izbrisani*) is cited as one of the worst human rights abuses in contemporary Europe. While engaging debates on the nation-state and neoliberalism, this article discusses the struggles of *Izbrisani* youth from 1992 to the present day through a consideration of the spatial effects of erasure, including trauma to families forced apart and young people locked in place. Theoretical insights are drawn from Agamben's ideas about bare life, Rancière's politicization of aesthetics, and Žižek's notion of radical ethical acts, which respectively provide lenses for understanding *Izbrisani* youth privations, awakenings, and transformations.

斯洛文尼亚抹除少数族裔人口 (Izbrisani) 的案例，被引用作为当代欧洲最为严重的侵犯人权案例之一。本文在涉入有关国族国家及新自由主义的辩论的同时，透过考量抹除的空间效应，包括家人被迫分离的创伤以及被禁闭的青年，探讨被抹除的青年自 1992 年至今的挣扎。本文引用阿冈本的 "裸命"、洪席耶的 "美学的政治化"，以及齐泽克的 "激进伦理行为" 概念之理论洞见，这些理论对于理解被抹除的青年的剥夺、觉醒与转变，提供了各别的分析视野。

El caso de las minorías desplazadas (*Izbrisani*) de Eslovenia se cita como uno de los peores abusos contra los derechos humanos en la Europa contemporánea. Al tiempo que se abocan debates sobre estado-nación y neoliberalismo, este artículo discute las luchas de la juventud *Izbrisani*, desde 1992 hasta el día de hoy, mediante una consideración de los efectos espaciales del desplazamiento, incluyendo los traumas infligidos a familias separadas a la fuerza y el encierro de gente joven en el lugar. Las bases teóricas provienen de las ideas de Agamben acerca de la vida al desnudo, la politización de la estética de Rancière y la noción de Žižek de los actos éticos radicales, que respectivamente proveen lentes para entender las privaciones de la juventud Izbrisani, despertares y transformaciones.

Young people's mobilities are integrally linked to allowed freedoms, and to the degree that those self-determinations are prescribed legally at the level of the state, they are often couched within human rights discourses. It is thus important to understand the interleaving of young people's mobilities and the contemporary nation-state. In a critical review of mobility studies, Söderström and his colleagues (2014) noted that "specific insights on inequality, domination, and constraint [are] provided by focusing on the interdependence of different mobilities and immobilities," and from the latter there arises a need to understand how engagements with the "state ... shape the possibilities and implications of different forms of mobility" (xv). It is imperative, they concluded, to look jointly at the experience of mobility and its institutional framing. This insight resonates with Massey's (2005) admonition that institutional frames "tame the spatial" into forms of "stasis and closure," which deny "heterogeneity, relationality, coevalness, ... liveliness indeed," and foreclose upon "a more challenging political landscape" (13, 61). In what follows, I discuss youth immobilities lodged within institutional frames that violated the rights of people who were removed from Slovenia's permanent residential register in 1992 and became, effectively, stateless. I elaborate some of the privations and hardships of young people from 1992 to the present, focusing on their immobilities and the ways they resisted and pushed back against the institutional frames through which they were locked in place.

Slovenia's declaration of independence in June 1991 was a move away from a socialist regime toward the increasingly neoliberal and transnational ideals of a borderless Europe. The contemporaneous expansion of the Schengen Agreement and the creation of the Eurozone pushed free movement of capital, people, and goods. If increased mobility and market fluidity were expectations of the nascent Slovenian government, their 1991 Alien Act in combination with the Citizenship of the Republic of Slovenia Act also established parameters for excluding certain residents.

Slovenia was the northern and most prosperous of the Yugoslav Socialist republics, and during the 1970s and 1980s it attracted workers from southern Balkan republics. Minorities from these republics who lived in Slovenia at the time of independence but had not applied for citizenship between June and December 1991 were removed from the permanent residential register. Official Slovenian statistics estimate 25,671 people lost permanent legal status, of whom 21 percent were under eighteen years of age (Kogovšek 2010). The *Izbrisani* (Slovenian for "erased," and first used disparagingly by former Secretary of State Andrej Šter) were mainly Croatians, Bosnians, and Serbians, but also some Roma, with legal status defined through parental ethnicity (Bajt 2010) and language (Petkovič 2010). Right-leaning nationalists saw the erasure as a move away from Slovenia's socialist past toward a neoliberalism that eschewed state welfare structures from which, they believed, ethnic minorities benefited unduly. Commenting provocatively on the formation of new states, Ong (2006) noted that certain people are "rendered excludable as citizens and subjects" and denied basic rights because they are "too complacent or lacking in neoliberal potential" (16). Although they were not the impresarios of the new Slovenian economy, *Izbrisani* youth were nonetheless potential cheap labor, whose "disposability [was] naturally and culturally scripted" (Wright 2006, 1).

A focus on Slovenian minority youth locked in place with no legal status offers an acute example of exclusion from basic human rights. Their subsequent recognition and push for citizenship is an example of mobile activism and transformative politics (Aitken 2014). In what follows, I use in-depth interviews with *Izbrisani* children and their families conducted between October 2013 and May 2014 to inform my theoretical analysis. The stories from a few young people do not generalize the consequences of being stuck in place, but they are in no sense unusual.[1] With this article, I offer partial insights—mere glimpses really—of suffering and exclusion. *Izbrisani* youth represent a unique case of denying mobility through administrative bordering; some are locked in place, living inconspicuously under fear of deportation, whereas others are locked out, unable to return to Slovenia. To understand the theoretical and political resonance of these issues, I draw on Agamben's (1995, 2001) ideas about bare life and states of exception. I then look at how *Izbrisani* privations push back against the system using Rancière's (2009, 1) "disruption of the sensible." The article ends with the transformative nature of this disruption, drawing on Žižek's (2010) suggestion that actions foment radical ethical acts when they not only change the person but also disrupt "a traditional ethic of common sense and common decency amongst ordinary people" (326).

Bare Life

Agamben (1995, 8) described bare life through the *homo sacer*, an obscure figure from Roman law who was killed with impunity rather than through honorable sacrifice. The life of the *homo sacer* was incorporated in the judicial order as exclusion: humans as animals (*zoē*) without political existence (*bios*). Agamben's (1995) project elaborates the function of bare life in modern politics through "the state of exception" (9), whereby citizens are protected by the sovereign rule of law and certain outsiders are purposively excluded. For Agamben (2001), the state of exception in modern politics is exemplified by the succession of camps from internment to concentration to extermination, which "represents a perfectly real filiation" (22). The Nazi camps were an extreme and inevitable form of a "political space in which we still live" (Agamben 2001, 36). Criticism is leveled at Agamben's unwillingness to distinguish between different forms of camps—or forms of exception not contained within camps—that could perhaps "dislodge the notion of a clear judicial filiation" (Owens 2009, 575). A loosening of Agamben's teleology enables consideration of the foundational relationship of the nation-state to other forms of exclusion. Ong (2006), for example, noted that the extraordinary malleability of neoliberalism as a form of governance enables the reengineering of political spaces and populations through exceptions beyond camps. Global market forces and the "neoliberal logic" of "emerging states," she argued, "reconfigures the territory of citizenship" through "new economic possibilities, spaces, and political constellations for governing the (national) population" (Ong 2006, 75).

In the early 1990s, the new state of Slovenia denunciated its Balkan identity through legal processes, distancing it from countries to the south. Slovenia escaped most of the Balkan War with its declaration of independence and the so-called ten-day war, after which the Serbian-controlled Yugoslav army focused its aggression on Croatia and then Bosnia. With independence, new laws defined who was a legitimate Slovenian citizen and who was not. Caught in the extremes of this legislation were young people of minority parents who had not applied for citizenship during the narrow window of opportunity offered in the second half of 1991. Faced with serious human rights abuses, these young *Izbrisani* hid for fear of deportation or, if stuck outside Slovenia, they fought (and some continue to fight) to return home.

"I Hid Down Like a Little Mouse"

Fear without hope. Uncertainty. Hiding from police. Officials everywhere only rejected me. They had power over me. I could not cross the border. I was undocumented [for] seven-and-a-half years. I lived in fear nonstop. However, I was young. I had more courage, more nerves. I tried to be inconspicuous. (Sonja Krupić, 31 January 2014)

Sonja Krupić was born in Slovenia when it was part of Yugoslavia. She was looked after by her grandmother in Bosnia following her father's death. At age five, Sonja returned to stay with her mother in a subsidized apartment in Fužine, a poor immigrant community in southeast Ljubljana. At seventeen, Sonja was left in charge of the apartment while her mother went to find work in Germany. In February 1992, with her mother still abroad, Sonja was identified as a citizen of a former Yugoslav republic who had not applied for Slovenian citizenship and lost her status as a permanent resident:

I never received any letter that I should arrange my citizenship. I called my mother [to see] if I have to arrange some papers. She didn't know what it was about. When I found out I was erased and they destroyed my Yugoslavian personal ID, I was forced to go to the Bosnian embassy, but this was the last thing I wanted to do: I don't belong there, I have nothing to do down there, I grew up here, in Slovenia; here I have my friends. (7 February 2014)

The policy of granting citizenship through bloodlines (*jus sanguinis*) preempted citizenship based on Sonja's Slovenian birth (*jus soli*). For *Izbisani*, the consequences of losing legal status included not just ineligibility for citizenship, but also removal from subsidized housing, and loss of legitimate employment and free health care. Sonja tried to get help from sympathetic friends and lawyers, but she was a teenager with little experience or influence:

I tried to arrange my status by calling to different offices, until someone started to yell at me that I should hide and be silent; if not, they are going to find me and deport me. I stopped calling and hid at home in fear that

someone will call me and deport me. I hid down like a little mouse. (7 February 2014)

For *Izbrisani* youth, the consequences of erasure showed up as strictures and rebukes, as detentions and expulsions, and as denial of access to bureaucratic processes that seemed at the whim of bureaucrats. Sonja avoided officials or people connected to the Slovenian government, including doctors:

I tried to survive on my own. In the meantime I was for some time without a job. They switched off my electricity. I was without money, without food. For two months I was literally starving. I lowered myself so much that I was begging in the city center. I never felt such humiliation before! It felt so terrible that I reached this point. (7 February 2014)

Young people are often resilient and flexible in the face of change but only to a degree (cf. Marshall 2013). Sonja's deprivation grew, and her freedom of movement was curtailed by fear of deportation to Bosnia, where war with Serbia continued during the 1990s:

I could not cross the border. They threatened me [with] deportation to a war zone. The horror. There I would be killed and raped. I too was raped by a friend of my mother's who was supposed to help me. [I did not report it] because I would be deported. (7 February 2014)

To the degree that the loss of legal rights was a stripping to bare life, young *Izbrisani* were particularly susceptible to ethnic and gender victimization. At a time in life when identity formation is paramount, these young people were told that they were nothing. In 2000, Sonja got temporary documentation from the Bosnian embassy in Ljubljana: "Six zeros at the end of my registration number. . . . Mine [used to be] 505 060 and now I have six zeros at the end. No birth-right. It was strange." The six zeros signified not only an erasure of legal rights, but a denial of parents and birth, placing the young person in a position that was less than human, without value.

"It Does Not Matter Who Is Really Guilty, I Am Always Blamed"

The mouthpieces of authority think that we are outcasts without value. I wish these mouthpieces could feel what we had to go through; then they would see. (Igor Fakaj, 27 February 2014)

While Sonja was locked in place in Slovenia, others were locked out as part of the erasure process. Fourteen-year-old Igor Fakaj was born in Germany. We interviewed Igor when he came to Ljubljana with his father and sister to seek residency. Igor's father, Amir, was deported to Albania in 1996 "simply because of my last name, I'd never been there before and have no family there" (Amir Fakaj, 26 February 2014). Located immediately south of the former Yugoslavia, Albania is perceived as an economically deprived, poor Balkan cousin. Amir escaped from an Albanian transitional home for foreigners and fled to Germany without documents. He was given temporary residence and met his wife, who gave birth to Igor a year later: "In Germany I live in fear. Every second the police can come and shoo me away to some municipality I have never been before." In 2005, Amir was threatened with deportation to Kosovo ("I've never been to Kosovo"), so he fled back to Slovenia with his wife and (now) four children, where he applied for asylum. Amir was granted asylum, but his children and wife were deported. According to his thirteen-year-old daughter:

> The police came and woke us up, and they told us we had to go back to Germany. I felt very bad. I was very frightened and did not want to go back. They forced us into a car and drove us to Germany. In Germany we live in a little room where everybody sleeps. They say ugly things; kids at school say they don't wish to have foreigners and that we must leave their country. (Alenka Fakaj, 27 February 2014)

Amir's asylum case was successful and he is now a Slovenian citizen, but he is unable to return with his family. Igor sums up his feelings in a quote that pointedly gets to the heart of bare life:

> In school I am both a German and a foreigner. There are different tensions between these groups, sometimes they fight each other and somehow, it does not matter who is really guilty, I am always blamed. I feel pushed away from school, from the surroundings, from everything. (27 February 2014)

The rights of those who are nothing but human (i.e., zoē) are not only voided as animals; bare life also creates them as scapegoats on which the ills of a society are heaped. Following Agamben, Rancière (2009, 132) referred to bare life as "identity fundamentalism" and argued that what is needed is the staging of dissent through disruption of the sensible.

Disrupting the Sensible

According to Dixon (2009), Rancière's politicization of aesthetics is "primal" because he sees space as a fundamental "*loci* for identity formation and the emergence of practices, and nodes in a regime of politics" (414). Aesthetics in this sense are primal because they come from bare life, but they also relate to Massey's (2005) ideas about countering the stasis of institutional frames when they "open space for deviations, modify the speeds, the trajectories, and the ways in which groups of people adhere to a condition" (Rancière 2007, 39). Rancière repoliticized aesthetics away from traditional ideas of beauty and art to a consideration of the "distribution of the sensible" (Rancière 2009, 1) in terms of relations "between what people do, what they see, what they hear, and what they know" (Rancière 2010, 15–17).

An archeology student, Aleksander Todorović, is widely credited as the initiator of the *Izbrisani* movement. In February 2002 he disrupted Slovenian sensibilities by chaining himself to a tree at the Ljubljana Zoo and beginning a hunger strike: "I don't know why in front of the zoo. I have no idea why right there. No, it is not that I don't know: I had this crazy idea about becoming an amoeba. And that's how it all began." At his wit's end after ten years of erasure, Todorović wanted to show that "something bad had happened here" (Todorović, cited in Zdravković 2010, 264). His allusion to "becoming an amoeba" suggests reduction to something primal, akin to bare life. In February 2003, he was part of the first Erased Week, which raised public awareness of the *Izbrisani* issue. Following this, disruptions continued every February. A series of hunger strikes in 2005 began at the Croatian border and then moved to the UNICEF headquarters in Ljubljana:

> We went like this, for some time walking and for some time driving. Where the roads were without settlements and towns, we were driving. And then we went through towns like Postojna and Logatec, we went by foot and came walking to Ljubljana. Everything was photographed. We were in every town, it was on TV. (Andro Duvnjak, 10 February 2014)

Joint themes of mobility and presence, which were long missing from *Izbrisani* lives, characterized the disruptions. Mobile activism focused attention on the implicit and explicit immobilities of erased people. In 2006, a thwarted attempt to occupy Slovenia's National Assembly turned into activists dressed in white lying on the road disrupting traffic flow to spell out *izbris* (canceled). Later that year, several children joined a group of *Izbrisani* activists in a journey to the European Court of Human Rights in Strasbourg in what was dubbed the Caravan of the Erased. Young *Izbrisani* were connecting and creating new sensibilities:

> Oh dear, when we started to talk … I felt like I got wings. As if a stone fell from my heart. Pains literally began to peel off my body. I could feel life, health, the future. I could see the light at the end of the tunnel. (Twenty-year-old Ismeta, cited in Zorn 2010, 23)

Žižek (2014) would describe these as events in transit, speaking truth to power: *Izbrisani* activism was, literally, a mobile protest.

Franjo Duvnjak (Andro's son) and Aleksandra Todorović (Aleksander's daughter) were two of the young *Izbrisani* who went to Strasbourg in the Caravan of the Erased. Aleksandra was twelve at the time and had participated with her father and mother in the attempt to occupy the National Assembly. She recalled this as a time of tension in her family because her father was dubbed an enemy of the people by the right-wing press:

> [Dad] never wanted to confess how much it hurt him, how much some things got to him. It is not a secret that he was a very depressed person. He had dark periods when he closed himself in the apartment and wouldn't go anywhere. (Aleksandra Todorović, 24 March 2014)

Activism brought *Izbrisani* out of hiding.

Franjo was fourteen, and the trip to Strasbourg was his first time out of the country. Franjo's mother, Mirna, became an activist in the early 2000s when he was still in elementary school. Franjo was candid that his trip to Strasbourg was mostly holiday: "I wanted to see how France looks like. I have never been so far, just in Slovenia. And I decided to go" (Franjo Duvnjak, 10 February 2014). Play is an important part of protest (Crossa 2013); Marshall (2013) argued that Rancière's relational and radical political aesthetics reside most potently among young people, and protest often arises subtly from less structured and more relational, serendipitous, playful, and mobile processes. In geography, ideas about young people's spontaneous disruption of the sensible can be traced in part to Katz's (2004) work on radical play and the ways children use things creatively and spontaneously. For Katz, play is identity making, but it can also be revolutionary and world making. In play, children might toy with social practices, but it is also where received meanings and relations are refused or reworked (Katz 2011). Marshall (2013) argued that Rancière highlights "how we might understand the present political moment through the lives of children and how children play a role in building alternative futures…how children both perform and transform the aesthetics of suffering" (54). Franjo focused on the fun aspects of his trip to Strasbourg but, importantly, later in the interview his mother made a pithy comment on some poignant and more subtle affects: "Like you have something and that something gives you energy. And this persistence I have inside, I move on that" (Mirna Duvnjak, 10 February 2014). The question that takes up the balance of this article is the degree to which internal energy of this kind can disrupt the status quo and change abusive regimes of power.

Toward Radical Ethical Acts

For Žižek (2008, 68), raging against abstract and anonymous sociopolitical systems is an "utterly senseless" form of protest. Most protests are not political acts, he argued, because nothing changes; they are acting out a spectacle that addresses the figure of the big Other but leaves it undisturbed. Žižek's big Other is an assemblage of symbolic (e.g., religious, capitalist) relations that are part of everybody's subjectivity and that cannot be appropriated by collective movements. For Žižek (2010), Lacan's (1992) *le passage à l'acte* is not acting out because it comprises a movement and an ethical act, which heralds the demise of the big Other. Radical ethical acts "open up the space of an authentic belief which sustains the act, a belief that is no longer transposed onto, sustained, or covered by some figure of the big Other" (Žižek 2010, 134). This resonates with Agamben (2001), who pointed out that sovereign power today does not hold any form of legitimation outside of fear and emergency, which we internalize in ways that enervate and disenfranchise. When sovereignty loses its precision and borders blur, though, "the bare life that dwelt there frees itself … and becomes both subject and object of the conflicts of the political order, the one place for both the organization of State power and emancipation from it" (Agamben 1995, 9).

Is *Izbrisani* activism changing Slovenia and Slovenians along lines that would meet with the approval of Žižek and Agamben? Did their bare life rise up to emancipate the *Izbrisani*? The answers to these questions are mixed. The Caravan of the Erased was successful to the extent that the European Court of Human Rights ruled in their favor.[2] Of the 25,671 erased people, between 10,000 and 11,000 have regulated their status in Slovenia either as citizens or permanent residents, but there are still more than 13,000 living outside of Slovenia who are potential applicants for legal status (Feffer 2013). More than 1,300 *Izbrisani* have died, which includes nearly 200 suicides (Aleksander Todorović took his own life in February 2014). The Slovenian government admits that a bureaucratic error was made in 1992, but there have been no formal apologies for the erasure, and the monetary reparations mandated in 2013 by the European Court are currently on appeal.

The Fakaj children are no closer to resettling with their father in Slovenia. Amnesty International is pushing the case, but there are as yet no signs of a settlement. Sonja Krupić's radical ethical act is through painting. Sonja continues to live in Fužine and uses art as therapy for the trauma she went through. She works in an elders' apartment complex where she shares her art. Sonja's paintings are stark, even brutal, portrayals of placid mountain and rural scenes. She is no longer interested in activism:

> I'm not a member of any party and never want to be: for me, all forms are small-minded customers. I only try to survive. I am sometimes angry; lies on media and some people believe them. I don't need that the government to say "sorry," but don't tell lies. (Sonja Krupić, 7 February 2014)

With the death of her father, Aleksandra Todorović took on the mantle of activism as a radical ethical act:

> I can see in my peers that they are against the violation of human rights, but they would never fight. They would never expose themselves, because they know that this can lead to bad things. The children of the erased know what this all means, but the society is not ready for this yet; it blindly follows some instructions and lies from politics. It is easy to support something quietly, but when it comes to doing something, it is a completely different story. Here we should achieve more. We lack open people, we really, really lack this. . . . This is something that needs change. (Aleksandra Todorović, 24 March 2014)

From watching the activism of her father, Aleksandra is aware of the importance of disrupting space and public sensibilities through protest, but she is also aware that his public struggle was at significant personal cost. Before she starts practicing her activism, Aleksandra wants to attain a university degree in social work or law.

The activism of Aleksandra and Sonja is more subdued, at times more despairing, than that of their forbearers, but I argue with Žižek that it is potentially more radical in its critique of politics because it is not about acting out but rather finds power from radical ethical acts as passages or movements.

Žižek, Rancière, and Agamben not only provide particular critiques of sovereign power and ways of understanding political opposition to that power, but they also offer critiques of (universal) human rights as unenforceable. To the degree that *Izbrisani* youth made successful appeals to human rights discourse and law, many have secured state citizenship, but where does that leave them? Harvey (2008) argued that "[t]he freedom to make and remake our cities [read spaces] and ourselves is . . . one of the most precious yet most neglected of our human rights" (23). For some *Izbrisani*, rights to space are rights to return to the country they consider home. For others, they are about deprivation of legal rights within political boundaries, which served to lock them in place. This has much to do with mobility because, as Lee and Pratt (2011) noted, tracing fixities in people's lives within the complexities of neoliberalism is important because the same values that revel in movement and freedom also work to constrain and disenfranchise certain people. In writing about Mexican youth mobility that speaks to issues of citizenship and rights to space in the United States, Torres and Carte (2014) noted that including marginalized and underrepresented young (and community) voices in mobility studies punctuates neoliberal policy interventions with differently imagined futures. The critical perspectives on youth rights and immobilities that I elaborate here reinforce the notion that although space is fluid and dynamic—that it is produced, palpable, and fully political—it is nonetheless problematically fixed by institutional and ideological frames. Isin (2008) argued that "processes of 'globalization,' 'neoliberalization' and 'post-modernization' . . . produce new, if not paradoxical, subjects of law and action, new subjectivities and identities, sites of struggle and new scales of identification" (16) through everyday acts of citizenship that foment out of mobilities in which the boundaries of the nation-state are increasingly irrelevant (Isin 2014) and different futures are imagined (Aitken 2014). Perhaps these futures will focus on rights to space rather than rights to citizenship.

Sonja's voice underscores the resilience and energy of young people surviving the structural violence of neoliberal frames, and her alternative future haltingly emerges from her art. Reworking space through play, resilience, and refusal is perhaps the most poignant of ethical transformations, and for this it could be argued that the passions of young people like Franjo and Aleksandra are Slovenia's best hope.

Acknowledgments

This work would not have been possible without my colleagues Anton Gosar and Stanko Pelk from the University of Primorska and Uršula Lipovec-Čebron and Jelka Zorn at the Ljubljana Peace Institute. I am particularly indebted to my translator and research assistant, Ines Hvala. Kate Swanson and three anonymous reviewers provided valuable comments on early drafts of the article.

Funding

This research was supported by a Fulbright Research Scholarship and the June Burnett Endowment.

Notes

1. More than twenty young people and various family members were interviewed by myself and my translator, Ines Hvala. I also interviewed Slovenian academics, news reporters, and filmmakers who followed the *Izbrisani* issue. The interviews were conducted in cafes, restaurants, and the homes of *Izbrisani*. In February 2014, we brought together several young *Izbrisani* in a symposium that got national media attention. With the exception of known public figures, the names used here are pseudonyms to protect participants facing ongoing legalities and discriminations. The fictitious surnames reflect interviewees' Serbian, Bosnian, or Croatian origins, thus highlighting a primary gateway for erasure.
2. *The European Court of Human Rights (Third Section), Kurić and others v. Slovenia* (application No. 26828/06), initiated in 2006, judgment in favor of plaintiffs in 2012.

References

Agamben, G. 1995. *Homo sacer: Sovereign power and bare life*. Stanford, CA: Stanford University Press.
———. 2001. *Means without end: Notes on politics*. Minneapolis: University of Minnesota Press.
Aitken, S. C. 2014. *The ethnopoetics of space and transformation: Young people's activism and aesthetics*. Farnham, UK: Ashgate.
Bajt, V. 2010. More than administratively created "foreigners": The erased people and a reflection of the nationalist construction of the other in the symbolic idea about "us." In *The scars of the erasure*, ed. N. Kogovšek, J. Zorn, S. Pistotnik, U. Lipovec-Čebron, V. Bajt, B. Petkovič, and L. Zdravkovic, 195–217. Ljubljana, Slovenia: Peace Institute.
Crossa, V. 2013. Play for protest, protest for play. *Antipode* 45 (4): 826–43.
Dixon, D. 2009. Creating the semi-living: On politics, aesthetics and the more-than-human. *Transactions of the Institute of British Geographers* 34:411–25.

Feffer, J. 2013. Restoring Slovenia's erased. *Institute for Policy Studies*. http://www.ips.dc.org/restoring_slovenias-erased (last accessed 11 November 2014).
Harvey, D. 2008. The right to the city. *New Left Review* 53 (September–October): 23–40.
Isin, E. F. 2008. Theorizing acts of citizenship. In *Acts of citizenship*, ed. E. F. Isin and G. M. Nielsen, 15–43. London: Zed Books.
———. 2014. *Citizens without frontiers*. London: Bloomsbury.
Katz, C. 2004. *Growing up global*. New York: Guilford.
———. 2011. Accumulation, excess, childhood: Towards a counter-topography of risk and waste. *Documents d'Anàlisi Geogràfica* 57 (1): 47–60.
Kogovšek, N. 2010. The erasure as a violation of legally protected human rights. In *The scars of the erasure*, ed. N. Kogovšek, J. Zorn, S. Pistotnik, U. Lipovec-Čebron, V. Bajt, B. Petkovič, and L. Zdravkovic, 83–140. Ljubljana, Slovenia: Peace Institute.
Lacan, J. 1992. *The seminars of Jacques Lacan: The ethics of psychoanalysis*, ed. J.-A. Miller, trans. D. Porter. London and New York: Routledge.
Lee, E., and G. Pratt. 2011. Migrant worker: Migrant stories. In *Geographies of mobilities: Practices, space, subjects*, ed. T. Cresswell and P. Merriman, 225–38. Farnham, UK: Ashgate.
Marshall, D. 2013. "All the beautiful things": Trauma, aesthetics and the politics of Palestinian childhood. *Space and Polity* 17 (1): 53–73.
Massey, D. 2005. *For space*. London: Sage.
Ong, A. 2006. *Neoliberalism as exception: Mutations in citizenship and sovereignty*. Durham, NC: Duke University Press.
Owens, P. 2009. Reclaiming "bare life"?: Against Agamben on refugees. *International Relations* 23:567–82.
Petkovič, B. 2010. The erased language. In *The scars of the erasure*, ed. N. Kogovšek, J. Zorn, S. Pistotnik, U. Lipovec-Čebron, V. Bajt, B. Petkovič, and L. Zdravkovic, 223–47. Ljubljana, Slovenia: Peace Institute.
Rancière, J. 2007. *The politics of aesthetics: The distribution of the sensible*. London: Continuum.
———. 2009. The aesthetic dimension: Aesthetics, politics, knowledge. *Critical Inquiry* Autumn:1–19.
———. 2010. The aesthetic heterotopia. *Philosophy Today* 54:15–25.
Söderström, O., S. Randeria, D. Ruedin, G. D'Amato, and F. Panese, eds. 2014. *Critical mobilities*. London and New York: Routledge.
Torres, R., and L. Carte. 2014. Community participation appraisal in migration research: Connecting neoliberalism, restructuring and mobility. *Transactions of the Institute of British Geographers* 39 (1): 140–54.
Wright, M. 2006. *Disposable women and other myths of global capitalism*. London and New York: Routledge.
Zdravković, L. 2010. The struggle against the denial of citizenship as a paradigm of emancipatory politics. In *The scars of the erasure*, ed. N. Kogovšek, J. Zorn, S. Pistotnik, U. Lipovec-Čebron, V. Bajt, B. Petkovič, and L. Zdravkovic, 257–77. Ljubljana, Slovenia: Peace Institute.
Žižek, S. 2008. *On violence*. London: Profile Books.

———. 2010. *Living in the end times*. New York: Verso.

———. 2014. *Event: Philosophy in transit*. London: Penguin.

Zorn, J. 2010. Registered as workers, erased as non-Slovenes: The transition period from the perspective of erased people. In *The scars of the erasure*, ed. N. Kogovšek, J. Zorn, S. Pistotnik, U. Lipovec-Čebron, V. Bajt, B. Petkovič, and L. Zdravkovic, 19–46. Ljubljana, Slovenia: Peace Institute.

Unintended Return: U.S. Deportations and the Fractious Politics of Mobility for Latinos

Marie Price and Derek Breese

Department of Geography, The George Washington University

A record-breaking 4.2 million people have been removed from the United States since 2000, with migrants from Latin America accounting for over 93 percent of all removals. The U.S. policy shift toward forced removals (commonly referred to as deportations) underscores many mobility politics and paradoxes that Latinos experience. Their determination to be mobile and leave their countries of origin often results in encountering various legal challenges in the United States that might limit their mobility within the United States and, sometimes, result in their involuntary mobility through forced return. This article is grounded in the politics of mobility literature interested in the frictions created within constellations of mobility that create unintended return. Drawing from administrative data produced by the Department of Homeland Security, Transactional Records Access Clearinghouse (TRAC) data, and the U.S. Census, this research (1) documents the scope and uneven practice of forced removal; (2) suggests how unintended return is affecting Mexico, Guatemala, and Honduras; and (3) develops the *unintended returnee* as an important mobility subject.

自从 2000 年以来, 有高达四百二十万的破纪录人次被美国递解出境, 在所有的递解者中, 拉丁美洲的移民佔了百分之九十三以上。美国政策转向强制递解 (一般被指称为"遣返") , 凸显出拉丁裔所经历的众多能动性政治与自相矛盾之事。他们流动且离开祖国的决定, 经常导致在美国面临各种法律的挑战, 并可能限制他们在美国的能动性, 且有时会导致他们因强迫遭返所面临的非自愿移动。本文植基于对创造非预期返回的能动性配置中生成的摩擦感兴趣的能动性政治之文献。本研究运用美国国土安全部所生产的行政数据, 政府档案纪录交换中心 (TRAC) 的数据, 以及美国人口调查, (1) 纪录强制递解出境的范畴与不均实践; (2) 指出非预期返回如何影响墨西哥、危地马拉与洪都拉斯; 以及 (3) 将非预期返回者建构成为重要的能动性主体。

Desde el año 2000 se ha removido de los Estados Unidos la cifra récord de 4.2 millones de personas, guarismo en el que los inmigrantes de América Latina representan más del 93 por ciento de la remoción. El cambio de política inmigratoria hacia la expulsión forzada (comúnmente referida como deportación) subraya muchas de las políticas sobre movilidad y paradojas que experimentan los latinos. Su determinación por hacerse móviles y abandonar sus países de origen a menudo resulta en la confrontación de varios retos legales en los Estados Unidos que podrían limitar su movilidad dentro de este país y, a veces, resultar en su involuntaria movilidad del regreso forzado. Este artículo está fundamentado en literatura sobre política de movilidad, interesada en las fricciones creadas dentro de las constelaciones de movilidad que determinan regreso no buscado. Con base en datos administrativos producidos por el Departamento de Seguridad Nacional, del Transactional Records Access Clearinghouse (TRAC) [Centro de Información sobre Acceso a Registros Transaccionales] y del Censo de los Estados Unidos, esta investigación, (1) documenta la forma y práctica inequitativa de la remoción forzada; (2) sugiere cómo el regreso no intencional está afectando a México, Guatemala y Honduras: y (3) convierte al retornado involuntario en un sujeto de movilidad importante.

Much of the U.S. discourse about immigration reform focuses on the fate of some 11 million undocumented individuals. Meanwhile, a formidable machinery of migration enforcement has been constructed over the last two decades that has led to 4.2 million removals since 2000 (Meissner et al. 2013; Department of Homeland Security [DHS] 2014). More removals took place in this period than in the entire 107-year period between 1892 and 1999 (DHS 2011, table 36). The vast majority of these removals are to Latin America, mostly Mexico, and 90 percent of those removed are men (even though women represent nearly half of the undocumented; Rosenblum and McCabe 2014). Hence, the gendered and racial removal of Latino immigrants is a politically charged and deeply personal issue for U.S. Latinos (Golash-Boza and Hondagneu-Sotelo 2013).

It is estimated that one in three Latinos in the United States knows someone who has been deported or detained in the last twelve months (Lopez, Morin, and Taylor 2010). The experience of removal from the United States and *unintended return* to a Latin American origin country is an increasingly familiar crisis for many Latino households. Yet the processes behind and consequences of these forced removals are less discussed in the English-language media, which instead focuses on the threat of unchecked Latino immigration (Chavez 2008; Golash-Boza 2012).

Since the 1996 Immigration Reform and Immigrant Responsibility Act, federal policy has concentrated on illegal immigration and securing national borders. Through the formation of a new federal agency (DHS), the application of enhanced technologies (biometrics and improved documents), and the expansion of new programs (Secure Communities and the 287(g) programs), the first thirteen years of the 2000s witnessed a surge in removals from the United States. Removals topped 200,000 for the first time in 2003, exceeded 300,000 in 2007, and topped 400,000 in 2012. In 2013, the last year for available data, 438,000 people were removed (DHS 2014).

Empirically, this article has three purposes. First, it examines the stunning rise in removals from the United States and the shifting terrain of court decisions at the state level with regard to enforcing or staying deportation orders. Second, an analysis of DHS administrative and U.S. census data shows the disproportionate impact removal has had on immigrants from three countries: Mexico, Honduras, and Guatemala. Third, the work introduces a new term, *unintended return*, and argues that it has conceptual and empirical merit. Theoretically, this work advances the politics of mobility literature (Coutin 2010; Cresswell 2010) by conceiving of unintended return as an outcome in constellations of mobility fraught with frictions and paradoxes.

Literature Review: Mobility Politics and Paradoxes

Geographers are actively engaged in a multidisciplinary effort to understand human mobility (Carling 2002; Cresswell 2006, 2014; Samers 2010; King 2012)

although much of the migration research does not explicitly engage the mobility literature. Carling (2002) raised the specter of involuntary immobility (people who would like to move but face barriers and experience unwanted fixity) when researching the mobility issues that Cape Verdeans experience. We believe that using a mobility perspective, rather than a migration-driven one, leads to consideration of the obstacles, stillnesses, and paradoxes embedded in mobility networks that include practices, representations, and physical movement (Cresswell 2010). In addition, this research is particularly concerned with the politics of mobility, especially the role that national borders, courts, and localities play in influencing mobility practices.

The mobility practices of migrants are heavily influenced by national governments that determine which migrants legally enter and under what terms. Yet many geographers argue that substate jurisdictions and transnational networks also influence the politics of immigrant selection, inclusion, and exclusion (Wright and Ellis 2000; Singer, Hardwick, and Brettell 2008; Varsanyi 2011; Leitner and Strunk 2014). U.S. states, counties, cities, and towns are the settings that shape how newcomers are received, through both inclusionary and exclusionary practices (Walker and Leitner 2011; Price 2014). The role of individual migrants and the social networks in which they are embedded also influence their mobility practices and destinations (Durand and Massey 2010).

Today, tolerance and intolerance in the United States toward immigrants, especially the undocumented, are in flux, given federal inaction on immigration reform. Geographers have been especially interested in places that exhibit hostile attitudes toward immigrants, with the literature suggesting that communities that undergo dramatic influxes in immigrant newcomers in a short period of time might be more inclined to see such groups as a threat (Winders 2006; Varsanyi 2008; Singer, Wilson, and DeRenzis 2009). An important study by Walker and Leitner (2011) maps 370 local governments that have proposed or implemented policies designed to address the undocumented, with the vast majority of these policies having been implemented since 2005. They found that local exclusionary ordinances outpaced inclusionary ones two to one. Regionally, the South was more exclusionary and the West was slightly more inclusive. Moreover, inner cities were more likely to promote inclusionary ordinances with regard to the undocumented, whereas the suburbs and rural areas were more likely to be exclusionary (Walker and Leitner 2011). This devolution of immigration enforcement policy and the expansion of exclusionary ordinances have "pushed the border inward" in ways that help to explain the rise and patterns of deportation within the United States (Walker and Leitner 2011, 156).

In terms of human mobility representations, De Genova (2010) noted that deportation is a critical locus for a theoretical elaboration of "the co-constituted problems of the state and its putative sovereignty, on the one hand, and the elementary precondition of human freedom which is the freedom of movement" (39). Although forced removal is represented as a state's ability to control and secure its territory, from the perspective of the detained and removed individual, forced removal represents a failure and even a crisis for individuals and households. Such tactics illustrate how migrants are "especially vulnerable to a 'rescaling of personhood' due to shifting conceptions of illegality, a rollback of rights, and a rise in deportation" (Varsanyi 2008, 882).

Given the rise in deportations from the United States, interest in the physical removal of people as mobility subjects is growing (Brabeck and Xu 2010; Coutin 2010; De Genova 2010; Golash-Boza 2012; Golash-Boza and Hondagneu-Sotelo 2013; Rosenblum and McCabe 2014). Today's deportees face additional constraints that populations deported from the United States in the twentieth century did not. The U.S.–Mexico border is more fortified, which makes illegal land entry more difficult and expensive (Meissner et al. 2013). Second, once someone has been removed, the possibilities of legal entry are postponed for at least ten years, which undermines family unification. Third, as the numbers of unintended returns grow, local jurisdictions in the United States must deal with more fragmented and needy households, especially when male income earners are removed, leaving behind women and often U.S. citizen children (Cardoso et al. 2014).

There have always been deportations, yet the fact that the United States expels so many Latinos underscores a mobility paradox in the age of migration that coexists within a deportation regime (De Genova 2010). Unintended return is a by-product of constellations of mobility grounded in particular historical and geographical settings. The practices, representations, and physical reality of mobility (or immobility) differ depending on the setting, country of origin, and gender of those targeted for removal. Golash-Boza and Hondagneu-Sotelo (2013) framed such forced mobilities as a "gendered racial removal program" driven by

male joblessness, the War on Terror, and the criminalization of Black and Latino men by police. The disproportionate removal of men is undisputed, but the reasons for it are complex and cannot be answered with the data used in this study.

Methodology and Definitions

This article relies on data from the DHS, the Transactional Records Access Clearinghouse (TRAC), and findings from the literature. There are two deportation classifications—returns and removals—used by DHS. *Returns*, sometimes termed voluntary departures, are identified as the deportation of undocumented individuals not based on a formal order of removal. They are mostly noncriminal apprehensions conducted by the Border Patrol at points of entry and along the Mexico–U.S. border. When an individual admits that his or her attempted entry was unauthorized, the person voluntarily forgoes any due-process procedures and is relocated outside the country. The number of returns hit an all-time high of 1.6 million in 2000, yet this figure has steadily declined and reached a forty-year low of 178,371 in 2013.[1]

Removals (including expedited removals) are the focus of this research. *Removals* generally pertain to those deportations ordered by a judge from the Executive Office of Immigration Review for criminal and noncriminal violations. Individuals subjected to this removal procedure are able to obtain legal counsel at their own expense and are provided with avenues to appeal a court decision. Over the past decade, however, Congress has granted the DHS new powers so that all deportees do not have to be presented with an official charge to immigration courts to be removed as an undocumented individual. Examples include when an undocumented person is convicted of a crime (even a minor one), DHS has the authority to administratively deport an individual without an immigration hearing. This deportation process is catalogued as an expedited removal and has accounted for nearly 1.2 million removals since 2003 (Rosenblum and McCabe 2014). Likewise, reinstatement of final removal orders, which are given to an immigrant who reenters the United States after a prior removal order, do not require a formal hearing or review for expulsion (DHS 2011).

Removals can happen anywhere in the United States, not just along the border. Individuals can be removed after entering into the country for only days or after several years. Most removal apprehensions near the border occur within the first two weeks of a migrant crossing into the United States, whereas the majority of the apprehensions in the interior occur among migrants who have entered the United States at least three years before (Rosenblum and McCabe 2014). The majority of removals are noncriminal, meaning that the only reason for the removal is that the person does not have the legal right to stay in the United States. In 2013 removals reached an all-time high at 438,421 individuals. At the national level, the decline in returns and the rise of removals is a major change in enforcement patterns. Fewer returns suggest that the number of people trying to enter the United States by illegally crossing the U.S.–Mexico border is declining. The rise in removals suggests that more immigrants (including the undocumented) are being removed from the United States than ever before.

To analyze variations in removal proceedings, we obtained information through the TRAC database maintained at Syracuse University. TRAC uses the Freedom of Information Act to obtain federal government records, which provides statistics by district court and state. More than fifty immigration courts exist throughout the country, each responsible for managing deportation proceedings in its specific geographic location, which could include multiple states. TRAC's ability to capture deportation orders at a subnational level is used to analyze the geography of removal (TRAC 2014).

From Removals to Unintended Return

More than 4.2 million removals occurred between 2000 and 2013 (DHS 2014). Latin American countries accounted for 94 percent of all removals from the United States between 2000 and 2013. Mexico dominates, accounting for seven out of every ten removals for a total of 3 million. Nearly 800,000 Central Americans were also removed between 2000 and 2013, with the incidence of removal increasing dramatically after 2006. Guatemala (296,110) and Honduras (275,113) produced the most Central American removals, followed by El Salvador (183,418; DHS 2014). By contrast, removals of South Americans totaled about 140,000; the frequency of removals for South Americans peaked in 2005 when 14,500 people were removed (Figure 1).

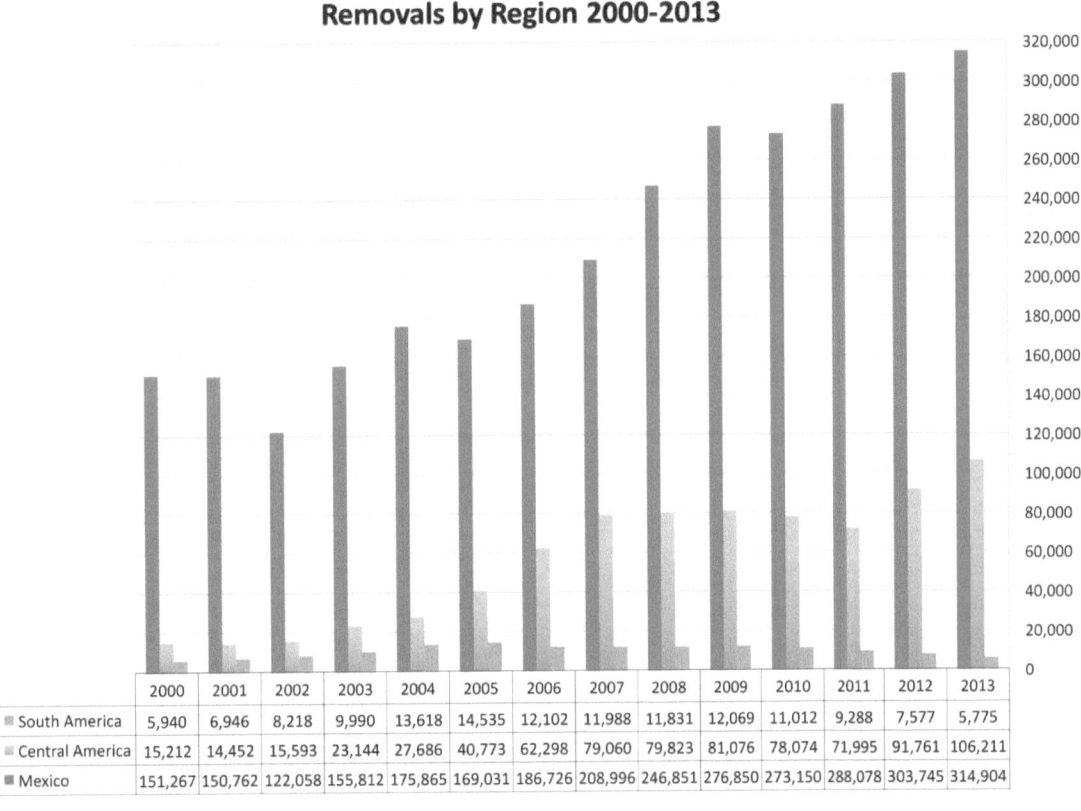

Removals by Region 2000-2013

	2000	2001	2002	2003	2004	2005	2006	2007	2008	2009	2010	2011	2012	2013
South America	5,940	6,946	8,218	9,990	13,618	14,535	12,102	11,988	11,831	12,069	11,012	9,288	7,577	5,775
Central America	15,212	14,452	15,593	23,144	27,686	40,773	62,298	79,060	79,823	81,076	78,074	71,995	91,761	106,211
Mexico	151,267	150,762	122,058	155,812	175,865	169,031	186,726	208,996	246,851	276,850	273,150	288,078	303,745	314,904

Figure 1. Latin American removals from the United States: 2000–2013. Removals to South America, Central America, and Mexico are shown in this graph. Removals increase substantially after 2006 for both Mexico and Central America. Reaching an all-time high in 2013, the last year data are available. The peak for South American removals, a much smaller number, occurs in 2005. *Source:* Department of Homeland Security (2014). (Color figure available online.)

When comparing the number of removals to Latin American countries in 2000 with 2013, there was an increase of 148 percent. Yet the increases for Central American countries were much greater: Guatemala (930 percent), Honduras (666 percent), and El Salvador (340 percent) experienced the sharpest rises in the number of removals from 2000 to 2013. In contrast, removals to Mexico increased by 108 percent.

Because many removals are undocumented individuals, the size and composition of the undocumented population in the United States should influence removal trends. Immigrants from Mexico and Central America account for over two thirds of the estimated 11.5 million unauthorized in the United States in 2011, yet they account for 91 percent of the removals (Hoefer, Rytina, and Baker 2012). Mexico (6.8 million), El Salvador (660,000), Guatemala (520,000), and Honduras (380,000) are the four nations with the largest estimated unauthorized populations. In addition, Ecuador's unauthorized population in the United States is ranked ninth. Yet other undocumented source countries such as China, the Philippines, India, Korea, and Vietnam have unauthorized estimates ranging from 170,000 to 280,000 people (Brick, Challinor, and Rosenblum 2011; Hoefer, Rytina, and Baker 2012). Their rates of removal, though, are proportionally much lower than those from Latin America.

For many years removals were reserved for immigrants with criminal convictions. The growth in noncriminal removals to Latin America since 2000 suggests that removal strategies are increasingly including the undocumented along with people who have committed serious crimes.[2] For Guatemala, 74.3 percent of removals are noncriminal, Honduras is 70.6 percent, and El Salvador is 63.2 percent. Overall, 61 percent of Latin American removals were in the noncriminal category from 2000 to 2013 (DHS 2014). Although criminal removals account for a growing share of removals, many of these are for nonviolent crimes, traffic crimes (including driving under the influence and other traffic offenses), and immigration crimes (Rosenblum and McCabe 2014).

The Shifting Constellation of Removal Practices

Removal rates vary by state within the United States. It is not surprising that the top deportation states in the United States also receive the most immigrants (Texas, California, Florida, and New York) and particularly the most Latino and unauthorized immigrants (Hoefer, Rytina, and Baker 2012). Even comparing these major destination states, however, federal courts in Texas are more likely to remove someone for an immigration charge than federal courts in California. California receives many more Latino immigrants and its estimated undocumented population in 2011 was 2.8 million versus 1.8 million for Texas (Hoefer, Rytina, and Baker 2012). Texas, however, had one third more removal proceedings than California from 2000 to 2010 (430,000 from Texas vs. 320,000 from California).

Analysis of the TRAC data demonstrates shifting trends in the application of prosecutorial discretion, which can result in a stay from removal, especially after new guidelines were introduced in June 2011 by U.S. Immigration and Customs Enforcement (ICE) Director John Morton.[3] In 2011 only 30 percent of cases brought to federal courts on immigration charges resulted in a person being allowed to stay in the country; the rest (70 percent) were removed. By 2014, with the new guidelines fully in place, 49 percent of cases were granted a stay from removal.

Federal courts adjudicate federal laws, yet the outcomes regarding stays from removal vary tremendously when federal courts are compared by the state in which they are located. Table 1 compares the percentage of individuals allowed to stay in twenty-seven states plus Puerto Rico in 2011 and 2014.[4] In 2011 the rates of stay from removal ranged from a high of 62 percent in New York to a low of 6 percent in Louisiana. In 2011, federal courts in New York exhibited ten times more tolerance than federal courts in Louisiana. The settings that exhibited more tolerance in 2011 included several major immigrant destination states such as New York, California, Massachusetts, Florida, and Hawaii along with Puerto Rico and Oregon. In contrast, federal courts in Louisiana, Georgia, Arizona, Utah, Texas, Minnesota, and Ohio were much more likely to remove people. The federal courts in Texas (16 percent) and Arizona (12 percent) seldom granted stays from removal in 2011.

By 2014, nearly half (49 percent) of all federal cases for immigration-only charges resulted in a stay from removal. This demonstrates the impact of the 2011 prosecutorial discretion guidelines with regard to handling removal proceedings. Courts in Oregon, New York, California, New Jersey, Puerto Rico, Virginia, and Massachusetts exhibited the most tolerance, with at least 59 percent of the cases resulting in a stay from removal. Federal courts in Georgia, Louisiana, Utah, Michigan, and Texas were at the bottom of the 2014

Table 1. The percentage of immigration-only charges stayed from removal by state for all origin groups, 2011 and 2014

State	2014 percentage stayed (immigration charge only) for all national groups	2011 percentage stayed (immigration charge only) for all national groups	Change (2011–2014) (%)
Oregon	73	55	19
New York	70	62	9
California	60	43	17
New Jersey	59	38	21
Puerto Rico	59	56	3
Virginia	59	39	20
Massachusetts	59	44	15
Arizona	55	12	43
Minnesota	55	17	38
Nebraska	55	19	36
Florida	54	45	9
Washington	54	24	30
Colorado	53	23	29
Hawaii	50	43	7
Entire U.S.	49	30	20
Pennsylvania	48	27	21
Tennessee	48	28	20
Nevada	42	27	15
Illinois	41	21	20
North Carolina	41	32	9
Maryland	40	47	−6
Connecticut	40	43	−3
Ohio	35	18	17
Missouri	35	19	16
Texas	34	16	18
Michigan	33	20	13
Utah	32	13	19
Louisiana	25	6	19
Georgia	23	9	13

Note: In 2011, 30 percent of cases resulted in stays from removal, but outcomes by state varied from a low of 6 percent to a high of 62 percent. By 2014, with the implementation of new prosecutorial discretion guidelines, 49 percent of cases on immigration charges only resulted in a stay from removal. Oregon courts granted stays from removal in 73 percent of cases, showing the most tolerance toward the undocumented. Georgia courts granted stays from removal in only 23 percent of cases, reflecting the least leniency toward the undocumented. *Source:* TRAC (2014).

rankings, with only 23 percent to 34 percent of cases resulting in a stay from removal. The most significant changes were in the federal courts in Arizona, Minnesota, and Nebraska. Federal courts in these states went from very low rates of stays from removal in 2011 to rates of 55 percent in 2014, well above the national average. Every state in Table 1 showed an increase in the percentage of people allowed to stay in 2014 with the exception of Maryland and Connecticut, which had declines of 6 percent and 3 percent, respectively.

When examining the same data set for Mexican nationals only (see Table 2), the rates of immigrant stays were far lower in 2011 at 13 percent but improved significantly in 2014 to 42 percent. This is still below

the national average of 49 percent for all country groups in 2014. Federal courts in Oregon, North Carolina, California, Arizona, Nebraska, Minnesota, and Tennessee issued stays from removal in more than half of all cases (ranging from 52 to 74 percent). What is notable about this list is that it includes two southern states, North Carolina and Tennessee, which as new destinations have been recognized for their more exclusionary practices toward immigrants and the undocumented (Winders 2005; Walker and Leitner 2011). Federal courts in the southern states of Louisiana and Georgia, however, had the lowest rates of staying removals. Arizona and Nebraska are also states that have engaged in more exclusionary practices, especially

Table 2. The percentage of immigration-only charges stayed from removal by state for Mexican nationals, 2011 and 2014

State	2014 percentage stayed (immigration charge only) for Mexican nationals	2011 percentage stayed (immigration charge only) for Mexican nationals	Change (2011–2014) (%)
Oregon	74	43	31
North Carolina	58	19	38
California	57	36	21
Arizona	54	9	45
Nebraska	53	10	42
Minnesota	52	7	46
Tennessee	52	13	39
Puerto Rico	50	50	0
Washington	49	14	35
Florida	48	18	30
Massachusetts	48	19	29
Colorado	44	18	27
Entire U.S.	42	13	29
New York	39	15	24
Virginia	38	13	25
Pennsylvania	36	6	30
Nevada	35	13	22
Missouri	35	11	24
New Jersey	34	10	24
Illinois	31	9	23
Texas	29	12	17
Connecticut	28	15	13
Utah	26	7	19
Maryland	25	19	6
Ohio	24	3	21
Hawaii	23	22	1
Michigan	21	6	15
Georgia	17	3	15
Louisiana	14	3	11

Note: The rates of immigrant stays from removal are far lower for Mexican nationals. In 2011 only 13 percent of immigration-only cases resulted in a stay from removal for Mexicans at the national level, with considerable variation by state location of federal courts. By 2014 a much higher percentage of Mexicans (42 percent) received stays from removal nationally, suggesting the impact of new prosecutorial discretion guidelines. *Source:* TRAC (2014).

Arizona's notorious SB1070 bill, but the TRAC data reflect a more inclusionary shift since 2014.

The data present a contradictory picture in which DHS and federal courts in particular states project different rates of tolerance or intolerance toward the undocumented. Nationally, more immigrants are being removed than ever. The DHS budget is for 400,000 removals per year, a target more or less maintained throughout the Obama administration by focusing on criminal removals but also using more expedited removals that bypass court proceedings. At the same time, immigration cases that make it to federal courts are being met with more stays from removals. Added to this mix are people with Deferred Action for Childhood Arrivals (DACA), created through executive action by President Obama in 2012. DACA has protected more than 700,000 undocumented youth from removal and given them work authorization and access to driver's licenses (Singer, Svajlenka, and Wilson 2015).[5] Yet, this is only a temporary protection and must be renewed every two years.

The TRAC data also show that when court location and country of origin are considered (in the case of Mexicans), there are major differences in how federal courts decide who should be removed. This suggests that local context matters, even in federal court proceedings. There is growing variation in how states and cities are addressing the undocumented. In October 2013, for example, California Governor Jerry Brown signed the Trust ACT, which limits local and state police from engaging in the removal of noncriminal unauthorized immigrants in California. Addressing public safety concerns, the undocumented can also get driver's licenses in California. Now, more than ever, the policies and practices toward the undocumented in the United States are decentralizing, which makes for even more complex constellations of mobility.

What Unintended Return Means for Latin Americans' Mobility

The unintended return of more than 4 million people from the United States to Latin America since 2000 is a major issue for Latinos and Latin American countries, especially Mexico. We conclude by suggesting what unintended return means for Latin America and why it requires more scholarly attention. Generally, return migrants in Latin America are a source of pride, even envy, in communities where they cluster; they are often constructed as positive agents for economic development and social change (Potter, Conway, and Phillips 2005; Jones 2011). Yet unintended return has far more negative connotations, shifting focus toward communities and countries of origin that receive those involuntarily removed.

Unintended return often begins suddenly and without warning. Like a lightning strike, a migrant can be stopped for a traffic violation or caught up in a workplace raid and suddenly be channeled back to his or her country of origin (often after months of detention and prolonged immobility). Rather than a triumphant return with new possessions, financial resources for investment, and even one's entire family, the unintended returnee often has little to show after expensive and risky border crossings and perhaps years of toil abroad. Coutin (2010) argued that deportations shift the meaning of one's home country from a place of familiarity and welcome to a "zone of confinement." The process of forced removal often results in financial calamity: Remittances are no longer sent, mortgages cannot be paid, and savings are rapidly depleted. There is an emerging literature that addresses what happens to those jurisdictions that produce deportees and receive unintended returnees (Hagan, Eschbach, and Rodriguez 2008; Hagan, Castro, and Rodriguez 2010; Hamann and Zúñiga 2011; Wheatley 2011; Hiemstra 2012), but more is needed given the scale of the phenomenon and its influence on mobility practices and representations.

Mexico eclipses all other countries by its sheer number of unintended returns, with some 3 million since 2000. Twelve million Mexican-born individuals live in the United States along with 35 million people of Mexican ancestry. The demographic and social ties between Mexico and the United States run deep. Thus, even in the case of removal from the U.S. territory, a return to the United States at some later date is possible, through legal or unauthorized means. When comparing the current size of the Mexican-born population in the United States with the number of removals, it is equivalent to nearly one in four foreign-born Mexicans in the United States who have been removed since 2000.

The research on unintended return from Mexico shows loss of income, family strain, social stigma, and the difficulties of reintegration, especially for young children and youth who have grown up in the United States and must now be integrated into Mexican schools (Hamann and Zúñiga 2011; Wheatley 2011). The number of deported men in their twenties and thirties who grew up in the United States and speak limited Spanish has grown so significantly that it has sparked a boom in English call centers in Mexico. Tijuana has thirty-five call centers employing nearly 10,000 workers; it is estimated that nearly half of these workers experienced unintended return (Spagat and Millan 2014). Mexico even has its own DREAMer population: youth raised in the United States who speak mostly English and are struggling to get their educational credentials recognized so they can work or continue with higher education (Anderson and Solis 2014).

Proportionally, the impact of unintended return might be greater for Honduras and Guatemala. Like Mexicans, Central Americans are leaving their countries in search of jobs, reuniting with family members, or seeking safety from rampant violence. Yet, their migratory roots in the United States are not as deep. Removals of Hondurans from 2000 to 2013 are equivalent to 55 percent of the Honduran population that currently lives in the United States. Similarly, since 2000, the removal of Guatemalans is equivalent to 36 percent of current foreign-born stock population in the United States. Removals of Salvadorans are numerically and proportionally less, equal to 15 percent of the 1.2 million foreign-born Salvadorans in the United States. Many Salvadorans receive temporary protected status (TPS), which provides a legal means to remain in the United States, shielding many from removals due to immigration violations alone. Scholarly and journalist research on the impact of unintended return on Central American communities is limited but growing (Hagan et al. 2008; Martinez 2013; Cardoso et al. 2014).

Each day chartered planes filled with deportees land in these Central American countries, as political leaders and nongovernmental organizations scramble to address the needs of growing numbers of unintended returnees. Guatemala has been a leader in responding to this crisis. In 2007 the Guatemalan Congress created CONAMIGUA (*Consejo Nacional de Atención al Migrante de Guatemala*) to coordinate among various government authorities responsible for protecting the rights of migrants and returnees. The newly returned are processed at a designated airport facility and are provided with a bag lunch, a phone call, and bus fare to their homes. In 2014 the government refurbished temporary shelters to house migrants who cannot immediately find family or return home (O'Keefe 2014). Guatemala also launched its *Quédate* (Stay) program, urging Guatemalans by radio, television, and the Internet not to leave. The same program built two technical centers for returned youth in need of vocational skills in 2015. How many of the unintended returnees remain in Central America or try to leave again is unknown.

Unintended return is one possible outcome in a friction-filled and politicized constellation of mobility actors and actions existing in the United States today. In practice, men are much more likely to end up as unintended returnees than women. Also, the U.S. state in which one resides, the court where one's case is heard, and one's country of origin are all factors that affect mobility outcomes with regard to forced removal. Due to new federal guidelines regarding prosecutorial discretion, more undocumented migrants have received stays from removal, yet removal figures are at an all-time high.

The U.S. policy shift toward forced removals underscores a mobility paradox for many Latinos: Their determination to be mobile and leave their countries of origin is driven by poverty, violence, wage differentials, employment opportunities in the United States, and social networks. Entering the United States without legal authorization quickly places them in a constellation of relationships that limits mobility and sometimes results in involuntary mobility. Unintended return is an expression of forced mobility that underscores the putative sovereignty of the state (in this case the United States) to enforce its territorial boundaries no matter the human or financial costs. At the same time it forces Latin American states to deal

with 4 million returnees, some of whom are disconnected from their homeland or traumatized by their dangerous journey to the north. Unless major immigration reform happens in the United States, offering some relief from the threat of removals, the human tide of unintended returnees will continue to pour into Mexico and Central America.

Acknowledgments

The authors gratefully acknowledge the helpful comments of Nancy Hiemstra, Audrey Singer, Jill Wilson, the editor, and three anonymous reviewers in shaping this work. Responsibility for the opinions expressed herein is solely that of the authors.

Notes

1. It is important to note that someone can be returned or removed more than once, and each count will be included in the *Yearbook of Immigration Statistics*.
2. Criminal removals vary in terms of seriousness. Removal of foreign-born individuals, regardless of legal status, convicted of homicide, sexual offenses, drug-related offenses, or driving under the influence regularly occurs. Yet the American Immigration Council argues that more of the criminal removals are individuals convicted of immigration crimes (e.g., illegal reentry) or traffic crimes (driving without a license).
3. The U.S. ICE Memorandum of 17 June 2011, Exercising Prosecutorial Discretion Consistent with the Civil Immigration Enforcement Priorities of the Agency for the Apprehension, Detention and Removal of Aliens, is available at www.ice.gov.
4. Not every state has a federal court where immigration charges are adjudicated; some states such as California, Texas, and New York have multiple courts.
5. A second executive order by President Obama in November 2014 expanded the individuals protected under DACA and created DAPA for the undocumented parents of U.S. citizen children. This executive order is currently held up in federal courts and has not been enacted. It has the potential to protect up to 4 million undocumented from removal.

References

Anderson, J., and N. Solis. 2014. *Los otros dreamers*. Mexico City: Authors.

Brabeck, K., and Q. Yu. 2010. The impact of detention and deportation of Latino immigrant children and families: A quantitative exploration. *Hispanic Journal of Behavioral Sciences* 32 (3): 341–61.

Brick, K., A. E. Challinor, and M. Rosenblum. 2011. *Mexican and Central American immigrants in the United States*. Washington, DC: Migration Policy Institute.

Cardoso, J. B., E. R. Hamilton, N. Rodriquez, K. Eschbach, and J. Hagan. 2014. Deporting fathers: Involuntary transnational families and intent to remigrate among Salvadoran deportees. *International Migration Review.* Advance online publication. doi: 10.1111/imre.12106.

Carling, J. 2002. Migration in the age of involuntary immobility: Theoretical reflections and Cape Verdean experiences. *Journal of Ethnic and Migration Studies* 28 (1): 5–42.

Chavez, L. R. 2008. *The Latino threat: Constructing immigrants, citizens and the nation.* Stanford, CA: Stanford University Press.

Coutin, S. B. 2010. Confined within: National territories as zones of confinement. *Political Geography* 29:200–08.

Cresswell, T. 2006. *On the move: Mobility in the modern western world.* London and New York: Routledge.

———. 2010. Towards a politics of mobility. *Environment and Planning D: Society and Space* 28:17–31.

———. 2014. Mobilities III: Moving on. *Progress in Human Geography* 38 (5): 712–21.

De Genova, N. 2010. The deportation regime: Sovereignty, space and the freedom of movement. In *The deportation regime: Sovereignty, space and the freedom of movement,* ed. N. De Genova and N. Peutz, 33–65. Durham, NC: Duke University Press.

Department of Homeland Security. 2011. *Yearbook of immigration statistics: 2010.* Washington, DC: Department of Homeland Security.

———. 2014. *Yearbook of immigration statistics: 2013.* Washington, DC: Department of Homeland Security.

Durand, J., and D. S. Massey. 2010. New world orders: Continuities and changes in Latin American migration. *The Annals of the American Academy of Political and Social Science* 630:6–52.

Golash-Boza, T. M. 2012. *Immigration nation: Raids, detentions, and deportations in post-9/11 America.* Boulder, CO: Paradigm.

Golash-Boza, T., and P. Hondagneu-Sotelo. 2013. Latino immigrant men and the deportation crisis: A gendered racial removal program. *Latino Studies* 11 (3): 271–92.

Hagan, J., B. Castro, and N. Rodriguez. 2010. The effects of US deportation policies on immigrant families and communities: Cross-border perspectives. *North Carolina Law Review* 88:1800–23.

Hagan, J., K. Eschbach, and N. Rodriguez. 2008. US deportation policy, family separation, and circulation migration. *International Migration Review* 42 (1): 64–88.

Hamann, E. T., and V. Zúñiga. 2011. Schooling and the everyday ruptures transnational children encounter in the United States and Mexico. In *Children, youth and migration in global perspective,* ed. C. Coe, R. R. Reynolds, D. A. Boehm, J. M. Hess, and H. Rae-Espinoza, 141–60. Nashville, TN: Vanderbilt University Press.

Hiemstra, N. 2012. Geopolitical reverberations of US migrant detention and deportation: The view from Ecuador. *Geopolitics* 17 (2): 293–311.

Hoefer, M., N. Rytina, and B. C. Baker. 2012. *Estimates of the unauthorized immigrant population residing in the United States: January 2011.* Washington, DC: Department of Homeland Security.

Jones, R. C. 2011. The local economic imprint of return migrants in Bolivia. *Population, Space and Place* 71:435–53.

King, R. 2012. Geography and migration studies: Retrospect and prospect. *Population, Space and Place* 18:134–53.

Leitner, H., and C. Strunk. 2014. Spaces of immigrant advocacy and liberal democratic citizenship. *Annals of the Association of American Geographers* 104 (2): 348–56.

Lopez, M. H., R. Morin, and P. Taylor. 2010. *Illegal immigration backlash worries, divides Latinos.* Washington, DC: Pew Hispanic Center.

Martinez, O. 2013. *The beast: Riding the rails and dodging narcos on the migrant rrail.* London: Verso.

Meissner, D., D. M. Kerwin, M. Chishti, and C. Bergeron. 2013. *Immigration enforcement in the United States: The rise of a formidable machinery.* Washington, DC: Migration Policy Institute.

O'Keefe, E. 2014. Ellis Island in reserve: Where deportees go when they get home to Guatemala. *Washington Post* 30 August 2014.

Potter, R. B., D. Conway, and J. Phillips, eds. 2005. *The experience of return migration: Caribbean perspectives.* Aldershot, UK: Ashgate.

Price, M. 2014. Cities welcoming immigrants: Local strategies to attract and retain immigrants in U.S. metropolitan areas. World Migration Report 2015, Background Paper for International Organization for Migration, Geneva, Switzerland.

Rosenblum, M. R., and K. McCabe. 2014. *Deportation and discretion: Reviewing the record and options for change.* Washington, DC: Migration Policy Institute.

Samers, M. 2010. *Migration.* London and New York: Routledge.

Singer, A., S. Hardwick, and C. Brettell. 2008. *Twenty-first century gateways: Immigrant incorporation in suburban America.* Washington, DC: The Brookings Institution.

Singer, A., N. P. Svajlenka, and J. H. Wilson. 2015. Local insights from DACA implementation. *Brookings Metropolitan Policy Program report, June 2015.* Washington, DC: The Brookings Institution.

Singer, A., J. H. Wilson, and B. DeRenzis. 2009. Immigrants, politics, and local response in suburban Washington. *Brookings Metropolitan Policy Program report, February 2009.* Washington, DC: The Brookings Institution.

Spagat, E., and O. Millan. 2014. Deported Mexicans find new life at call centers. Associated Press 22 August 2014.

Transactional Records Access Clearinghouse (TRAC). 2014. U.S. deportation outcomes by charge: Completed cases in immigration courts. In Transactional Records Access Clearinghouse and Syracuse University [database online]. http://www.trac.syr.edu/phptools/immigration/charges/deport_filing_charge.php (last accessed 2 June 2015).

Varsanyi, M. W. 2008. Rescaling the "alien," rescaling personhood: Neoliberalism, immigration and the state. *Annals of the Association of American Geographers* 98 (4): 877–96.

———. 2011. Neoliberalism and nativism: Local anti-immigrant policy activism and an emerging politics of scale. *International Journal of Urban and Regional Research* 35 (2): 295–311.

Walker, K. E., and H. Leitner. 2011. The variegated landscape of local immigration policies in the United States. *Urban Geography* 32 (2): 156–78.

Wheatley, C. 2011. Push back: U.S. deportation policy and the reincorporation of involuntary return migrants in Mexico. *The Latin Americanist* 55 (4): 35–60.

Winders, J. 2005. Changing politics of race and region: Latino migration to the US South. *Progress in Human Geography* 29 (6): 683–99.

Wright, R., and M. Ellis. 2000. Race, region, and the territorial politics of immigration in the U.S. *International Journal of Population Geography* 6:197–211.

Circulations and the Entanglements of Citizenship Formation

Lynn A. Staeheli, David J. Marshall, and Naomi Maynard

Department of Geography, Durham University

Citizenship is given form, meaning, and power through the transactions and circulations that constitute it. Our focus in this article is the ways in which circulations through networks and institutions that extend beyond nation-states are enacted and encouraged through pedagogies and practices that moor habits of citizenship in daily lives. Although there has been significant attention to those practices at national and local levels, there has been relatively little attention to the ways that floating sites of citizenship formation are entwined with, but also seem to be suspended above, other sites. There are at least three ways in which circulations both construct those sites and are entwined in citizenship formation: They are the reason that the seeming contradiction between cosmopolitanism and efforts to moor citizens to place becomes unremarkable; they enable and shape the modes of interaction that conjoin politics and emotional geographies; and they are part of the way in which a common understanding of active citizenship is accepted almost without question. We use the examples of two international conferences for young citizen-activists to illustrate our arguments regarding the circulations of ideas, norms, and practice that are central to citizenship formation.

公民权透过组构自身的实施与循环，被赋予形态、意义和权力。我们于本文中的焦点是，透过超越国族国家的网络与制度的循环，藉由日常生活中定着公民权惯习的教学法和实践被激活并受到鼓励的方式。尽管对国家和地方层级的实践有着大量的关注，但相对而言，却鲜少有研究关注公民权形成的浮动场域和其他场域相互交织、却似乎同时被悬置于其上的方式。循环同时建构那些场域，并与公民权形构相互交织的方式至少有三种：它们是让寰宇主义和将公民定着一地这两造看似冲突的意图变得不显着的原因；它们赋予并形塑结合政治与情感地理的互动模式；它们是积极公民权的普遍理解被毋庸置疑地接受的方式中的一部分。我们运用两个青年公民行动者的国际会议之案例，描绘我们有关公民权形塑的核心概念、规范和实践的循环之主张。

La ciudadanía está otorgando forma, sentido y poder por medio de las transacciones y circulaciones que la constituyen. Nuestro punto focal en este artículo son las maneras como las circulaciones a través de redes e instituciones que se extienden más allá de los estados-naciones son habilitadas y estimuladas por medio de pedagogías y prácticas que anclan los hábitos de ciudadanía en vidas cotidianas. Aunque se ha dado una atención significativa a esas prácticas a niveles nacionales y locales, poca ha sido la atención prestada a las maneras como sitios flotantes de formación ciudadana se entrelazan con otros sitios, pero que parecen también suspendidos sobre éstos. Existen por lo menos tres maneras como las circulaciones construyen esos sitios y son involucradas en formación de ciudadanía: Ellas son la razón para que la aparente contradicción entre el cosmopolitismo y los esfuerzos para atar los ciudadanos al lugar se vuelva inocua; ellas habilitan y configuran los modos de interacción que conjugan la política y las geografías emocionales; y son parte del modo como un común entendimiento de ciudadanía activa se acepta casi sin preguntar. Usamos los ejemplos de dos conferencias internacionales de ciudadanos jóvenes–activistas para ilustrar nuestros argumentos en relación con la circulación de ideas, normas y prácticas que son centrales para la formación de ciudadanía.

Citizenship is freighted with many, sometimes contradictory meanings. It is a status conferred by a nation-state. It is a marker of belonging and inclusion, even as it creates exclusions. It conveys expectations of how subjects should behave. It is a Western category that is treated as though it is universal. It guarantees rights. It obligates subjects to serve the state. It is conditioned by local, everyday relationships and practices. It represents global, cosmopolitan ideals. Collectively, the academic literature on citizenship reveals it as a complex, multivalent concept.[1]

Our intervention in this wide-ranging literature focuses on the ways in which citizenship is formed through an intimacy-geopolitics of circulation. As we explain, such circulations simultaneously attach citizenship—or at least the practices and behaviors undertaken by citizens—in localities and communities, even as it is encouraged and performed through sites and relations that are seemingly detached from those very same places, communities, and nations.

The kernel of our argument is as follows. Citizenship is constructed through a complex set of relationships among qualities, norms, interactions, and positionings with respect to a collective, a collective that itself could be undergoing transformation. This conceptualization has at least two implications. First, numerous embodied, institutional, and affective agencies are involved in citizenship formation. Second, the processes of entangling and ordering imply the circulation of ideas and norms through multiple means. Our particular focus is with the ways in which circulations through networks and institutions that extend beyond nation-states are enacted and encouraged through pedagogies and practices that moor habits of citizenship in daily lives. Although there has been significant attention to those practices at national and local levels, there has been relatively little attention to the ways in which floating sites of citizenship formation are entwined with, but also seem to float above, other sites. We use the examples of two international conferences for young citizen-activists to illustrate our arguments regarding the circulations of ideas, norms, and practices that are central to citizenship formation.

Intimacy-Geopolitics, Circulation, and Citizenship Formation

The term *intimacy-geopolitics* highlights the inseparability of, and tensions between, intimacy and geopolitics (Pain and Staeheli 2014). A growing literature has pointed to the ways that intimacy is important to geopolitics, often arguing that it is necessary to recognize the ways in which actions and relations at multiple scales condition geopolitical relationships; this literature is often concerned with the spatial relationships that entangle near and distant places, such that the presumed binary between them is dissolved (e.g., Mountz and Hyndman 2006; Pratt and Rosner 2012). In so doing, this literature often argues for the importance of recognizing the political and politicized nature of intimacy (which is, in itself, ambiguous and complex) and its roles in shaping geopolitics and relationships, such as through the invocation of gender-based violence as a rationale for war (e.g., Fluri 2011)

or the role of gender in development policy and practices (e.g., Nagar et al. 2002).

In defining intimacy-geopolitics, however, Pain and Staeheli (2014) argued for more than the importance of intimacy to geopolitics and instead argued that they are inseparable from each other and are mutually constituted, rather than being prefigured. We build on this argument to suggest that citizenship, as an instantiation of intimacy-geopolitics, is given form, meaning, and power through the transactions and circulations that constitute it. We use the term *circulation* rather than the more common *mobility* advisedly. There is, for instance, a burgeoning literature on policy mobilities that might have been called on (e.g., McCann and Ward 2011). Likewise, mobility features prominently in the literature on children and young people's geographies (e.g., Barker et al. 2009). In both such instances, however, mobility is used somewhat generically to refer to a broad array of phenomena ranging from the dissemination of policy through global governance networks, to young people's experiences of transnational migration (e.g., Hopkins and Alexander 2010) and the everyday movement of young people to and from home, school, and elsewhere (e.g., Harker 2009; Skelton 2013; Horton et al. 2014). Such research has been fruitfully informed by a new mobilities paradigm that emphasizes the relational character of mobility and immobility (Adey 2006; Hannam, Sheller, and Urry 2006; Sheller and Urry 2006). This approach has been useful in challenging idealized notions of unencumbered movement and circulation of goods, ideas, people, and capital conjured by terms like *mobility*, *flow*, and *networks*, emphasizing the blockages to and unevenness of mobility. Yet in many analyses, the term mobility tends to still be used in a binary fashion[2] (i.e., people, things, or policies are either mobile or not; Salter 2013); this has the potential to obscure more complex power relations that condition mobilities (Cresswell 2012).

Often missing in this notion of relative im/mobilities is the shape that movement takes beyond stop and go. What kinds of movement are encouraged or discouraged by various social and institutional norms and moorings? Rather than simply being overlooked in analyses, Salter (2013) argued that the very concept of mobility does not lend itself easily to the *dispositif* implied in circulation, which, as we see later, is important to the ways that citizenship formation proceeds. By referring specifically to circulation, we seek to emphasize a particular, circular movement of ideas and people that organizers of international conferences typically envision, as well as the messy entanglements that come about in practice as a result of the multiply scaled political contestations and improvisations that take place in such settings. For example, at the 2014 international youth conference described later, it was clear that many of the participants were engaged in a back-and-forth movement between international conferences and activism in local or regional politics in their home countries. Many delegates were veterans of an international youth conference circuit, having attended numerous international and regional conferences that are held in different cities around the globe. Such conferences are meant to serve as sites where skills and ideas can be exchanged among circulating delegates before returning home to be practiced in place. These circulations, which are part of intimacy-geopolitics and shape citizenship formation, are not easily anticipated or described in a straightforward manner, however, and their outcomes are not easily predicted. Rather, such circulations are shaped by complex and long-standing relationships and sudden disruptions, operating across multiple spatial and temporal scales.

As the preceding comments imply, citizenship is more than a status but instead involves relationships that condition individuals' positioning, capacities, and agencies with respect to a collective. That collective is commonly assumed to be a state, but it need not be. Indeed, in many formulations and in some circumstances, citizenship is held to operate outwith the state, either as in some calls for cosmopolitan or global citizenship or in some civic formulations of citizenship in which civil society and communities stand as the collectivity (Staeheli 2011). This is not to say that the state is irrelevant but rather that citizenship is forged, developed, experienced, and practiced in sites and institutions beyond those defined or contained by the state. Citizenship—as distinct from the legal status of citizen—is thus formed in and through the relationships and circulations we describe in terms of intimacy-geopolitics.

There are at least three ways in which circulations are important to citizenship formation. First, they sustain the spatial relationships that entangle proximate and distant spaces. In the example we develop, they are the reason that the seeming contradiction involved in entwining cosmopolitanism—which commonly implies transcendence of the nation and the particular—with efforts to moor action by citizens to place and as national citizens becomes unremarkable. Second, they enable and shape the modes of interaction that conjoin politics and emotional geographies, as in the feelings and obligations of belonging as citizens; the circulation of a common understanding of citizenship as both feeling and status is one means by which this occurs (Osler and Starkey 2005). Finally, they are part of the way that a common understanding of active citizenship—or commitments to certain practices as citizens—is accepted almost without question, seeming to emerge as commonsensical, without an apparent source or genealogy.

Reading the Circulations of Youth Citizenship Formation

We illustrate the argument outlined earlier by drawing from a larger study of citizenship formation in divided societies. The study is primarily concerned with efforts to encourage behaviors, attitudes, and practices among young people. One component of the research attends to the efforts of an ensemble of organizations and agents—nongovernmental organizations (NGOs), governments, foundations, international organizations, and activists—that attempt to intervene in processes of citizenship formation to encourage qualities that are seen as conducive to stability, security, and reconciliation in countries marked by deep division. In this article, we focus on efforts of international organizations to encourage certain practices of citizenship and, in particular, on the use of international conferences that bring young people together to debate common issues, to be seen and heard as active participants in decision making, and to provide a forum in which skills and expectations of active citizenship can be imparted.[3] Imaginatively, the conferences float above the fray created by national and local conditions, politics, and conflicts, removing the youth from the distractions of daily life and the real-world, nitty-gritty encounters that seem to corrupt or impede political action taken as citizens.

These conferences are part of a larger infrastructure or organizational apparatus that has been constructed to encourage particular kinds of young citizens.[4] Young people are often seen as paradoxical with regard to citizenship. They are lauded as having great potential but are also seen as security threats. They are sometimes represented as only loosely bound by existing norms and institutions, but they are also the focus of state efforts to forge national identities. They are seen as malleable but also as resistant to norms and expectations. Due to their uncertain, even unstable relationships with communities, nations, and social norms, there is often considerable effort to shape the identities, behaviors, and values of young people as citizens (Pykett 2010; Staeheli and Hammett 2010).

These efforts are linked by agents who work in international organizations, government institutions, civil society organizations, religious organizations, schools, and NGOs. In the mobilities literature, the relationships among these organizations might be described as providing an infrastructure for citizenship formation (Hannam, Sheller, and Urry 2006), whereas others might describe them as forming a network or an assemblage (Salter 2013). From our perspective, the language of infrastructure or assemblage is less important than the ways in which ideas, practices, and bodies flow between them and become entangled. We focus on international conferences because they seem to float above local and national efforts to form young citizens, collecting influences and ideas from multiple sources, even as they encourage youth to immerse themselves in actions to address problems in their communities and countries. There is a pervasive assumption that removing young people from their everyday environments might expand their worldviews but also remove them from the pernicious influences that might be found "at home." It thus might impart a kind of cosmopolitanism to those who attend, even if it is temporary, intermittent, or blended with other citizenship values and practices on return (Baillie Smith and Jenkins 2011; Diprose 2012; Baillie Smith et al. 2013). Consistent with our conceptualization as citizenship being formed through circulations that we analyze in terms of intimacy-geopolitics, we read the conferences in terms of the ways that proximate and distant are entangled (i.e., in terms of spatial relations), the encouragement of commitments to action in civil society for the good of self and others (i.e., conjoining politics and emotions), and cementing the hegemony of active citizenship (i.e., the practices of citizenship).

We focus on two conferences: the 1970 United Nations (UN)-sponsored World Assembly of Youth and the 2014 World Conference on Youth. These conferences bookend our larger study of international efforts at citizenship promotion as they are entwined with national and local organizations and social activists. Information about the 1970 Assembly is drawn from files in the UN Archives and Records Management Section, and the information about the 2014 conference draws primarily on participant observation. We are also informed by a small set of interviews with people who were involved in the conferences as participants or in the organizations that supported them. We do not claim that the conferences are representative of all such events. Instead, we use the conferences to illustrate our conceptual argument about the role of circulation in the intimacy-geopolitics of citizenship formation.

On the surface, the two conferences might seem rather different. The 1970 World Youth Assembly was a late addition—an afterthought of sorts—to the celebrations of the twenty-fifth anniversary of the United Nations. The Assembly drew approximately 750 delegates from member states and from thirteen international youth organizations. The theme of the Assembly was "Peace, Progress and International Co-operation," and the stated objectives included enrolling young people in supporting UN efforts to address problems facing the world and member states and drawing attention to the roles that youth could play. What was intended to be something of a feel-good gathering, with a long roster of social and cultural events in New York City that delegates could attend, quickly became contentious. The U.S. government refused to provide funds for the Assembly, so ad hoc committees were created to solicit funds from corporations, foundations, and the general public; judging by the "thank you" notes, the latter were typically in the range of $5 to $10. After the UN-led organizing committee was joined by representatives of the thirteen international youth organizations, the *New York Post* reported that the diplomats were outmaneuvered by the youth, who ranged "from the Boy Scouts to Communist-dominated organizations" and who won the right to select about 20 percent of the delegates. There were concerns that these delegates would be uncontrollable (Berlin 1970). Indeed, officials commented in their post-assembly review that they were surprised at how seriously youth delegates took the conferences, eschewing cultural events for meetings with officials and with other delegates and rejecting stances taken by the UN on contentious topics (World Youth Assembly 1970b). Reflecting the tumultuous politics of the time, the latter happened frequently. The organizing committee had established commissions on World Peace, Development, Education, and the Environment and wrote draft reports for each (apparently with little to no input from youth or youth organizations). At the Assembly, delegates ripped apart the prepared report and inserted a far more radical agenda for change. The final report of the World Peace Commission called for the end of imperialism and colonialism, called for the right to self-determination (most notably for Palestine and Puerto Rico), condemned aggression on the part of the United States and other Western powers, and called for the end of the blockade of Cuba. It also called on young people to demonstrate solidarity with oppressed peoples around the world (World Youth Assembly 1970a). U.S. officials, without any apparent irony, noted that this was the inevitable outcome of allowing governments and youth to select the delegates, as they would bring ideological commitments to the Assembly. Officials were particularly concerned by delegates from Soviet-aligned countries, claiming they were too old and too entrenched in party politics to be free from the influence of government propaganda (World Youth Assembly 1970b).

By 2014, the machinery for international conferences had become well-oiled and there were few opportunities for the disruptive activities that marked the 1970 conference. The World Conference on Youth was one of more than 100 international youth conferences held in 2014 that addressed citizenship in some way. It was attended by nearly 1,000 delegates, including representatives of youth organizations, youth leaders who applied to the organizing committee, delegates selected by national governments, facilitators, social media fellows, and 100 youth leaders from Sri Lanka, which hosted the conference. The stated goal of the conference was to mainstream youth into the UN post-2015 development agenda, but some observers believed that it was also a ploy to promote the national and international standing of the Sri Lankan president. The conference ran over several days and involved a mix of plenary sessions, focused discussions on substantive issues related to the Millennium Development Goals, and training and leadership workshops. Meanwhile, officials of national governments finalized the Columbo Declaration on Youth at the conference. Youth delegates lobbied representatives of their national governments separately to make changes to the declaration, committed themselves to hold their governments to account, and were then sent back to their homes to organize communities in support of the policies advocated in the Declaration. Although youth delegates debated topics and disagreed with each other, their influence on the actual Declaration is not clear, as much of it was worked out in regional and national conferences at which they were not typically present (see Riles 2000). Although youth delegates might have had some influence, there was no such dramatic rewriting of the declaration as happened in 1970. Central planks in the declaration included the need for inclusive and participatory youth policies in member states and the integration of young people into democratic processes in a "meaningful way at local, national, regional and international levels"; volunteering programs were specifically mentioned (World Conference on Youth 2014).

Side events and training workshops allowed more direct involvement of young people than did the working group finalizing the Declaration. At some of these events, peer education projects were discussed where information was shared about how to spread good practice for youth participation in their localities and civil society, as well as in national politics. Other events talked about ways to enhance global awareness among marginalized youth who might not be aware of the broader contexts in which their marginality was enforced. Similar themes were addressed in sessions aimed at young people involved in conflict resolution. Cosmopolitanism and global citizenship were often presented as means of overcoming internal, communitarian conflict. For instance, a young woman from Moldova claimed, "I was a citizen of my city or neighborhood, but now I am a citizen of the world. We have to get outside our internal conflict mentality and achieve a global awareness." At other sessions, the importance of holding governments to account was discussed and strategies for encouraging good practice were disseminated. One representative of a national youth council spoke of the ideals of citizenship and the need to activate the notion of "values-based leadership." In these sessions, civil society was argued to be important as a site from which to hold governments to account, but some delegates also spoke of the need to create civic and political spaces of their own. These were not necessarily spaces of confrontation, however, and a representative of a youth organization reminded delegates of the values of empathy. Although leaders often patronize young people, she argued that youth should exercise empathy with leaders, noting that they were usually good people who really want to help and who are also frustrated by the narrow confines of their own position.

Circulations and the Entanglements of Citizenship Formation

The comments of the representative just noted served as a reminder to the delegates that they did not act in a vacuum—that even if they were acting locally, there were influences and constraints on the actions of other agents. Although she would never have used this language, we interpret it in terms of circulations, intimacy-geopolitics, and the entanglements of citizenship formation. She reminded the delegates that when they returned home, they would be back in the morass of relationships and constraints that affect all agents, not just youth. Even though they might act locally (while perhaps thinking globally), they were interacting with others whose range of actions were also constrained and shaped in complicated ways. We briefly illustrate these issues in terms of the circulations that link near and distant, the ways in which politics and affective feelings are intertwined, and the practices and practicalities of acting as citizens.

The conferences themselves are an attempt to lift activists out of the day-to-day of their lives and to link them with agents and knowledges that come from other places and contexts. To facilitate learning across differences—but also to create a common basis for acting as citizens—international organizations, foundations, and governments develop training materials that conferences delegates can take home.[5] Although there are differences in specific materials, there is convergence around commitments to active citizenship and in many instances to some form of cosmopolitanism or globalism, such as discussions of human rights, the interconnectedness of people and places, and the necessity to work as citizens irrespective of nationality on issues of global concern. Metaphors of boats—as in "we are all in the same boat"—are common in these materials. Furthermore, active citizenship, as presented by organizations such as UNESCO and the European Commission, requires that actors be knowledgeable of others and be willing to engage in constructive and accountable ways, no matter where they are located or with whom they interact (Basok and Ilcan 2006; Skelton 2007). Training materials encourage youth to look beyond parochial concerns of their own group and their own location and to interact more broadly and with more respect for—and even a stake in—the perspectives of other people and places. These interventions in what might be thought of as topological and topographical spatial relations also have implications for affective and political relations and for the kinds of practices that are constructed as normal and legitimate for citizens.

Such circulations, however, can be made more difficult by blockages and disruptions that limit the movement of delegates. These might be geopolitical, such as the problems some delegates faced in obtaining visas to travel to countries. Although one arm of a government or an international organization operating within a country might welcome delegates, visa and passport regimes of those same countries can block such movements (Neumayer 2006). Even if visas are not an issue, travel is never "free" and the costs of

attendance were a challenge for many delegates, particularly to the 1970 World Youth Assembly; the archival record is full of pleas for money or for expedited approval of visas for the attendees (World Youth Assembly 1970b). In such cases, international efforts to rise above geopolitics and the conditions attendees faced at home were entangled with the real politics and economics of international travel and the support governments offer to each other.

National and local contexts affected the long-term impacts of the conferences, as well. Delegates to the 2014 conference questioned the value of encouraging participation at international conferences when opportunities for participation locally were nearly absent and when the circulation of ideas was limited to the small number of people who attended. Several delegates struggled with the feeling that conferences provided a veneer of youth inclusion in ways that seemed to co-opt and tame their political agendas. One delegate at the 2014 conference complained, "I mean, why this fancy conference hall and fancy hotels? It is like we're just acting. I feel like they are just preparing me to be like them. That's what they mean by training and participation."

Delegates at both the 1970 and 2014 gatherings argued that concerns for democracy and citizenship were not evident in the actions of governments and organizations such as the UN, or at least that the actions had multiple political valences that complicated—entangled and confused—their politics. In the 1970 World Peace Commission report, for example, proclamations about democracy were interlaced with denunciations of imperialism and colonialism by superpowers. Furthermore, delegates questioned the meaning and politics of cosmopolitanism and the supposed universalism of concepts such as rights. In discussions at both gatherings, delegates debated how to make "universal rights" interpretable and meaningful in their local and regional contexts and in ways that served—rather than obscured—their political goals. Yet critical and skeptical as delegates might have been, there was also a sense of possibility and commitment on their part that was fostered by meeting other young people who shared commitments to making a difference in their communities, nation, and world.

Conclusion

The preceding examples point to the complex ways that spatial relations, politics, affective agency, and practice are entangled in the circulations that are part of citizenship formation. The conferences we discussed are merely illustrations but are nevertheless suggestive of both the efforts to construct citizenship as floating above yet still moored to place(s) and communities and profoundly conditioned by geopolitical, social, and economic relations.

The circulations and movements of ideas, practices, and people—as well as the disruptions to them—entangle local contexts, political goals, feelings of power, activism, national politics, and broad economic and political relationships. These are all evidence of the intimacy-geopolitics of citizenship formation. Approaching citizenship formation as an example of intimacy-geopolitics enabled in and through movement and circulations allows us to recast—and perhaps ultimately discard—several canards about citizenship. The idea that citizenship is created by and primarily relevant to nation-states should finally and decisively be put aside, as should claims that global and cosmopolitan citizenship somehow transcend nation-states or make them less relevant. Circulations of ideas, values, and bodies are critical to the ways in which near and distant are co-constituted, as well as to the ways in which the intimacy-geopolitics of citizenship formation become evident. Rather than attempting to locate citizenship in specific sites or scales, our attention is directed to the relationships through which citizenship is constructed, enacted, and given meaning. In these relationships, we can see circulations, citizenship formation, and intimacy-geopolitics as providing the resources and rationales for contestation and activism in which new qualities of citizens, new collectivities, and new ways of being political might emerge.

Acknowledgments

The UN Archives and Records Management Section in New York provided assistance with the research on the 1970 World Assembly of Youth conference, for which we are very grateful. We also thank the reviewers and editor for their helpful comments on the article.

Funding

This research was supported by grants from the European Research Council (grant ES/J500082/1). This support is gratefully acknowledged.

Notes

1. See Ehrkamp and Jacobson (2015) for an excellent recent review. See also Staeheli (2011) and Kofman (2003).
2. Bissell and Fuller's (2011) edited collection *Stillness in a Mobile World* provides a notable exception to this, using the concept of stillness to challenge the (over)attention to the dialectic between stasis and movement within mobility studies.
3. See Basok and Ilcan (2006), Skelton (2007), and Diprose (2012) as other examples of the effort by international and transnational organizations to train citizens and, in particular, young citizens. These are efforts with uncertain and inconsistent outcomes, and we make no claims as to their "real" effects.
4. *Youth* and *young people* are used interchangeably, and perhaps loosely, in this article. There is an academic literature that debates the boundaries of youth and even the utility of the category. For our purposes, however, these debates seem less relevant, as the countries that send delegates to conferences set their own definitions, which are quite varied. In this article, our focus is on the circulations and idea of floating sites of citizenship formation using young people as an example, rather than on the boundaries of the category or on definitions of youth.
5. As an example, see the Junior Chamber International's (2013) materials on active citizenship. This framework was used in presentations at the World Conference on Youth.

References

Adey, P. 2006. If mobility is everything then it is nothing: Towards a relational politics of (im)mobilities. *Mobilities* 1 (1): 75–94.

Baillie Smith, M., and K. Jenkins. 2011. Disconnections and exclusions: Professionalization, cosmopolitanism and (global?) civil society. *Global Networks* 22 (2): 160–79.

Baillie Smith, M., N. Laurie, P. Hopkins, and E. Olson. 2013. International volunteering, faith and subjectivity: Negotiating cosmopolitanism, citizenship and development. *Geoforum* 45:126–35.

Barker, J., P. Kraftl, J. Horton, and F. Tucker. 2009. The road less travelled—New directions in children's and young people's mobility. *Mobilities* 4 (1): 1–10.

Basok, T., and S. Ilcan. 2006. In the name of human rights: Global organizations and participating citizens. *Citizenship Studies* 10 (3): 309–27.

Berlin, M. 1970. UN due for rambunctious youth parley. *New York Post* 13 April 1970. (UN Archives and Records Management Section, S-0249-0086, United Nations, New York.)

Bissell, D., and G. Fuller. 2011. *Stillness in a mobile world*. London and New York: Routledge.

Cresswell, T. 2012. Mobilities II: Still. *Progress in Human Geography* 36 (5): 645–53.

Diprose, K. 2012. Critical distance: Doing development education through international volunteering. *Area* 44 (2): 186–92.

Ehrkamp, P., and M. Jacobsen. 2015. Citizenship. In *The Wiley Blackwell companion to political geography*, ed. J. Agnew, V. Mamadouh, J. Sharp, and A. Secor, 152–64. Oxford, UK: Wiley Blackwell.

Fluri, J. 2011. Bodies, bombs, and barricades: Gendered geographies of (in)security. *Transactions of the Institute of British Geographers* 36 (3): 280–96.

Hannam, K., M. Sheller, and J. Urry. 2006. Mobilities, immobilities and moorings. *Mobilities* 1 (1): 1–22.

Harker, C. 2009. Student im/mobility in Birzeit, Palestine. *Mobilities* 4 (1): 11–35.

Hopkins, P., and C. Alexander. 2010. Politics, mobility and nationhood: Upscaling young people's geographies: Introduction to special section. *Area* 42 (2): 142–44.

Horton, J., P. Christensen, P. Kraftl, and S. Hadfield-Hill. 2014. "Walking . . . just walking": How children and young people's everyday pedestrian practices matter. *Social and Cultural Geography* 15 (1): 94–115.

Junior Chamber International. 2013. Home page. http://www.jci.cc/about/whatwedo (last accessed 25 June 2015).

Kofman, E. 2003. Rights and citizenship. In *A companion to political geography*, ed. J. Agnew, K. Cox, and K. Mitchell, 393–407. Malden, MA: Blackwell.

McCann, E., and K. Ward. 2011. *Mobile urbanism: Cities and policymaking in the global age*. Minneapolis: University of Minnesota Press.

Mountz, A., and J. Hyndman. 2006. Feminist approaches to the global intimate. *Women's Studies Quarterly* 34:446–63.

Nagar, R., V. Lawson, L. McDowell, and S. Hanson. 2002. Locating globalization: Feminist (re)readings of the subjects and spaces of globalization. *Economic Geography* 78:257–84.

Neumayer, E. 2006. Unequal access to foreign spaces: How states use visa restrictions to regulate mobility in a globalized world. *Transactions, Institute of British Geographers* 31:72–84.

Osler, A., and H. Starkey. 2005. *Changing citizenship: Democracy and inclusion in education*. Maidenhead, UK: Open University Press.

Pain, R., and L. Staeheli. 2014. Introduction: Intimacy-geopolitics and violence. *Area* 46 (4): 344–47.

Pratt, G., and V. Rosner. 2012. Introduction: The global and the intimate. In *The global and the intimate: Feminism in our time*, ed. G. Pratt and V. Rosner, 1–27. New York: Columbia University Press.

Pykett, J. 2010. Citizenship education and narratives of pedagogy. *Citizenship Studies* 14 (6): 621–35.

Riles, A. 2000. *The network inside out*. Ann Arbor: University of Michigan Press.

Salter, M. 2013. To make move and let stop: Mobility and the assembly of circulation. *Mobilities* 8 (1): 7–19.

Sheller, M., and J. Urry. 2006. The new mobilities paradigm. *Environment and Planning A* 38:207–26.

Skelton, T. 2007. Children, young people, UNICEF, and participation. *Children's Geographies* 5:165–81.

———. 2013. Young people's urban im/mobilities: Relationality and identity formation. *Urban Studies* 50 (3): 467–83.

Staeheli, L. 2011. Political geography: Where's citizenship? *Progress in Human Geography* 35 (3): 393–400.

Staeheli, L., and D. Hammett. 2010. Educating the new national citizen: Education, political subjectivity, and divided societies. *Citizenship Studies* 14 (6): 667–80.

World Conference on Youth. 2014. Colombo Declaration on Youth: Mainstreaming youth in the post-2015 development agenda. http://wcy2014.com/pdf/colombo-declaration-on-youth-final.pdf (last accessed 25 May 2015).

World Youth Assembly. 1970a. Report of the Commission on World Peace. UN Archives and Records Management Section, S-0249-00806, United Nations, New York.

———. 1970b. World Youth Assembly—Summary reports. UN Archives and Records Management Section, S-0241-002-10, United Nations, New York.

The Geopolitics of Tourism: Mobilities, Territory, and Protest in China, Taiwan, and Hong Kong

Ian Rowen

Department of Geography, University of Colorado Boulder

This article analyzes outbound tourism from mainland China to Hong Kong and Taiwan, two territories claimed by the People's Republic of China, to unpack the geopolitics of the state and the everyday, to theorize the mutual constitution of the tourist and the nation-state, and to explore the role of tourism in new forms of protest and resistance. Based on ethnographies of tourism practices and spaces of resistance conducted between 2012 and 2015 and supported by ethnographic content analysis, this article demonstrates that tourism mobilities are entangled with shifting forms of sovereignty, territoriality, and bordering. The case of China, the world's fastest growing tourism market, is exemplary. Tourism is profoundly affecting spatial, social, political, and economic order throughout the wider region, reconfiguring leisure spaces and economies, transportation infrastructure, popular political discourse, and geopolitical imaginaries. At the same time that tourism is being used to project Chinese state authority over Taiwan and consolidate control over Tibet and Xinjiang, it has also triggered popular protest in Hong Kong (including the pro-democracy Umbrella Movement and its aftermath), and international protest over the territorially contested South China Sea. This article argues that embodied, everyday practices such as tourism cannot be divorced from state-scale geopolitics and that future research should pay closer attention to its unpredictable political instrumentalities and chaotic effects. In dialogue with both mobilities research and borders studies, it sheds light not only on the vivid particularities of the region but on the cultural politics and geopolitics of tourism in general.

本文分析从中国大陆到香港与台湾——两处皆被中华人民共和国宣称为其领土之地——的观光, 以拆解国家的地缘政治和每日生活, 理论化观光客与国族国家之间的相互建构, 并探讨观光在新的抗议及抵抗形式中的角色。本文根据在 2012 年与 2015 年间所进行的观光实践与抵抗空间的民族志研究, 并由民族志内容分析加以支持, 证明观光能动性与转变中的主权、领域性及边界划定形式相互交缠。作为全世界增长最为快速的观光市场, 中国的案例可作为示范。观光对于更广大区域的空间、社会、政治与经济秩序影响深切, 重构了休憩空间与经济, 交通建设, 流行政治论述和地缘政治想像。于此同时, 观光被用来投射中国政府对台湾的权威, 并巩固对西藏和新疆的控制, 此外更引发了香港的群众抗议 (包含捍卫民主的雨伞运动及其馀波), 以及针对具有领土争议的南中国海的国际抗议。本文主张, 如同观光一般的身体化的每日生活实践, 无法与国家尺度的地缘政治分离, 而未来的研究应更加关注其无法预测的政治工具性与混乱效应。本研究同时与能动性研究和边界研究进行对话, 不仅对于该区域显见的特殊性、亦对一般的观光文化政治与地缘政治提出洞见。

Este artículo analiza el turismo de orientación externa desde la China continental a Hong Kong y Taiwán, dos territorios reclamados por la República Popular China, para descargar la geopolítica del estado y de lo cotidiano, para teorizar la constitución mutua del turista y del estado-nación, y para explorar el papel del turismo en las nuevas formas de protesta y resistencia. Con base en etnografías de las prácticas turísticas y espacios de resistencia conducidas entre 2012 y 2015 y con el apoyo de análisis del contenido etnográfico, este artículo demuestra que las movilidades del turismo se hallan enredadas con formas cambiantes de soberanía, territorialidad y demarcación fronteriza. El caso de China, mercado turístico de más rápido crecimiento en el mundo, es un buen ejemplo al respecto. El turismo está afectando profundamente el orden espacial, social, político y económico a través de la región de mayor más amplitud, reconfigurando los espacios y economías del ocio, la infraestructura del transporte, el discurso político popular y los imaginarios geopolíticos. Al propio tiempo que el turismo se utiliza para proyectar la autoridad estatal china sobre Taiwán y consolida su poder sobre el Tíbet y Xinjiang, eso también ha desencadenado la protesta popular en Hong Kong (incluso el prodemocrático Movimiento Sombrilla y sus secuelas) y la protesta internacional en relación con el Mar Meridional de la China, objeto de disputa territorial. Este artículo sostiene que las prácticas personificadas y cotidianas, como el turismo, no pueden divorciarse de las geopolíticas a escala de estado, y que la investigación futura debería poner mayor atención a sus impredecibles instrumentalidades políticas y caóticos efectos. Alternando con la investigación de movilidades y estudios fronterizos, el artículo arroja luz no solo sobre las vívidas particularidades de la región sino sobre la política cultural y sobre la geopolítica del turismo en general.

Tourism is no mere leisure activity, as the case of "Greater China" makes clear. In the complicated sovereign and territorial topology of this "contingent state" (Callahan 2004), tourism is political instrument, provocation to protest, and stage of high-stakes struggle over ethnic identity, national borders, and state territory. This article analyzes outbound tourism from mainland China to Hong Kong and Taiwan, two territories claimed by the People's Republic of China (PRC), to unpack such geopolitics of the state and the everyday, to theorize the mutual constitution of the national tourist and the nation-state, and to explore the role of tourism in new forms of dissent and resistance. Examination of this case sheds light not only on the vivid particularities of the region but on the cultural politics and geopolitics of tourism in general.

Mobilities and borders are increasingly recognized as inseparable domains (Cresswell 2010; Richardson 2013; Salter 2013). Indeed, "to theorize mobilities and networks is at the same time to theorize borders" (Rumford 2006, 155). Cultural and political geographers have conducted insightful studies on the role of tourism in domestic nation-building and modernization projects (Oakes 1998; Johnson 1999; Light 2001). The political implications and instrumentalities of tourism mobilities between and at the edges of national territory demand deeper attention, however. Tourists, a particular kind of mobile subject, traverse a bordered world, and their movements affect and are affected by the construction and performance of those borders.

Although much recent mobilities literature relates migration to state sovereignty and the performance of borders and state territory (Parsley 2003; Dauvergne 2004; Salter 2006, 2008; Wonders 2006), tourism has received insufficient analysis. There have indeed been some examinations of the role of borders in encouraging or restricting tourism (Timothy 1995, 2004; Sofield 2006), the potential instrumentality of tourism for achieving world peace or for reconciliation or unification between nation-states (D'Amore 1988; Jafari 1989; Guo et al. 2006; Seongseop, Timothy, and Han 2007), and the use of tourism as an instrument of foreign policy (Richter 1983; Arlt 2006), but tourism has rarely been treated as a bordering or territorializing process in its own right.

Within the subfield of tourism geography and the broader interdisciplinary realm of tourism studies, recent themes of embodiment (Gibson 2009), physicality and performance (Edensor 2001), and performativity have led researchers in interesting regional and methodological directions (Gibson 2008) but have also tended to shift the discussion farther away from state-scale politics. This article responds by arguing that the geopolitics and the everyday embodied encounters of tourism articulate together and should be researched in tandem.

The case of China, the world's fastest growing tourism market, is exemplary. Tourism is profoundly affecting spatial, social, political, and economic order throughout the region, reconfiguring leisure spaces and economies, transportation infrastructure, popular political discourse, and geopolitical imaginaries. Outbound tourism from the PRC has been used as an economic lever for extracting political concessions not only in nearby Taiwan but as far away as Canada. At the same time that tourism is being used to consolidate state authority in Tibet and Xinjiang, it has also triggered wide popular protest in semiautonomous Hong Kong and international criticism over the territorially contested South China Sea, where the PRC began cruise ship tourism in 2013, and for which the Communist Party's United Front Work Department declared explicitly that "tourism will have an important function" to "pledge and protect our nation's sovereignty" (United Front Work Department 2015; author's translation). This wide range of reactions underscores the political stakes and sites of tourism, which touch on territorial extent and definition, bordering technologies, sovereign claims, and the rights and lived experiences of mobile subjects.

China is remarkable for not only its rapidly growing outbound tourism but also its rise in global geopolitical prominence and its experiments in new forms of sovereignty. Ong (2004) argued that the PRC uses "variegated sovereignty" as a "technology of governance" designed to exert influence and integrate its territorial claims over Hong Kong, Taiwan, and elsewhere by flexibly allowing for different techniques of rule. By variegated sovereignty, Ong was referring to "differential powers of autonomy and social orders that are allowed by the Chinese state" in different but connected economic and political zones, designed instrumentally for "incremental but eventual political integration" (83). Here, I highlight how tourism mobilities are a fragile component of this fraught project.

The background section briefly introduces theoretical concepts and empirical cases useful for analyzing tourism in general and Chinese tourism in particular. Hong Kong and Taiwan are the foci of the empirical

144

section. These two territories on China's historical periphery have two different and complicated sovereign regimes, but they are both territories where tourism has been deployed by the Chinese Communist Party (CCP) for similar geopolitical aims of greater political, economic, and cultural integration with mainland China as directed from Beijing. An examination of these cases will demonstrate how tourism produces national subjects and national borders—or, in other words, functions as a technology of state territorialization (Rowen 2014). I further suggest that embodiment, both in practice and in representation, is key to this process.

The following discussion is based on multisited, mobile ethnography (Marcus 1995; Buscher and Urry 2009) and ethnographic content analysis (Altheide 1987) conducted between 2012 and 2015 in Taiwan, Hong Kong, and the PRC. This fieldwork included fourteen months of participant observation of Chinese tourism within Taiwan, two months of participant observation within both the Taiwan Sunflower Movement and Hong Kong Umbrella Movement occupations, sixty interviews with Chinese tourists in Taiwan, thirty-six interviews with political activists or protest-site visitors (including both Hong Kong and Taiwan), and twenty interviews with Taiwanese tourist site staff and vendors. Based on respondent availability, interviews ranged from ten to seventy-five minutes. Most interviews were conducted on site, and others took place in nearby parks or cafes. Concurrent and later research included extensive analysis of regional print, radio, TV, and online popular and social media.

The Geopolitics and Cultural Politics of Tourism

Tourism is more than the aggregate of human flows through a world traced by package tours and guidebooks. Rather, tourist bodies, sites, the state apparatus that manages them, and regulatory devices such as visas or passports constitute a "hybrid assemblage" with a wide range of effects (Franklin 2004; Salter 2013). In this ontology, tourism can be treated as an "active ordering of modernity" that produces nationalized subjects and spaces through ideological regimes, site management and design, and mobility regulation. These effects extend beyond bodies and spaces nominally recognized as touristic.

The tourist moves as a stage on which national or racial values are not only inscribed but performed domestically and abroad. State actors project, inscribe, and proscribe moral values onto the bodies and

representations of mobile subjects (Sun 2002; Nyíri 2010). For example, in China, even if tourism is usually portrayed as a recreational activity, tourists' behavior has affected the perception of the nation more widely. Chio (2010) observed that "[negative] stories of the Chinese tourist abroad have put a damper on this upbeat association between travel, individual character, and national character" (14). In response, China's leadership has launched multiple campaigns to promote "civilized tourism," portraying its tourists as ambassadors both at home and abroad, enrolling tourists into this national project.

Such moral values and national education campaigns are inscribed not only on bodies but on sites designated and bounded specifically for tourist experience. The cultural authority exerted via the construction and management of such sites is an important component of national self-definition, as demonstrated by an insightful analysis of nation-building narratives at a Vietnamese war museum (Gillen 2014). Such tactics are also well documented in the case of China, where the state "sponsors a discursive regime in which scenic spots and their state-endorsed hierarchy are tools of patriotic education and modernization, and in which the state has the ultimate authority to determine the meaning of the landscape" (Nyíri 2006, 75). The organizational conditions that enable this regime include deep institutional and personal overlaps among state regulatory agencies, tour operators, and site developers and management.

Normative conceptions of national territory are also inscribed in mobility regulation devices such as passports and visas. In the case of transnational or border-crossing tourism, use of these devices enrolls not only tourists, but other actors in the "global mobility regime" (Salter 2006) of mobility regulation, including embassies, consulates, and customs and immigration officials. These devices rely on consistent citation of the extent and division of sovereign territory. Their instrumentality can also make them subject to contestation.

China's so-called Passport War of 2012 is illustrative. In May 2012, the PRC released a new passport that not only includes images of Taiwan but also includes maps that cover disputed territories including parts of Kashmir (administered by India), most of the South China Sea (claimed by several countries, including Vietnam and the Philippines), and the Senkaku/Diaoyu Islands (claimed by Japan; Tharoor 2012). The passport quickly drew objections from India, Vietnam, and the Philippines, whose foreign ministries directed their immigration officers to not stamp the new passports for fear of legitimizing the PRC's territorial claims. Their solution was the creation of another device: entry stamps on specially issued, separate forms. Indian authorities even began issuing visas to PRC nationals that include a map of India claiming the disputed territories.

Caught in the act of border crossing, tourist bodies collide with contradictory ethno-national and territorial claims. Between liminal spaces of contested sovereignty and identity, as in Taiwan or Hong Kong, such encounters are punctuated with "material moments" that reveal the complexity and fluidity of national identity (Zhang 2013). Tourism's wide range of political instrumentalities can also produce "retrenchment of identity in a territory" (Park 2005, 110), however, and fuel territorial conflict (Rowen 2014).

China's construction and deployment of Approved Destination Status (ADS) is another example of tourism's political instrumentality. Outbound tourism has, since 1995, been regulated by a system that confers ADS on countries that have signed bilateral agreements with China. ADS allows outbound group tourists to apply for visas through travel agencies, saving them a trip to the consulate. It also encourages greater marketing of group tours. ADS is therefore a highly desirable designation for countries that are eager to boost inbound tourism revenue.

The initial purpose of the ADS system was to limit Chinese nationals from bringing hard currency abroad (Arlt 2006). ADS later became a tool to exert other forms of political pressure. A primary criterion is that "the country should have a favorable political relationship with China" (Kim, Guo, and Agrusa 2005, 212). This includes not maintaining official diplomatic relations with Taiwan. Taiwan and Hong Kong are involved in several other promotional schemes in addition to ADS, and they coordinate with PRC agencies accustomed to using the economic benefits of tourism as political tools. However, as I will argue below, these political programs can be subverted by the actual practices and effects of tourism.

Chinese Tourists as "Locusts" Raiding a Restive Hong Kong

The cultural and political integration of Hong Kong and Taiwan is a fraught and complicated project of the CCP leadership. Taiwan floats in a kind of liminal space between state and nonstate and Chinese and non-Chinese (Corcuff 2012), as does Hong Kong. The

unusual legal and administrative status of these territories has required both polities to employ various regimes and devices for regulating mobility, not all of which can be negotiated on an equal basis with Beijing.

The Hong Kong Special Administrative Region (SAR) was created in 1997 following the British colonial "handover" of the territory to China. The SAR ostensibly functions on a one country, two systems (OCTS) principle, in which both Hong Kong and mainland China belong to the same country but are meant to operate according to two different systems of governance until 2047. The OCTS scheme was initially formulated as a model for the annexation and integration of Taiwan, and its implementation has been watched closely by Taiwanese observers (Cooney 1997). Some vestiges of the British system remain, including an independent judiciary and partially elected legislature, but the territory's chief executive is appointed by the PRC.

Hong Kong's cultural integration has been challenging for the PRC. Anson Chan, former chief secretary for administration, observed that "the real transition is about identity and not sovereignty" (Chan 1998), and the CCP position is generally represented as one that knows that it has won the territory but not the "hearts" of the people. Tourism has further problematized this project.

In 2003, Hong Kong's economy appeared imperiled following the outbreak of the infectious disease SARS. Ostensibly to improve the financial outlook, China raised its caps on outbound tourists by implementing the Individual Visit Scheme. In just over ten years, the annual number of mainland Chinese tourist arrivals rose from 8.5 million to 40 million (Chiu, Ho, and Osawa 2014) in a territory of just 7 million people. Their spending has been significant but so has the corresponding rise in commodity prices. This is due not only to spending by leisure tourists but also to the rise of the "water goods" (*shui huo*) trade, in which day visitors cross from China to purchase essentials that are either cheaper due to Hong Kong's lower taxes or perceived to be of higher quality due to China's relatively lax food safety enforcement. This trade has precipitated a backlash from Hong Kong people who fear rising prices and food shortages.

Tourism from mainland China has accelerated the development of a distinct Hong Kong subjectivity defined in part by difference from China. Popular and social media long reflected widespread discontent with the behavior of Chinese tourists, which reached a boiling point with the "anti-locust" (*fan huang-chong*) protests in early 2014. These widely publicized demonstrations actually drew only a few hundred activists but reflected an incipient nativism that has been aggravated no less by widely reported damages and social ruptures of tourism than by Beijing's policy interventions and public statements (Garrett and Ho 2014).

The animality of the terms used to deride Chinese tourists conflates the physical with the geopolitical. *Locust* has been in common use at least since 2012, when a full-page ad, paid for via crowd-sourced funds, appeared in the popular daily newspaper *Apple Daily*, featuring an image of Chinese tourists as locusts flying over Lion Rock, an iconic Hong Kong site. The term is particularly directed at tourists who visit primarily to buy goods to bring back for use or sale in China—they are said to scour the shops and leave nothing affordable for local people. Another term, *pigs* (*zhu*), has likewise been directed at tourists and recalls the same epithet used by Taiwanese to insult unwelcome arrivals from China in the 1940s (Kerr 1965).

Driving much of the tension have been depictions of the supposedly uncouth and unhygienic practices of Chinese tourists. Blogs that document public urination and defecation, spitting, shoving, and other forms of behavior unacceptable in Hong Kong have proliferated rapidly. Public urination in Hong Kong and Taiwan is presented by area netizens not only as an annoyance but as an act of geopolitical provocation. Perhaps the most spectacular example is the viral YouTube video, "Locust World," released in 2011 and since seen by more than 1.4 million viewers, which includes the following lyrics (translated from Cantonese by Bad Canto 2011):

Locust come out from nowhere, overwhelm everywhere
Shouting, screaming, yelling like no one could hear
Ever feel shame to yourself?
Smoke like breathing in hellAnd your fucking son who shit right in the mall
See this country? countrymen expert in stealing, cheating, deceiving, lying
"I'm Chinese!" scares the piss out of everyone
Locust nation named "Cina"—disgusted by the whole of East Asia
Everyday trying to naturalise us with Mandarin
Invading across the Hong Kong border and taking over our land—that's your speciality
Parasitic until your citizenship is recognised
Big-belly locust like aliens; pregnant and not stopped by immigration . . .

. . . Locust eggs hatch in hospitals—taking over beds and not paying bills
We thought we've seen the worst, but . . . doing your toilet business on the streets?
There's no shame—jumping queues, spitting in public . . .
. . . we witness and condemn these acts everyday
Inch by inch, Hong Kong is now being taken over by these pests
Those glittering days are now long gone
While our citizens are bleeding, the locusts buy out all our food
How can we retake our homeland?

The imagery accompanying the song is a carefully crafted pastiche of real-life scenes from Chinese tourist sites, including crowded shopping centers, queue jumping, shoving, and of course, public urination.

Tensions between Hong Kong and mainland China rose spectacularly during the Umbrella Movement of late 2014, in which hundreds of thousands of young people flooded the streets to protest Beijing's policies. Although the rallying cry of this movement was for "genuine universal suffrage," the long-promised right for Hong Kong people to elect a leader of their own choosing, in fact the zones around the several occupation sites presented a panoply of identity politics and civic passions, some of which was anti-China and anti-Chinese.

For one month during the Umbrella Movement, I regularly conducted participant observation, in-depth interviews, and ad hoc focus groups with both Hong Kong demonstrators and Chinese visitors to the site. Although many were careful to articulate their demands in the terms of demands for electoral reforms, cultural and embodied difference was still a persistent theme. "It's nice to be here with each other with just Hong Kong people. I don't think I've heard so much pure Cantonese in weeks," said a twenty-six-year-old journalist. "This is like the Hong Kong of my youth," said a forty-five-year-old salon worker. She clarified that she was referring not only to the high proportion of "locals" but also to the general everyday qualities of civility, order, and hygiene that she did not associate with China.

Attention to such qualities of embodiment helps illuminate the origins and the bordering effects of Hong Kong protest and discontent. Both the small anti-locust protests, as well as the mass mobilizations of the Umbrella Movement, were driven by ambivalence about Hong Kong's integration with China and the perceived disappearance of Hong Kong's autonomy. The large influx of Chinese tourist "ambassadors" increased such tension. In this already tense mix, a child's public urination—something I frequently witnessed within walking distance from the protest encampments in which I resided—became tantamount to geopolitical provocation. The political potency of the act—in itself banal, particularly in China—could not be understood without reference to Hong Kong's incomplete social, cultural, and geopolitical reterritorialization as a part of China. Put another way, the affective characteristics of the protests were inexorably inflected by the relational difference in the embodied behavior of locals and mainland tourists, augmenting widespread discontent about deferred democratization and subverted local identities.

Yet, Chinese tourists were frequently evident on site during the Movement and were in fact often cited by activists as the strategic target of the smaller Causeway Bay commercial district occupation (see Figure 1). "It's important that we stay

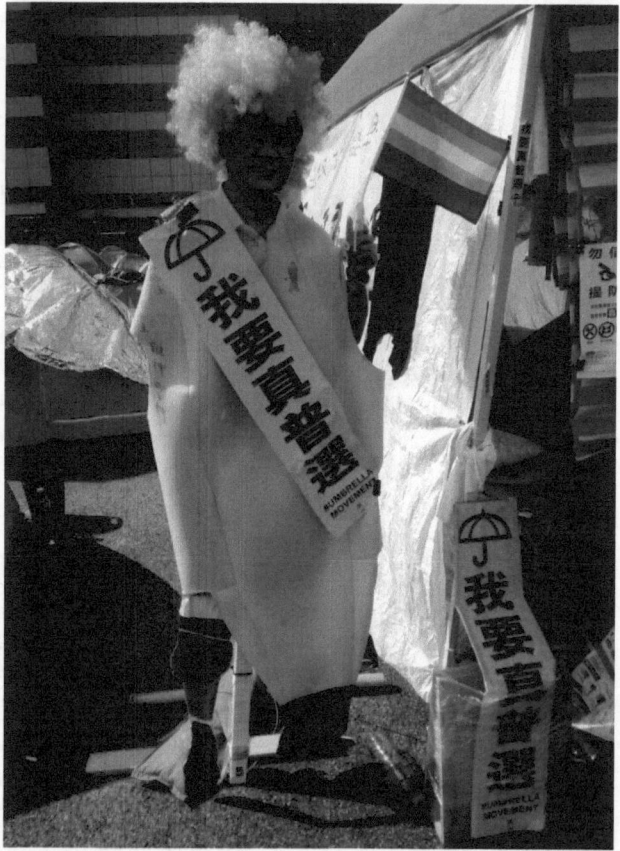

Figure 1. Remixed imagery of Chinese leader Xi Jinping calling for genuine universal suffrage gained the attention of many Chinese visitors to Hong Kong Umbrella Movement protest sites. Admiralty, 17 November 2015. (Color figure available online.)

here to sway their hearts and minds, since they'll go back to China afterwards," said a twenty-four-year-old philosophy graduate student. Tourism, in this case, was doubly problematic for the CCP's territorial program—not only did it spark protest, but it threatened and sometimes even succeeded in incorporating tourists into those very protests.

Chinese Tourists and the "Taiwan Question"

Taiwan's sovereign regime is dramatically different from that of Hong Kong. Although only officially recognized by twenty-two other states, Taiwan functions as a de facto independent democratic state with its own military and directly elected president. Taiwan's state administration includes its own Ministry of Foreign Affairs, Department of Immigration, as well as a Mainland Affairs Council, an agency under the executive branch tasked with conducting official coordination with its counterpart in the PRC, the Taiwan Affairs Office. Although the PRC claims both Hong Kong and Taiwan as its sovereign territory and officially groups them together as outbound destinations with the same nominal status, Taiwan's leadership has far greater capacity to control its own borders and conduct negotiations than does Hong Kong.

An agreement to receive direct tourist arrivals from China was not made until 2008, after the election of President Ma Ying-jeou of the pro-unification Chinese Nationalist Party (KMT). Eager to trumpet political breakthroughs and economic gains, the Ma administration acceded to the PRC's demand that it accept "entry/exit permits" for Chinese tourists, as does Hong Kong, instead of requiring passports and visas, which would imply that Taiwan was a formally independent country. By 2014, annual tourist arrivals had risen to nearly 3 million and were often presented as a showcase example of Ma's "successful" cross-strait policies (see Figure 2). Yet, Ma's China policies were panned by the electorate, later earning him approval ratings as low as 9 percent and sparking the March 2014 Sunflower Movement, when thousands of student and civic activists occupied the area inside and around the Legislative Yuan (parliament) to protest a trade deal that included provisions that would liberalize the tourism industry (Rowen 2015). The KMT's landslide defeat in the November 2014 local elections was widely portrayed as a referendum on Ma's China's policies (Harrison 2014).

Figure 2. Taiwan independence demonstrators, Falun Gong practitioners, and pro-PRC demonstrators overlap at a site popular with Chinese tourists. Taipei 101, 25 May 2014. (Color figure available online.)

Although it would be unjustifiable to draw a direct causal arrow between the parallel growth in inbound tourism from China and popular protest against China policy, their tandem acceleration deserves analysis. Tourism has frequently been presented by the ruling KMT as a boon to the economy. This has stoked opposition from a variety of actors: independence advocates eager to reduce Taiwan's reliance on China, populists who complain that the benefits of cross-strait trade have been felt only by people with KMT or PRC connections, and activists who claim that the costs are therefore displaced onto the Taiwanese public. A characteristic example follows:

> They [Chinese] create their own market—they fly their own airlines, they hire their own buses, eat and live at their own hotels—but they are using our land and our scenery, to make money. Our scenic hotspots such as Sun Moon Lake and Kenting are now filled with Chinese. We are left with their trash. Allowing Chinese tourists into the country costs more than we gain. ("Bohmann von Formosa" quoted in Tsai and Chung 2014)

There are few reliable data about Taiwan's actual economic gains from tourism. Tourism Bureau figures, both published online and reconfirmed to me in my interviews with officials, are an estimate based on tourist self-reported guesses of per day spending multiplied by total arrivals, instead of analysis of actual revenues. Although economic benefits are therefore unclear, unseen, and immaterial for the vast majority of Taiwanese, analysis of my interview and media data suggests that it is precisely the representations of tourist embodiment that imbue them with geopolitical salience.

As in Hong Kong, Chinese tourists are frequently depicted as rude, loutish, noisy, smelly, and unhygienic. Reports both on social media and in the popular press include tourists defacing plants on the east coast (Fauna 2012), tourists bathing in their underwear in the popular southern beach town of Kenting (Tsai and Chung 2014), and public urination (Ramzy 2014). Similar sentiment was expressed by a colleague: "I don't go to the beach at Kenting anymore. There are too many mainlanders there now. It's like going to China."

Although there is an element of "othering" at play here, arguably with racist or discriminatory overtones, this reaction is situated in an uncomfortable historical context. For many in Taiwan, tourism from China recapitulates a kind of geopolitical invasion: its occupation by the KMT in the late 1940s, when the same word now used for today's mainland Chinese tourists, "mainland guest" (*luke*), referred to incoming waves of

KMT soldiers. Like tourists, they were also widely perceived by local residents as uncouth, unhygienic, and abusive (Kerr 1965). Tourism in Taiwan is therefore part of an ongoing, highly politicized saga of mobility, identity, bordering, and territorialization.

Conclusion

Tourism mobilities constitute national subjects and nation-states and reproduce and undermine borders and territories. As a political technology, tourism is part and parcel of state geopolitical programs. These effects articulate not just via state-scale visa and passport regimes but through the messy outcomes of everyday embodied behavior. Far from being a reliable tool of peacemaking, rapprochement, or even territorial claim-making, tourism can also aggravate alienation and precipitate protest.

In China, authorities have used tourism as a tool of foreign policy and a tactic of territorial projects. In Hong Kong and Taiwan, Chinese tourists have become issues in electoral and protest politics. This has produced contradictions between the territorial and cultural programs of the different state administrations in all three territories. These contradictions emerge through changing mobility regimes and conflicting sovereign programs, as well as through representations of tourists and tourist spaces that proliferate beyond the bounds of state control.

The practices of individual mobile subjects, or of aggregated tourist flows, are only partially determined by state policy and programs. State projects themselves might be impacted by the unexpected outcomes of tourist practice. This is due to tourism's imbrication with wider issues of national identity, territory, and geopolitical order. Future mobilities and borders research, whether in this region or beyond, would be well served by closer attention to such unpredictable political instrumentalities and chaotic effects of tourist practice.

Acknowledgments

Partial research and writing support for this article was provided by the Fulbright Commission, the National Science Foundation, the Foundation for Scholarly Exchange, the Taiwan Foundation for Democracy, and the University of Colorado.

References

Altheide, D. L. 1987. Reflections: Ethnographic content analysis. *Qualitative Sociology* 10 (1): 65–77.

Arlt, W. G. 2006. *China's outbound tourism*. London and New York: Routledge.

Bad Canto. 2011. 蝗蟲 Hong Kong "Locust." *Bad Canto*. https://badcanto.wordpress.com/2011/03/03/蝗蟲/ (last accessed 10 November 2015).

Buscher, M., and J. Urry. 2009. Mobile methods and the empirical. *European Journal of Social Theory* 12 (1): 99–116.

Callahan, W. A. 2004. *Contingent states: Greater China and transnational relations*. Minneapolis: University of Minnesota Press.

Chan, A. 1998. Hong Kong: Riding out the Asian storm. http://www.info.gov.hk/gia/general/199806/12/0612097.htm (last accessed 10 November 2015).

Chio, J. 2010. China's campaign for civilized tourism: What to do when tourists behave badly. *Anthropology News* November:14–15.

Chiu, J., P. Ho, and J. Osawa. 2014. China travel-permit suspension weighs on Hong Kong tourism. *Wall Street Journal* 2 October. http://www.wsj.com/articles/china-travel-permit-suspension-weighs-on-hong-kong-tourism-1412258672 (last accessed 10 November 2015).

Cooney, S. 1997. Why Taiwan is not Hong Kong: A review of the PRC's "One Country Two Systems" model for reunification with Taiwan. *Pacific Rim Law & Policy Association* 6 (3): 497–548.

Corcuff, S. 2012. The liminality of Taiwan: A case-study in geopolitics. *Taiwan in Comparative Perspective* 4 (December): 34–64.

Cresswell, T. 2010. Towards a politics of mobility. *Environment and Planning D: Society and Space* 28 (1): 17–31. http://www.envplan.com/abstract.cgi?id=d11407 (last accessed 21 October 2013).

D'Amore, L. J. 1988. Tourism: A vital force for peace. *Annals of Tourism Research* 15:269–83.

Dauvergne, C. 2004. Sovereignty, migration and the rule of law in global times. *The Modern Law Review* 67 (4): 588–615.

Edensor, T. 2001. Performing tourism, staging tourism: (Re)producing tourist space and practice. *Tourist Studies* 2001 (1): 59–81.

Fauna. 2012. Mainland Chinese tourists deface plants in Taitung, Taiwan. *chinaSMACK*. http://www.chinasmack.com/2012/stories/mainland-chinese-tourists-deface-plants-in-taitung-taiwan.html (last accessed 10 November 2015).

Franklin, A. 2004. Tourism as an ordering: Towards a new ontology of tourism. *Tourist Studies* 4 (3): 277–301.

Garrett, D., and W. Ho. 2014. Hong Kong at the brink: Emerging forms of political participation in the new social movement. In *New trends of political participation in Hong Kong*, 347–83. Hong Kong: City University of Hong Kong Press.

Gibson, C. 2008. Locating geographies of tourism. *Progress in Human Geography* 32 (3): 407–22.

———. 2009. Geographies of tourism: (Un)ethical encounters. *Progress in Human Geography* 34 (4): 521–27.

Gillen, J. 2014. Tourism and nation building at the War Remnants Museum in Ho Chi Minh City, Vietnam. *Annals of the Association of American Geographers* 104 (6): 1307–21.

Guo, Y., S. S. Kim, D. J. Timothy, and K.-C. Wang. 2006. Tourism and reconciliation between Mainland China and Taiwan. *Tourism Management* 27 (5): 997–1005.

Harrison, M. 2014. Cross-straits relations: An era of uncertainty. *The China Story.* http://www.thechinastory.org/2014/12/cross-straits-relations-an-era-of-uncertainty/ (last accessed 10 November 2015).

Jafari, J. 1989. Tourism and peace. *Annals of Tourism Research* 16 (3): 439–43.

Johnson, N. C. 1999. Framing the past: Time, space and the politics of heritage tourism in Ireland. *Political Geography* 18 (2): 187–207.

Kerr, G. 1965. *Formosa betrayed.* Boston: Houghton Mifflin.

Kim, S. S., Y. Guo, and J. Agrusa. 2005. Preference and positioning analyses of overseas destinations by mainland Chinese outbound pleasure tourists. *Journal of Travel Research* 44 (2): 212–20.

Light, D. 2001. "Facing the future": Tourism and identity-building in post-socialist Romania. *Political Geography* 20 (8): 1053–74.

Marcus, G. E. 1995. Ethnography in/of the world system: The emergence of multi-sited ethnography. *Annual Review of Anthropology* 24:95–117.

Nyíri, P. 2006. *Scenic spots: Chinese tourism, the state, and cultural authority.* Seattle: University of Washington Press.

———. 2010. *Mobility and cultural authority in contemporary China.* Seattle: University of Washington Press.

Oakes, T. 1998. *Tourism and modernity in China.* London and New York: Routledge.

Ong, A. 2004. The Chinese axis: Zoning technologies and variegated sovereignty. *Journal of East Asian Studies* 4:69–96.

Park, C. J. 2005. Politics of Geumgansan tourism: Sovereignty in contestation. *Korean Studies* 8 (3): 113–35.

Parsley, C. 2003. Performing the border: Australia's judgment of "unauthorised arrivals" at the airport. *Australian Feminist Law Journal* 18 (1): 55–75.

Ramzy, A. 2014. Tempest over a chamber pot as Taiwan scrutinizes Chinese tourists' manners. *New York Times* 22 October. http://sinosphere.blogs.nytimes.com/2014/10/22/tempest-over-a-chamber-pot-as-taiwan-scrutinizes-chinese-tourists-manners/ (last accessed 10 November 2015).

Richardson, T. 2013. Borders and mobilities: Introduction to the special issue. *Mobilities* 8 (1): 1–6.

Richter, L. K. 1983. Tourism politics and political science: A case of not so benign neglect. *Annals of Tourism Research* 10 (3): 313–35.

Rowen, I. 2014. Tourism as a territorial strategy: The case of China and Taiwan. *Annals of Tourism Research* 46:62–74.

———. 2015. Inside Taiwan's Sunflower Movement: Twenty-four days in a student-occupied Parliament, and the future of the region. *The Journal of Asian Studies* 74 (1): 5–21.

Rumford, C. 2006. Theorizing borders. *European Journal of Social Theory* 9 (2): 155–69.

Salter, M. B. 2006. The global visa regime and the political technologies of the international self: Borders, bodies, biopolitics. *Alternatives: Global, Local, Political* 31:167–89.

———. 2008. When the exception becomes the rule: Borders, sovereignty, and citizenship. *Citizenship Studies* 12 (4): 365–80.

———. 2013. To make move and let stop: Mobility and the assemblage of circulation. *Mobilities* 8 (1): 7–19.

Seongseop, S., D. J. Timothy, and H. Han. 2007. Tourism and political ideologies: A case of tourism in North Korea. *Tourism Management* 28:1031–43.

Sofield, T. H. B. 2006. Border tourism and border communities: An overview. *Tourism Geographies* 8 (2): 102–21.

Sun, W. 2002. *Leaving China: Media, migration, and transnational imagination.* Lanham, MD: Rowman & Littlefield.

Tharoor, I. 2012. Asia's passport wars: Chinese map triggers diplomatic firestorm. *TIME World* 27 November. http://world.time.com/2012/11/27/asias-passport-wars-chinese-map-triggers-diplomatic-firestorm/ (last accessed 10 November 2015).

Timothy, D. J. 1995. Political boundaries and tourism: Borders as tourist attractions. *Tourism Management* 16 (7): 525–32.

———. 2004. *Tourism and political boundaries.* London and New York: Routledge.

Tsai, T., and J. Chung. 2014. Chinese tourists shock swimmers. *Taipei Times* 9 April. http://www.taipeitimes.com/News/taiwan/archives/2014/04/09/2003587637 (last accessed 10 November 2015).

United Front Work Department of the CPC Central Committee. 2015. 关于加强三沙旅游安全保障的建议 [Recommendations for the strengthening of Sansha tourism security]. 21 May. http://www.zytzb.gov.cn/tzb2010/jcjyxd/201505/c999d08b36a14b93aae8bbff263155bd.shtml (last accessed 20 August 2015).

Wonders, N. A. 2006. Global flows, semi-permeable borders and new channels of inequality: Border crossers and border performativity. In *Borders, mobility and technologies of control,* ed. S. Pickering and L. Weber, 63–86. Amsterdam: Springer.

Zhang, J. J. 2013. Borders on the move: Cross-strait tourists' material moments on "the other side" in the midst of rapprochement between China and Taiwan. *Geoforum* 48:94–101.

Micropolitics of Mobility: Public Transport Commuting and Everyday Encounters with Forces of Enablement and Constraint

David Bissell

Research School of Social Sciences, The Australian National University

Politics in geographical research on mobilities evaluates the nature of power and control of mobility and considers how people are differently enabled and constrained by these processes. Politics is usually approached along subject-centered lines where the task is to identify who is enabled and who is constrained and subsequently to account for the hidden mechanisms of power behind this unevenness. This article argues that what these subject-centered analyses can risk underplaying are the very transformations that mobility practices such as commuting themselves actually give rise to. This article draws on qualitative fieldwork during an evening train commute between Sydney and Wollongong in Australia to argue that the politics of mobilities needs to attend to ongoing processes of "micropolitical" transformation that take place through events and encounters, changing relations of enablement and constraint in the process. My argument is that we need to expand our understanding of what constitutes mobility politics to understand the nature and reach of the multiple forces that are at play, affecting and transforming life in this zone. This potentially enables us to more sensitively evaluate questions of responsibility and intervention.

地理学对能动性的研究之政治,评价权力的本质与能动性的控制,并考量人们如何不一而足地被这些过程赋予能力或受其限制。政治经常以聚焦主体的方式进行探讨,而该项工作旨在辨别谁被赋予能力以及谁受到限制,从而解释此般不均背后所隐藏的权力机制。本文主张,这些以主体为核心的分析所可能导致的不充分之处,正是在于诸如通勤本身的能动性实践实际上能够带来的改变。本文运用在澳大利亚悉尼与卧龙岗之间的夜间火车通勤之质性田野工作,主张能动性的政治,必须观照透过事件与偶遇而发生的"微政治"改变的持续过程,并在过程中改变了赋予能力与限制的关系。我主张,我们必须扩张对于何者构成了能动性的政治之理解,以了解正在上演的多重驱力的本质与范围,而它们影响并改变了此区内的生活。这麼做,能够潜在地让我们更加敏感地评估有关责任和介入的问题。

La política en investigación geográfica sobre movilidades evalúa la naturaleza del poder y el control de la movilidad, y considera cómo la gente es diferentemente capacidad y constreñida por estos procesos. Usualmente la política es aproximada a lo largo de líneas centradas en sujeto donde la tarea consiste en identificar quién está habilitado y quién constreñido, y subsiguientemente tomar en cuenta los mecanismos de poder ocultos detrás de esta desigualdad. Este artículo arguye que lo que estos análisis centrados en sujeto pueden arriesgar por intervenir poco son las propias transformaciones a que dan lugar las prácticas de movilidad, tales como el viaje al trabajo. El artículo se basa en trabajo de campo cualitativo durante una noche de viaje en tren entre Sídney y Wollongong, en Australia, para plantear que las políticas de movilidades deben atender los procesos corrientes de transformación "micropolítica" que tienen lugar a través de eventos y encuentros, cambiando durante el proceso las relaciones de habilitación y restricción. Mi argumento es que necesitamos ampliar nuestro entendimiento de lo que constituye política de movilidad para entender la naturaleza y alcance de las múltiples fuerzas que entran en juego, afectando y transformando la vida en esta zona. Esto potencialmente nos capacita para evaluar más sensiblemente cuestiones de responsabilidad e intervención.

"GOLD-DIGGERS! Look at them!" sneers a teenage boy loudly, swaying from a yellow pole. From his stage at the end of this packed double-deck train carriage, his accusation spears the placid atmosphere, addressing no one in particular but startling everyone. "I can see he is!" the boy grins, wagging his finger knowingly at a man in a white shirt sitting near to him, as if to expose his ruse.

The man refuses to look at him, his eyes trained on some point distant to this scene. The boy's eyes casually scan the carriage. "She isn't!" This time I can't see who his target is. One of the other three boys sniggers. The choice must be similarly condescending. Standing facing the passengers downstairs in the carriage who are all seated facing him, one of the other boys pipes up: "It's all about the money, money, money!" "MONEY, MONEY, MONEY!" chimes another, this time more provocatively, holding both hands out as he says it, rubbing his thumbs against his fingers to emphasize the sentiment. The first boy laughs.

The journeys that people make to and from work are one of the most significant contemporary mobility practices (Lyons and Chatterjee 2008; Edensor 2010). In many cities people are commuting further than ever before (Aldred 2014). In the context of rapid rates of urbanization worldwide, understanding the impacts of these practices on urban life is of paramount importance. The aim of this article is to develop our understanding of the politics of commuting mobilities.

Politics in mobilities research evaluates the nature of power and control of mobility and considers how people are differently enabled and constrained by these processes (Adey 2010b). Politics is usually approached along subject-centered lines where the task is to identify who is enabled, and who is constrained, and subsequently to account for the hidden mechanisms of power behind this unevenness (Morley 2000; Cresswell 2006). Much attention has quite rightly been devoted to identifying which groups of people benefit most from the infrastructures and services that are delivered by governments and transport providers and, conversely, which groups of people lose out from such decisions (Graham and Marvin 2001). Although such analyses are helpful for diagnosing situations of enablement and constraint, what these subject-centered analyses can risk underplaying are the very transformations that mobility practices themselves actually create, giving rise to a different sort of unevenness. This is significant for a politics of mobility, because if transformations are happening, then relations of enablement and constraint must also be changing in the process. My argument is that we need to expand our understanding of what constitutes mobility politics to appreciate the nature and reach of the multiple forces that are at play, transforming life in this zone. This enables us to more sensitively evaluate questions of responsibility and intervention—questions that are usually evaluated in terms of government transport and infrastructure provision.

The article focuses on some events that happened during one late afternoon commuter train journey between Sydney and Wollongong in Australia to argue that the politics of commuting mobilities cannot be adequately accounted for through a subject-

centered understanding of politics alone, where some individuals, groups, or institutions are understood to have more power, whereas others have less. I argue that these macro political analyses of larger scaled and seemingly longer durational social formations such as subject identities and institutions must be considered hand-in-hand with the more micro events and encounters that take place during the commute. The micropolitics of mobility can be understood as the ongoing processes of transformation that take place through events and encounters on the move. These events matter because they have transformative powers in and of themselves. Crucially, they cannot be understood as just the playing out of macropolitical forces such as institutional control. Rather, they can help us to appreciate the affective forces that are created by the actual practices of commuting themselves and the uneven enablements and constraints that these forces produce.

To help appreciate these ongoing transformations that mobilities generate, this article draws succour from nonrepresentational theories in geography.[1] These theories are useful because they open up the "very definition of 'politics' as a creative space for the contingent emergence of the new" (Roberts 2011, 2526). They help to show how practices of mobility can be understood as constitutive of subjects and milieus, rather than just derivative from practical necessity (Hansen 2004). What this ultimately means is that mobility systems are actively changing people, rather than just passively transporting them.

Illawarra Commuting Research Methods

Until the moment at the top of this article, the soporific roar of the air conditioning has smothered the train carriage. White noise to lull tired bodies. It is well past five o'clock and many people in this carriage have finished their paid working day. Foreheads are gleaming. It will be a while, though, before they are home. Many people in this carriage live in the Illawarra region, over sixty kilometers to the south of Sydney. This is Australia's busiest commuting corridor. Twenty thousand people (out of a workforce of 81,000 residents) travel from the Illawarra to Sydney every day (Bureau of Infrastructure Transport and Regional Economics [BITRE] 2012). Around 15 percent of these people take the train. Many commuters are from Wollongong, the region's main city.[2] The journey from Sydney to Wollongong takes an hour and thirty-five minutes.[3] On this late afternoon in early summer 2013, the atmosphere in the below-ground station at Redfern, close to the city center, is fuggy. I board the front "quiet" carriage, go upstairs, and find one of the few remaining seats. The feel of air conditioning is quite beautiful, relieving the humidity outside.

This is my penultimate journey back to Wollongong from Sydney. I have been doing participant observation on this line for the past week as part of a project that is investigating the impacts of commuting on people's lives in Sydney. I have been leaving Wollongong on the 6:56 a.m. train and then returning on the 5:07 p.m. train from Redfern, two of the busiest services. I have been using a notepad on my mobile phone as a way to inconspicuously make brief notes of happenings during each journey. The narrative vignettes in this article are drawn from these notes. They were typed up as soon as I returned to my accommodation in Wollongong that evening while the intensities of the events were still fresh in my memory.

There were many other encounters that I witnessed over the course of the week, but my choice to reenact this specific journey here is based on how it affected me powerfully at the time and how it continued to resonate for me long after the journey itself. Appreciating how our capacities to perceive and sense things are shaped by our previous experiences, my re-enactment of these events in this account is affected by the other fieldwork that I have conducted in Sydney as well as reading that I have undertaken. Nine months prior to this "week in the life" experiment, I had conducted in-depth interviews in Sydney with fifty-three commuters who identified with the concept of stressful commutes. Subsequent close reading and analysis of these transcripts helped to attune me to some of the enablements and constraints voiced by commuters in Sydney (cf. Vannini and Taggart 2013). The reflections in these interview encounters (Bissell 2014a) sharpened my apprehension of the way that the events reenacted here might be experienced by others present.

The Macropolitics of Longer Durational Social Formations

"Do you think they're fake eyelashes?" one of the boys quips at a woman, turning back to the more intimate vestibule area between the upper and lower deck. A dozen

or so passengers are sitting side-on to the direction of travel facing inwards towards the boys. They have resumed their swinging around the yellow pole in the centre of the carriage. One is holding onto the yellow ceiling grab-rails, lifting himself and straining. Their unsuspecting audience looks unenthused. But the boys continue their jibing unperturbed. "Do you laaaiyke ma haircut?" the boy in the white top rhetorically asks the vestibule, drawling "like" into almost two syllables and flashing a wide smile. His eyes dart between the two rows of faces that are doing well to suppress a reaction. "A dirty mullet?" he adds, as if to both pre-empt his audience and mark himself out as different from them. "A dirty mullet!" "A dirty mullet!" he repeats, dropping the final t, each time a little slower than the last, nodding calmly. An over-cheery recorded voice relieves the scene: "This train will stop at Hurstville." One of the women in the vestibule stands to leave and others follow. The automatic doors close tightly again. I feel a twinge of panic that the four boys are going to come upstairs to where I'm sitting. I contemplate changing carriages but it's quiet for the moment. The air conditioning roar provides reassuring comfort, broken suddenly by laughter. A woman enters the vestibule from downstairs. "You don't have to run away!" one of the boys call to her. "I've been sitting in an office chair all day, that's alright," she says back, sounding nonplussed and tired. More laughter.

As Cresswell (2006) argues, "The way people are enabled or constrained in terms of their mobile practices differs markedly according to their position in social hierarchies" (199). Reflecting this important contention, research on the politics of mobility has examined how different axes of social inequality such as class (Ohnmacht et al. 2009), gender (Uteng and Cresswell 2008), race and ethnicity (Silvey 2005), and sexuality (Nash and Gorman-Murray 2014) can enable or constrain people's mobility. In this vein, one way of interpreting the politics of the preceding event would be to interpret some of the longer durational social formations that might be discernible here.

First, there is classed inequality. One boy refers to a haircut (mullet) that is associated with a specific working-class Australian identity (see Pini, McDonald, and Mayes 2012), which serves to mark them out as different from the city workers. Furthermore, the boys' term "gold-diggers" in the opening vignette and financial greed that is implied could be interpreted as a reflection on their own economic and political disenfranchisement. This could be read as symptomatic of the problem of youth unemployment in the Illawarra region (Burrows 2010). Second, there are social inequalities among the commuters themselves. These worker hierarchies are diagnosed by the boys from their postures, comportments, and clothing ("He is/She isn't"). The greater number of males in this train carriage points to gender inequalities. Law (1999) argues that where male journeys are often planned for, women have disproportionately more responsibility to undertake the ad hoc "triangular" journeys associated with child care and household management. Third, there are inequalities between different modes of transport along this commuting corridor. Infrastructure investment has prioritized roads over public transport in New South Wales over the past century (Thomas 2014). Therefore, those who cannot afford toll roads and parking in Sydney, or those who do not drive, have longer commutes and have to withstand confrontations such as these. This aligns with O. Jensen's (2013) argument that policymakers can intensify social inequalities by perpetuating certain interests "from above." Fourth, there are inequalities between these commuters and people who live closer to their workplace. This form of inequality is described by Lefebvre (1984), who argues that because the commute constrains workers' time, it demonstrates the ongoing power of capitalism over workers. Bærenholdt's (2013) concept of "governmobility" (29), which describes how the regulation of mobilities occurs through the ways that they are internalized through practices, is clearly pertinent here because it adds weight to the contention that mobilities "constitute a significant stratifying force through which unequal life chances are being continuously reproduced" (Manderscheid 2009, 7).[4]

The Micropolitics of Barely Perceived Transitions in Power

These axes of enablement and constraint are a powerful way of evaluating the politics of mobility. They are important because, as Wolff (1993) argues, the "suggestion of free and equal mobility is itself a deception, since we don't all have the same access to the road" (253). They reinforce Massey's (1993) contention that the enhanced mobility of some comes at the expense of the reduced mobility of others (see also Cresswell 2010). This "macropolitical" interpretation based on longer durational social formations is only one way of interpreting politics in this unfolding scene, however.

The events themselves and the specific way in which they unfold can complicate these apparently more rigid social formations. This is not about redistributing power among the people in the carriage; for instance, suggesting that the boys actually have more power than a macropolitical interpretation might imply. Rather, it is about interpreting the power of the event itself in a way that appreciates the transformative impacts of the multiple forces at play. These forces include the sound of the boys' playful intonation and timbre of voice, alternations of silence and noise, and the repetition of phrases over and over.[5] They include the gestures of swinging and circling around the pole. They include the action of the opening and closing of doors. They include the juxtapositions of the heat of the boys' taunts and the cool automatic announcements. They include the fatigue in the woman's retort. Each of these elements, and more, contributes to the *in situ* enablements and constraints that are created by this unique event. Each of these elements builds on and combines with each other. There is no script.

Micropolitics refers to the barely perceived transitions in power that occur in and through situated encounters. This makes it an ideal concept to think about how this is an event that might have powerful consequences through the way that it transforms relations of enablement and constraint. For our purposes here, an instructive understanding of micropolitics comes from Deleuze's (1990) interpretations of the writings of Spinoza. The key point that I draw from them is that every encounter a body has increases or decreases its capacities to do things and to sense things. This observation has some important implications for our understanding of the bodies that are involved in this event. First, a body becomes understood in terms of its capacities to do things and to sense things, rather than just by its more rigid lines of identity. Second, if a body's capacities to do and sense are changed by the encounters that it has, rather than understanding these passengers as self-contained entities that are being displaced through space by the train, bodies are always transforming, however subtly, in relation to what is happening around them.

Different qualities of encounter do different things. Some encounters are constructive and might enhance a body's capacity to act, which Spinoza called joyful encounters. Others are destructive and diminish its capacity to act, which Spinoza termed sorrowful. What this means is that encounters are not innately good or bad according to a predetermined logic. Rather, a body's power, understood here in terms of its changing capacities to do and sense, depends on the exact unfolding of the event itself. Each encounter, however big or small, is therefore a recomposition of the capacities of those involved. These encounters do not have to be monumental. As Massumi (2009) says, "It doesn't have to be a drama. It's really more about microshocks of the kind that populate every moment of our lives. For example a change in focus, or a rustle at the periphery of vision that draws the gaze toward it" (2). Indeed, the vast majority of encounters that our bodies have are not perceived by our consciousness. Yet this does not stop them from affecting us at the subconscious level. These encounters are microperceptions whose effect might only become consciously registered much later on. Furthermore, every encounter is full of these microperceptions that hit us before we are able to consciously make sense of them through our habitual modes of interpretation. It is these microperceptions that cut in and change our capacities to act, priming future moments by creating new potentials (Massumi 2009).

Rather than taking the more rigid identities of subjects as the starting point for analysis, the concept of micropolitics helps us to appreciate how multiple forces are actually creating these subjects here and now. For example, the space itself alters bodily capacities to do and sense things. As the carriage doors close at Hurstville station, this space becomes a sealed capsule with little escape. This potentially transforms the way that the unfolding events are sensed, heightening capacities for anxiety. So it is not just the boys' actions that have power here but how their performances become heightened by the enclosure of the carriage itself. The performativity of the space becomes intensified by the boys' actions. Undertaking this research project invariably altered my capacities to sense this event through a mixture of anxiety and curiosity. Correspondingly, different commuters' capacities will be affected by the stream of previous encounters that have led up to this moment.[6] Most commuters in this carriage are returning from a day at work in the city. Faces are etched with signs of tiredness. For one person, this event was the final straw, a body affected by a teeming mass of microperceptions that pushed a tolerance threshold to its breaking point.

"BOYS!" roars a man's voice from downstairs. "IF YOU WANT TO TALK, F**K OFF!" The surface tension has broken. There is a momentary drawn-out silence from where it's hard to imagine what will happen next. "F**k off!" one of the boys stammers back to him, somewhat clumsily, as if taken aback and still processing this unanticipated response. Following his lead the others join in taking turns and the downstairs carriage erupts into a volatile zone of choppy, overlapping scatter-gunned obscenities. "You f**king pommy c***s!" "You dirty f**king dog mate!" "You faggot!" The older man sitting next to me, whose gaze has been through the window until this point as if lost in thought, turns his head very slowly, looks straight ahead, and with wearied sarcasm says, "You're really smart guys," raising his eyebrows. My heart is racing. It occurs to me that they might come upstairs and start on us. "YOU in the glasses! You DOG!," one of them shouts. "DOG!" "Daaaawg!" "Daaaawg!" "I'll break your jaw," another shouts. "Yeah, yeah everyone's looking now." "Wait 'til my brother comes to get me at the train station at North Wollongong." "Why are you recording us?" "You f**king paedo!" "Let's steal his iPhone." The recorded voice, oddly indifferent to this onslaught, optimistically announces that "This train will stop at Sutherland." Upstairs, a man sitting across from me answers his ringing phone softly. "Yep, carriage O-D-six-nine-two-four. Just coming into Sutherland. Yep, four boys." A young woman sitting adjacent to me smiles at him weakly. "Yes. ... I can't really. ... I didn't see their faces." The shouting continues downstairs. "Get a haircut you dreadlocked c***." The train doors open at Sutherland and a man with glasses in a crisp, white shirt, perhaps the man who lost it with the boys, steps onto the platform, takes a deep breath, waits, and is met by two guards who have walked down to this end of the train. The four boys step onto the platform, turn to the left and nonchalantly walk away. The man gives a card to the train guard. The train guard is holding a translucent purple plastic toy gun.

Analyzing the politics of this event according to longer durational social formations such as identity alone fails to account for the micropolitical transitions that are actually taking place in this moment. This heady mix of expressive forces is affecting bodies in the here and now. The anger and frustration in the volume of the man's yell startles the boys, momentarily constraining them. But this momentary incapacity seems to agitate them further, the transition in their capacities palpably felt through the newfound excitability and resentment of the boys' razor-sharp profanities. At the same time, the volatility of this event creates new constraints for others in the train carriage, unsure about how to respond, perhaps intensifying a sense of claustrophobia. It also creates new capacities for other people in the carriage to reach out. It enables consolation between me and the older man sitting next to me, as he turns to me to express his weariness. It enables an intimacy between the man on the phone and the woman adjacent who smiles at him. Witnessing the purple toy gun at the end, which adds a sense of barminess to this strange drama, changes my capacities to grasp what is actually going on. It is in this event's precise unfolding that relations of enablement and constraint are transforming.

Micropolitics refocuses attention from pregiven differences to the moment-to-moment transitions in power that give rise to difference. This invites us to appreciate the emergence of differences that mobility practices such as commuting create, rather than orienting analysis around the different identities of the people within the carriage. Empirically, these differences might be difficult to discern. If the people in the carriage look like the same people after this event, however, it is only the image of those same bodies (and myself) that persists. The capacities that make up the bodies in this scene are changed, however subtly, by the encounters that they have had.

Encounters are what make fieldwork in these spaces so vital, because fieldwork attends to the uniqueness and specificity of an event. Encounters that might appear to be the same play out differently every time because the world has moved on, relations between bodies and their surrounds have changed. Different potentials resonate and interfere, changing what happens, however minutely. Although commuting journeys might appear to be the same, day in, day out, every journey is unique, overlain with the difference that every previous experience makes. Bodies are, of course, prepopulated by the lived memory of the past through habits, skills, and inclinations (Dewsbury 2011; Bissell 2014b). These inclinations will be reconstituted, however slightly, by the encounters that those bodies have.

The Entanglement of Micropolitics with Macropolitics

In policy debates, it is commuting's macropolitical dimensions that receive the most attention (Shaw and Docherty 2014). The assumption is that political action, properly understood, takes place through the maneuverings of national-, state-, and city-level institutions that have the capacity to build infrastructures and manage transport operations. Indeed, there has been some criticism that mobilities research has not taken enough interest in macropolitical issues. As Bærenholdt (2013) laments, "There is a dominating interest in, if not even fascination with, the microsociology and phenomenology of mobile practices rather than macro issues" (20). His claim is that although "there is a common, interdisciplinary, awareness and attention to the concrete mobile social practices of people in the world ... more fundamental political issues about the making and governing of societies are absent. ... When dealing with issues of power, hegemony and social order, mobility studies are rather vague" (21).

The vital point to make here, however, is that the macropolitics of larger scaled and seemingly longer durational social formations is absolutely inseparable from the micropolitics of barely perceived transitions in power. Parnet's discussion with Deleuze provides an explanation of why this is the case. They discuss how individuals can be thought of as made up of different sorts of lines, of which two are important for this discussion. On the one hand, there are the "more visible lines" (Deleuze and Parnet 2002, 124), which are the rigid "molar" differences that organize us in different ways. These are the segments that often come to define our identity, such as occupation, age, and gender. On the other hand, there are lines that are "molecular." These lines "trace out modifications, they make detours, they sketch out rises and falls. ... But rather than molar lines with segments, they are molecular fluxes with thresholds" (124). These are the rises and falls that are reenacted through the scenes in this article.

Deleuze, however, argues that although these "supple" lines are "the connections, the attractions and repulsions, which do not coincide with the segments," it is vital to realize that they "nevertheless relate to [them]" (Deleuze and Parnet 2002, 125). This is a point he reaffirms in his writing with Guattari when they emphasize that "every politics is simultaneously a macropolitics and a micropolitics" (Deleuze and Guattari 1992, 213). It is the entanglement of the rigid with the more supple lines that is important here. In the opening scene of this article, for instance, it is a specific molar identity of classed privilege that is being expressed by the boys ("Gold-diggers!"). But it is through the precise playing out of this expression—for instance, the gesturing and the threatening tone of voices within an enclosed space—that molecular, micropolitical transitions in power happen, where "postures, attitudes, perceptions and expectations" (Deleuze and Guattari 1992, 215) are shaped. Therefore, although it might be tempting to focus our attention on just the seemingly larger scaled and longer durational social formations associated with the more rigid lines of social identities and institutional powers, the significance of a micropolitics of mobility comes from the fact that these barely perceived transitions push at the limits of the ways in which the everyday is held consistent by these more rigid lines.

Although Bærenholdt (2013) contends that microsociologies of mobile practices avoid "more fundamental political issues about the making and governing of societies" (21), micropolitical moment-to-moment shifts in capacities are not politics on a smaller scale. Macropolitical decision making is also concerned with changing capacities. As Massumi (2009) succinctly points out, "It would be naïve to think that [micropolitics] is separate from that kind of macro-activity. Anything that augments powers of existence creates conditions for micropolitical flourishings" (19). Investment in new public transport infrastructures is vital, but as Massumi reminds us, "Success at the macropolitical level is at best partial without a complementary micropolitical flourishing" (19). As the transitions that take place through the encounters in this article show, macropolitical decision making is always complicated by virtue of there being a much greater range of forces at play.

Macropolitical decisions regarding quiet train carriages, police and transit officers, closed circuit television, and the playing of announcements—all of which were at play in the commute described in this article—will always have micropolitical dimensions through the ways that they give rise to ongoing barely perceived transitions that enlarge or restrict capacities to do and sense things. This entanglement of the macropolitical and the micropolitical could, for instance, be traced through the history of quiet carriage provision on commuter trains in New South Wales. Collectively experienced frustration about noise on long-distance commuter trains reached a tipping point that mobilized a commuter-led grassroots campaign in 2012 to try quiet carriages on these trains (see Hughes 2014). The decision by Transport for New South Wales to formally instigate quiet carriages across the entire long-distance commuter train network in 2013 consequently changed the conditions for how encounters are sensed in these carriages.

Although it is often tempting to diagnose how events might be the product of dominant powers, Adey (2010b) warns that there is "a danger that we see mobility strangled, sorted and denied, or overdetermined as an outcome of some other practice, regulation or policy" (119). From this top-down perspective, events such as those in this article can only ever be understood as small openings of resistance that chip away at a dominant powers but only after power has had its way (Hynes 2013). A micropolitics of mobility refocuses attention away from the lures of dominant powers (e.g., architects, designers, planners, and policymakers) toward an appreciation of the more complex, emergent distribution of forces that play out in unpredictable ways from moment to moment, altering powers of existence. A micropolitics of mobility is concerned with how an event's openness comes from multiple forces that are immanent to the event itself. So rather than orienting analysis around rigid lines of longer durational social formations, a micropolitical focus encourages us to evaluate how the moment-to-moment transitions in power can disrupt, or indeed strengthen, these more rigid lines.

Conclusion

Because the twenty-first century will be the first properly urban century, understanding the significance and dynamics of commuting practices is a vital challenge of our time (Lin 2012). In Sydney, as is the case in many other cities, debates on addressing commuting problems tend to revolve around formal government responses to commuting problems, such as the construction of new transport infrastructures. Acknowledging the simple but significant fact that "to move is to be political" (Adey 2010b, 131), however, mobilities researchers have shown that such debates often draw on a restricted understanding of politics. Expanding the scope of the politics of mobility, mobilities research has examined how mobilities are socially differentiated and therefore how systems of mobility are responsible for both intensifying and alleviating different forms of inequality (Cresswell 2006). Recognizing that the movement of some people comes at the expense of the movement of other people is essential if we are to avoid the pitfalls of abstract and universalizing analysis.

I have argued that we can push our understanding of mobility politics even further in light of theoretical moves within the discipline that have questioned our long-held image of the durability and stability of the bodies caught up within these mobilities. Such theories provide us with a more indeterminate understanding of bodies, defined instead by their changing capacities to do and sense things, rather than just their rather more rigid molar identities. This shifts our attention from just those pregiven differences to the actual events and encounters that give rise to difference itself. Capacities are transformed, however subtly, by the experiences that bodies have and the spaces in which these experiences take place. This more immanent way of understanding the location and operation of power in mobilities provides an exciting response to "the still very open question of the role of power in mobility" (A. Jensen 2011, 255). This immanent sense of politics is beautifully expressed through Stewart's (2014) description of emergent systems. She describes how "its elements are thrown together not through the conspiracy of a state power, or a pre-existing common ground or ideal, but through events of articulation, histories of use, unintended consequences, and experiments that register. Its lines of force propel forward, spread laterally, and diverge into distinct trajectories" (550).[7]

The significance of micropolitical transitions are, of course, not limited to public transport (see Bissell 2015). The distinctive spaces of public transport, where different people are encapsulated in close proximity, do introduce a specific set of concerns (Adey et al. 2012). For example, in the context of concerns about security on public transport, encounters such as these might do little to assuage such anxieties. The intention here, however, is not to point to a security deficit that needs fixing. Rather, it is to suggest how such events cannot necessarily be anticipated because they emerge in unique ways in response to specific *in situ* conditions. Retrospective analysis could attempt to identify some of the *in situ* triggers that shaped the event.[8] Yet such suggestions would remain inadequate because they could never account for the sheer multiplicity of forces acting together in the event, giving it both momentum and volatility. Where a macropolitical reading of this event might interpret it as a clash of privileged and disenfranchised bodies, such demarcations fail to account for the micropolitical transformations that were taking place there and then, changing relations of enablement and constraint in the process.

The point is that we need to be wary about imagining that it is through macropolitical action where mobility politics is really determined. A decision in 2012 to replace traveling transit officers in New South Wales with spot police checks on board trains no doubt subtly altered the conditions in which this event took place, but this would be one dimension of the event among many others. It is an event that continues to affect in peculiar ways, folded through other experiences, even as I write this now.

The train is now deep into the Royal National Park (or "Nasho," as it is known in these parts), the vast tract of coastal heathland and littoral rainforest between Sydney and the Illawarra. The sun is now much lower in the sky and the eucalypts next to the track are creating a chiaroscuro effect in the carriage. Did the boys get back on the train? Their destination North Wollongong is still to come. The train stops at Engadine and through the twittering of rainbow lorikeets I hear police sirens on the freeway parallel to the railway. Are they coming for the train? Will they stop at Heathcote? A rabbit scurries to safety as we stop. As we pull out from Waterfall station I'm still feeling a bit shaken. I fantasise about what I might have done if I was sitting next to the man in glasses and whether I have confronted the boys. A compendium of much-too-late retorts dart across my mind. At Helensburgh the man who called the police gets up and walks down the carriage. I nod slightly unexpectedly to him. The lady adjacent to me resumes her Flight Centre careers quiz. I notice that she has to link landmarks with countries. The binder announces: "Training for a career that will take you places." The woman next to her is playing Bejewelled on her phone.

Acknowledgments

I am grateful for the comments and kind suggestions offered by audiences at the University of Western Australia, the University of Western Sydney, the University of Tasmania, and the University of Waikato. I would like to thank Elaine Stratford for giving me the opportunity to present this paper as a public lecture in Hobart, and Peter Thomas for providing comments on a draft version. I would also like to thank Mei-Po Kwan and Tim Schwanen for their editorial support, and the three reviewers for their supportive recommendations.

Funding

This research is part of a larger project titled "Stressed Mobilities: Understanding the Significance of the Commute for City-Workers" (http://commutinglife.com), funded by an Australian Research Council Discovery Early Career Award (DE120102279).

Notes

1. Nonrepresentational theories are a diverse body of geographical thought that do not prioritize the role of representation when making claims about the composition and enactment of social life. They offer ways of

thinking about how our lifeworlds are made up of all kinds of different actors, agents, and forces with different capacities, expanding what counts as both the social and the human. It is largely through focusing on embodied mobile practices that nonrepresentational theories have been particularly inspirational for geographical research on mobility. Through very different contexts, McCormack's (2014) research on dance, Vannini's (2012) research on ferry passengering, Adey's (2010a) research on aeromobilities, and Sharpe's (2013) research on parkour each develop nonrepresentational theories to provide new understandings of the power and significance of mobility for our lifeworlds.

2. Around 2.3 percent of the whole of Sydney's workforce is now from Wollongong (BITRE 2012), making it a large dormitory suburb (Waitt and Gibson 2009).

3. This is well in excess of the average one-way commuting time in Sydney of just under thirty-five minutes (BITRE 2012).

4. Such inequalities could read through the processes of economic restructuring that have occurred in this region over the past century. Since the 1930s, Wollongong and the Illawarra was a center of heavy industry, most famed for its steel manufacturing. Global economic restructuring during the 1980s resulted in mass deindustrialization that saw the unemployment rate increase. Much of the Illawarra's heavy industry has since closed down, although the city retains a significant blue-collar workforce. Although the health and education sectors have since grown significantly, the region continues to be influenced by its proximity to the more diverse and cosmopolitan economy of Sydney. The effects of this economic restructuring on commuting practices have been complex. On one hand, there has been a significant in-migration to the area, where creative workers and professionals are seeking more affordable housing and a lifestyle change in a picturesque setting between the Illawarra escarpment and the Tasman Sea (Waitt and Gibson 2009). On the other hand, there has been an increase in people leaving the area each day to work in Sydney. These broad processes of economic restructuring are complicated by the distinctive local politics that this railway line transects (see also Dorling 2013). It traverses through places known for their relative affluence (Loftus) and poverty (Redfern) and areas that are known for their distinctive cultural contexts (e.g., Sutherland Shire, or "The Shire" as it is known locally).

5. It is important to acknowledge the specific role of sound within this event. As Simpson (2009) shows, sound both signifies and affects. Let's take the boy's line, "A dirty mullet," from the unfolding event here. There is certainly a discursive, representational dimension to this line that we could interpret as a reference to his own haircut. To focus on the symbolic dimensions, though, would overlook the complex affective dimension created by the materiality of the sound itself. This is about the boy's timbre of voice and volume within the specific milieu of this carriage at this time. These symbolic and affective dimensions are, of course, entwined. Voiced over and over in the way it is here in the context of this event, "A dirty mullet" is not just a reflexive reference to a haircut. I experienced it as a complex torsion of taunt, irony, disparagement, and humor. It is these dimensions together that give the sound its capacity to resonate in particular ways.

6. In a discussion on how to think about the collectivity of an event, Manning (2013) reminds us of the importance of appreciating that different bodies will, of course, carry different tendencies and capacities. She said, "The point is that the same macro-event creates different bodyings in different ecologies co-constituted by different emerging milieus" (27). What this means for the present argument is that people "will be *responding differently together*, as inhabitants of the same affective environment" (Massumi 2014, 108, italics added).

7. So what this immanent understanding of politics invites, then, is a "shift away from the question of agency per se, towards an ontology in which the capacities to act and to be acted upon are modulated by the relations afforded by a particular milieu across a multiplicity of scales" (Roberts 2011, 2516).

8. For instance, the boys' frustrations of not being able to find four seats to sit together on a packed train after walking the entire length of the eight-car train while it was in motion (I realized later on that they boarded the rear carriage of the train at Redfern Station—microperceptions of the luminous orange top worn by one of the boys); the carriage that they arrived at was designated a quiet carriage; the boredom of having a long journey ahead of them; the high temperatures that afternoon; of being a minority in the carriage in terms of dress and age; and the fatigue of the commuters at the end of the day.

References

Adey, P. 2010a. *Aerial life: Spaces, mobilities, affects.* Oxford, UK: Wiley Blackwell.

———. 2010b. *Mobility.* London and New York: Routledge.

Adey, P., D. Bissell, D. McCormack, and P. Merriman. 2012. Profiling the passenger: Mobilities, identities, embodiments. *Cultural Geographies* 19 (2): 169–93.

Aldred, R. 2014. The commute. In *The Routledge handbook of mobilities*, ed. P. Adey, D. Bissell, K. Hannam, P. Merriman, and M. Sheller, 214–24. London and New York: Routledge.

Bærenholdt, J. 2013. Governmobility: The powers of mobility. *Mobilities* 8 (1): 20–34.

Bissell, D. 2014a. Encountering stressed bodies: Slow creep transformations and tipping points of commuting mobilities. *Geoforum* 51 (1): 191–201.

———. 2014b. Transforming commuting mobilities: The memory of practice. *Environment and Planning A* 46 (8): 1946–65.

———. 2015. Virtual infrastructures of habit: The changing intensities of habit through gracefulness, restlessness and clumsiness. *Cultural Geographies* 22 (1): 127–46.

Bureau of Infrastructure Transport and Regional Economics. 2012. Population growth, jobs growth and commuting flows in Sydney. Research Report 132, Australian Government Department of Infrastructure and Transport, Canberra, Australia.

Burrows, S. 2010. Youth unemployment in the Illawarra region. *The Journal of Australian Political Economy* 65:88–105.

Cresswell, T. 2006. *On the move: Mobility in the modern western world.* London and New York: Routledge.

———. 2010. Towards a politics of mobility. *Environment and Planning D: Society and Space* 28 (1): 17–31.

Deleuze, G. 1990. *Expressionism in philosophy: Spinoza.* New York: Zone Books.

Deleuze, G., and F. Guattari. 1992. *A thousand plateaus: Capitalism and schizophrenia.* London: Continuum.

Deleuze, G., and C. Parnet. 2002. *Dialogues II.* London: Continuum.

Dewsbury, J.-D. 2011. The Deleuze–Guattarian assemblage: Plastic habits. *Area* 43 (2): 148–53.

Dorling, D. 2013. *The 32 stops: Lives on London's Central Line.* London: Penguin.

Edensor, T. 2010. Commuter: Mobility, rhythm, and commuting. In *Geographies of mobilities: Practices, space, subjects,* ed. T. Cresswell and P. Merriman, 198–204. Aldershot, UK: Ashgate.

Graham, S., and S. Marvin. 2001. *Splintering urbanism.* London and New York: Routledge.

Hansen, S. 2004. The context of urban travel: Concepts and recent trends. In *The geography of urban transportation,* ed. S. Hanson and G. Giuliano, 3–29. New York: Guilford.

Hughes, A. 2014. "Super simple stuff?": Crafting quiet in trains between Newcastle and Sydney. Unpublished Honours Thesis, University of Newcastle, Callaghan, Australia.

Hynes, M. 2013. Reconceptualizing resistance: Sociology and the affective dimension of resistance. *The British Journal of Sociology* 64 (4): 559–77.

Jensen, A. 2011. Mobility, space and power: On the multiplicities of seeing mobility. *Mobilities* 6 (2): 255–71.

Jensen, O. 2013. *Staging mobilities.* London and New York: Routledge.

Law, R. 1999. Beyond "women and transport": Towards new geographies of gender and daily mobility. *Progress in Human Geography* 23 (4): 567–88.

Lefebvre, H. 1984. *Everyday life in the modern world.* London: Transaction.

Lin, W. 2012. Wasting time? The differentiation of travel time in urban transport. *Environment and Planning A* 44 (10): 2477–92.

Lyons, G., and K. Chatterjee. 2008. A human perspective on the daily commute: Costs, benefits and trade-offs. *Transport Reviews* 28 (2): 181–98.

Manderscheid, K. 2009. Integrating space and mobilities into the analysis of social inequality. *Distinktion: Scandinavian Journal of Social Theory* 10 (1): 8–27.

Manning, E. 2013. *Always more than one: Individuation's dance.* Durham, NC: Duke University Press.

Massey, D. 1993. Power-geometry and a progressive sense of place. In *Mapping the futures: Local cultures, global change,* ed. J. Bird, B. Curtis, T. Putnam, and L. Tickner, 59–69. London and New York: Routledge.

Massumi, B. 2009. Of microperception and micropolitics. *Inflexions: A Journal for Research-Creation* 3:1–20.

———. 2014. *The power at the end of the economy.* Durham, NC: Duke University Press.

McCormack, D. 2014. *Refrains for moving bodies: Experience and experiment in affective spaces.* Durham, NC: Duke University Press.

Morley, D. 2000. *Home territories: Media, mobility and identity.* London and New York: Routledge.

Nash, C., and A. Gorman-Murray. 2014. LGBT neighbourhoods and "new mobilities": Towards understanding transformations in sexual and gendered urban landscapes. *International Journal of Urban and Regional Research* 38 (3): 756–72.

Ohnmacht, T., H. Maksim, and M. Bergman, eds. 2009. *Mobilities and inequality.* Aldershot, UK: Ashgate.

Pini, B., P. McDonald, and R. Mayes. 2012. Class contestations and Australia's resource boom: The emergence of the "cashed-up bogan." *Sociology* 46 (1): 142–58.

Roberts, T. 2011. From "new materialism" to "machinic assemblage": Agency and affect in IKEA. *Environment and Planning A* 44 (10): 2512–29.

Sharpe, S. 2013. The aesthetics of urban movement: Habits, mobility, and resistance. *Geographical Research* 51 (2): 166–72.

Shaw, J., and I. Docherty. 2014. *The transport debate.* Cambridge, UK: Polity.

Silvey, R. 2005. Transnational Islam: Indonesian migrant domestic workers in Saudi Arabia. In *Geographies of Muslim women: Gender, religion, and space,* ed. G. Falah and C. Nagel, 127–46. New York: Guilford.

Simpson, P. 2009. Falling on deaf ears: A post-phenomenology of sonorous presence. *Environment and Planning A* 41 (11): 2556–75.

Stewart, K. 2014. Road registers. *Cultural Geographies* 21 (4): 543–47.

Thomas, P. 2014. Railways. In *The Routledge handbook of mobilities,* ed. P. Adey, D. Bissell, K., Hannam, P. Merriman, and M. Sheller, 214–24. London and New York: Routledge.

Uteng, T., and T. Cresswell. 2008. *Gendered mobilities.* Aldershot, UK: Ashgate.

Vannini, P. 2012. *Ferry tales: Mobility, place, and time on Canada's west coast.* London and New York: Routledge.

Vannini, P., and J. Taggart. 2013. Doing islandness: A non-representational approach to an island's sense of place. *Cultural Geographies* 20 (2): 225–42.

Waitt, G., and C. Gibson. 2009. Creative small cities: Rethinking the creative economy in place. *Urban Studies* 46 (5–6): 1223–46.

Wolff, J. 1993. On the road again: Metaphors of travel in cultural criticism. *Cultural Studies* 7 (2): 224–39.

Mobility Disadvantage and Livelihood Opportunities of Marginalized Widowed Women in Rural Uganda

Deborah Naybor,* Jessie P. H. Poon,† and Irene Casas‡

*Department of Environmental Studies, Paul Smiths College
†Department of Geography, University at Buffalo
‡School of History and Social Sciences, Louisiana Tech University

The adverse effect of mobility restrictions on the livelihood of economically marginalized women in rural Africa is considerable. This study investigates the space–time paths of twenty-seven widowed women in rural Uganda through methodological pluralism that integrates multiple sources of quantitative and qualitative data collected from Global Positioning System tracking, in-depth interviews, and participant observation. Geographic information systems mapping of activity space suggests that mobility patterns are characterized by frequent short repetitive trips and less flexible space–time budgets. In turn, this reduces opportunities to pursue diversified sources of income that enhance livelihood. Statistical regressions and qualitative interviews also show, however, that access to use of motorized vehicles such as cars and motorcycle taxis significantly strengthens livelihood by reducing time poverty, rendering time as a resource for pursuing income opportunities.

能动性的限制, 对于生活在非洲农村经济上边缘化的女性生计而言, 具有不利的影响。本研究透过整合从全球定位系统追踪、深度访谈和参与式观察取得的各种量化与质化数据来源的多重方法论, 探讨乌干达农村二十七位寡妇的时空路径。地理信息系统对于活动空间的地图绘制显示, 她们的能动性模式以经常性的短程反覆旅次和较缺乏弹性的时空预算为特征。这种情形随之减少了追求能够促进生计的多样化收入来源的机会。但统计回归与质化访谈亦同时显示, 拥有利用诸如汽车与摩托出租车的机动车辆之管道, 透过降低时间贫穷, 显着地增加了生计, 而时间则成为追求所得机会的资源。

Es considerable el efecto adverso que tienen las restricciones de movilidad sobre la subsistencia de mujeres económicamente marginadas del África rural. Este estudio investiga las trayectorias espacio-temporales de veintisiete viudas en la Uganda rural mediante un pluralismo metodológico que integra múltiples fuentes de datos cuantitativos y cualitativos obtenidos por rastreo del Sistema de Posicionamiento Global, por entrevistas a profundidad y observación participativa. El mapeo de la actividad espacial por sistemas de información geográfica sugiere que los patrones de movilidad están caracterizados por la frecuencia de viajes cortos repetidos y por presupuestos menos flexibles del espacio-tiempo. A su turno, esto reduce las oportunidades para buscar otras fuentes de ingreso que mejoren la subsistencia. Las regresiones estadísticas y las entrevistas cualitativas muestran también, sin embargo, que el acceso al uso de vehículos motorizados, tales como automóviles y mototaxis, fortalece significativamente la subsistencia al reducir la pobreza del tiempo, convirtiendo al tiempo en un recurso para buscar oportunidades de ingreso.

Transportation and mobility are relevant influences on livelihood opportunities among marginalized populations in rural Africa, whose most common mode of transportation is walking. Lack of infrastructure and accessible affordable transportation in sub-Saharan Africa creates a widening mobility gap. The divide between those who can afford and access transportation and "Africans stranded in rural villages where mobility deprivation is acute" (Pirie 2009, 22) limits livelihood strategies that affect economic survivability adversely. This article examines time use and mobility of a sample of marginalized Ugandan women in the context of their economic activities. We attempt to understand the relationship between mobility and livelihood by engaging in methodological triangulation of geographic information systems (GIS), quantitative data, and qualitative data.

The subjects of this study are widowed women in remote rural Uganda. These women are extremely poor and most rely on subsistence agriculture for survival in difficult conditions. Ellis (2000) theorized that livelihood, the activities required for survivability and satisfaction of basic human needs, is less uncertain among the poor if they pursue a diversified range of income sources besides farming. Mobility constraints for pursuing diversified income are also formidable, however. Using methodological pluralism that favors multiple sources of data and quantitative as well as qualitative approaches, we analyze the patterns of everyday mobility among the women and the influence of mobility on their livelihoods.

In the next section, we frame mobility within the time geography literature, highlighting constraints on human movement. We then detail the research methodology, including the recruitment of women and collection of Global Positioning System (GPS), survey, and qualitative data. This is followed by an elaboration of the findings before concluding with implications of the analysis.

Space–Time and Mobility

Hagerstrand's (1970) time geography focused on the limitations of human activity and the influence of fixed and flexible activities on time use and spatial location (Pred 1977). The framework offers a way to trace daily routines as specific locations of time and space (Miller 2005). The space–time path, which illustrates how people choose their course within their spatial–temporal environment (Corbett 2001), has proven to have greater significance in exposing patterns of human behavior through the use of new communication and information technology (Kwan and Webber 2003; Couclelis 2009). The number of tasks that each person can achieve depends on movement through a web of paths between stations where tasks are performed. Thrift (2005) observed that as individuals allocate certain spaces to fit certain needs, they use "time as a resource, allocating particular intervals to particular uses" (4).

Time use in space is a relevant framework for understanding the relationship between mobility and livelihood in rural sub-Saharan Africa, where lack of ability to pay for transport adversely affects mobility, livelihood, and health care (Porter 2002; Jaramillo, Grindlay, and Lizarraga 2012). Spatial capacity for livelihood opportunities depends on resource endowment, economic opportunity, and social capital (Bebbington 2009). Yet to take full advantage of these opportunities, the poor have to be able to negotiate the distance to access resources and convert them into productive assets. Access to affordable transportation could be limited by many factors, including poverty, poor road conditions, and cultural restrictions, especially in regard to gender (e.g., women riding a bicycle is not culturally acceptable in this

village). For the rural poor in Africa, affordable transportation is usually confined to the bicycle, and motorized vehicle use is limited due to cost. It is important to note, however, that motorized transportation such as the car tends in general to be used less by women even in industrialized countries, as a result of gendered differences in economic power and patriarchal social structures (Scheiner and Holz-Rau 2012).

In Kenya, Oluoko-Odingo (2011) found that better transport networks improved farmers' health through access to better medical care and also helped to stabilize food prices. Temporal constraints on spatial freedom involve choice and access, influenced by a wide variety of actors within a social framework. For example, many of the women in our study expressed a desire to work at a shop but could not be away from their home for hours each day due to reproductive responsibilities. The women have learned to work within spatiotemporal confines by developing survival strategies that influence livelihood under harsh physical conditions.

Walker and Vajjhala (2009) found that gender dimensions of social marginalization and vulnerability can be evaluated in terms of transport using GIS data. Yamano and Kijima (2011) used GIS data from 894 rural households in Uganda and found that perishable and high-income crops are more at risk of damage on long, rough roads to illustrate the importance of transportation and mobility on the ability of rural farmers to increase their livelihood levels. Taken together, the "mobility of poverty" (McQuoid and Dijst 2012, 27) is a pressing problem that demands scholarly attention if impoverished women in Africa are to improve their livelihood.

Study Area, Data, and Methodology

Shoval (2008) maintained that data collected on time and travel are often distorted because of the laborious process of logging time diaries. For this reason, we use GPS tracking to more accurately track our subjects' space–time paths and to understand the impact of mobility patterns on livelihood. United Nations Economic Commission for Africa studies found that high illiteracy rates in Africa call for face-to-face interviews rather than the use of diaries. Even collecting minimal accurate time diary information can be difficult, though, due to budget limitations and a "general lack of adherence to rigid time schedules by African people, particularly in rural areas" (Abdourahman 2010, 27).

163

Twenty-seven widowed women, age twenty-three to seventy-three, volunteered for the study. They carried GPS tracking devices over a period of three months, for an average of twenty days per participant. The storage of data is triggered by motion and stored on-board until downloaded. To assure ethical human subject compliance, the women controlled which days they collected data and for how long. This ensures privacy as participants were reluctant to disclose trips, for example, to an HIV/AIDS clinic or illegal alcohol vendor.

The women were recruited through local community leaders and were compensated at the rate of $1 for every day that they collected data (equivalent to their daily average earnings) and $2 plus a provided meal for participation in interviews. In addition to GPS tracking, participant observation provided data on cycles of activities, social structure, and temporal patterns during three two-week intervals within a one-year period. The home and farm of each woman were visited several times, as well as the local marketplace, churches, clinics, schools, and water sources. A two-hour semistructured interview with each participant gathered both quantitative and qualitative data on education, social life, family, livelihood, health, transportation, and time use. Livelihood factors included questions that allowed for assessment of the individual's level of subsistence and market agriculture and other methods of earning income. Triangulation of multiple sources of data, both quantitative and qualitative, enhances methodological pluralism that has come to underscore much geographic research (Poon 2005).

Women who have been widowed were chosen because they are among the most marginalized groups in Uganda (International Fund for Agricultural Development 2012). The Uganda Bureau of Statistics (2010) reported that over 79 percent of widows are dependent on subsistence farming and lack physical and financial resources, increasing their poverty vulnerability. In addition, the subjects are more willing than married women to provide information on their activity spaces and livelihoods and act more independent of male-dominated influence on time use and decision making. Evers and Walters (2000) found that men in sub-Saharan Africa dominate women's inputs in cash crops. Although a small group had remarried or found male companions since being widowed, they indicated that they had entered into the relationship with more decision-making power and economic independence.

The research was conducted in the village of Nakagongo in south central Uganda. It is a rural community that is highly dependent on agriculture and subsistence farming. The nearest paved road is approximately eight miles from the village in the town of Kyotera, which has a population of about 8,000. There are small informal and formal shops at scattered market areas in nearby villages, with the nearest small town, Bethlehem (population estimated at less than 1,000), acting as a trading and community center approximately three miles southwest of Nakagongo. One third of the study participants live near the local market, and the rest are located off-road. Porter (2002, 291) aptly noted that "to live off-road is to be invisible." The Uganda National Household Survey 2009–2010 reported that only 19.6 percent of Ugandan communities have paved roads, and 18 percent of towns or villages have a bus stop (Uganda Bureau of Statistics 2010). No public transportation is available in the area, but there are infrequent minibuses for hire in Bethlehem and larger buses from Kyotera for hire for longer trips. Bicycle or motorcycle taxis operated by local men are available in the village area but the fees are $0.50 to $1.00, almost an average day's wage for many of the participants.

Space–Time Analysis of Rural Ugandan Widows

Space–time paths were created from ArcGIS using the widows' GPS tracking data. The paths were constructed by connecting the sorted temporal sequence of all the locations visited by a subject. Given that GPS tracking data were collected for different periods of time, we used the longest period of consecutive data of each subject to calculate the space–time path. Time budget was examined by analyzing time spent in various economic, household, and social activities. Specific attention is paid to productive (income producing), reproductive (household) work, and patterns of spatial fixity. As the geographic literature has shown, women's reproductive work influences their daily patterns of mobility (Hanson and Pratt 1995; Schwanen, Kwan, and Ren 2008; Kwan 2012; McQuoid and Dijst 2012) and thereby impacts livelihood. The women farm and sell excess crops, and they raise livestock such as goats, chicken, and pigs. They also engage in informal work (e.g., crafts and sewing) and waged work in the rural labor market. A diversified range of activities decreases livelihood uncertainty: The more diversified the workers' sources of income, the better their survivability.

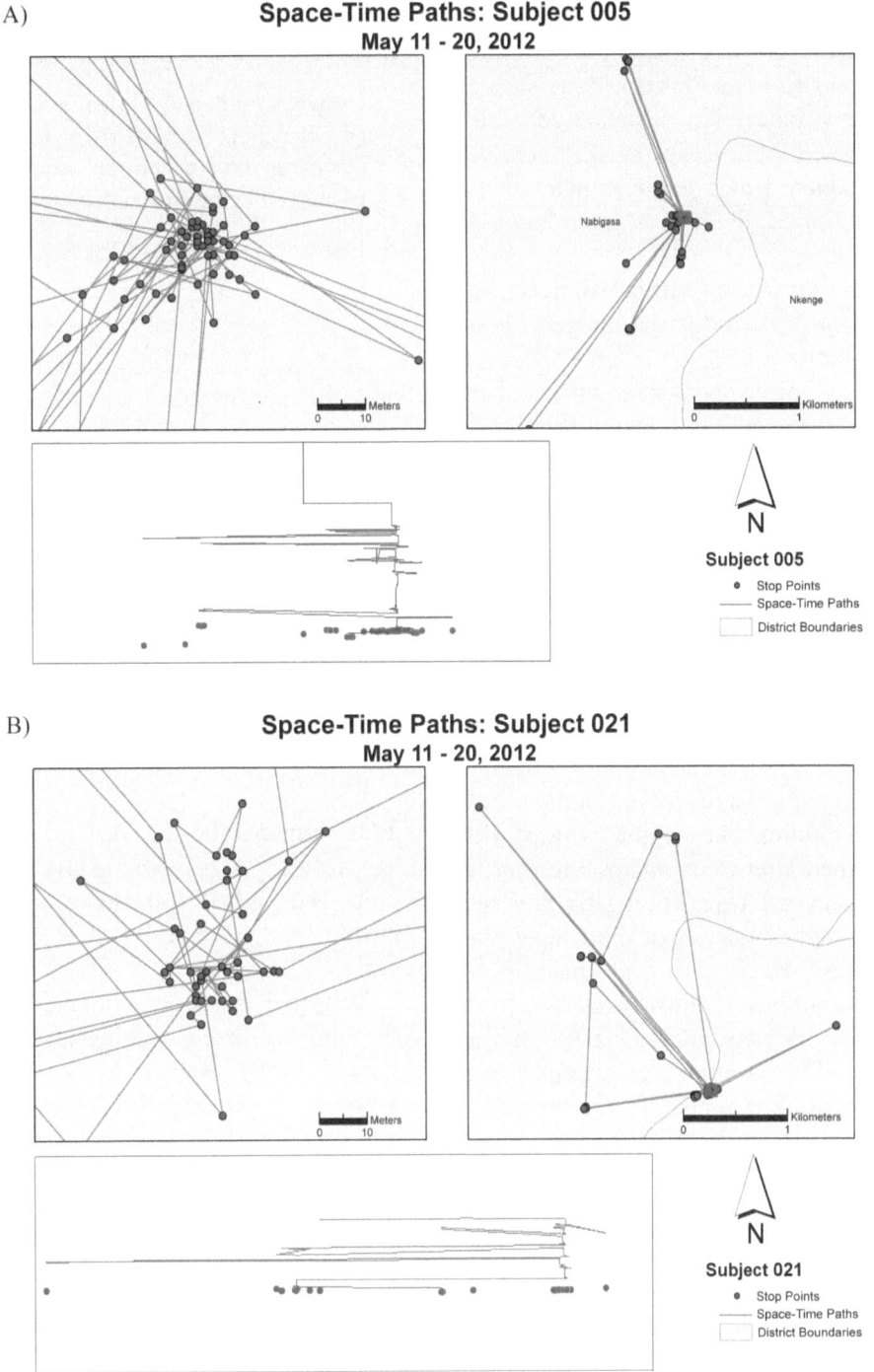

Figure 1. (A) Less diversified livelihood. (B) More diversified livelihood. (Color figure available online.)

In Figure 1, we present space–time paths of two women. Figure 1 represents the average of their "less diversified" and "more diversified" livelihood groups: The first, Subject 5, is characterized by fewer livelihood activities, whereas the second, Subject 21, has a wider range of income opportunities. The most frequently visited point is each woman's home, with multiple trips in and out of their home base, but Subject 5 is also characterized by a higher spatial fixity. The upper right side of each graphic shows a wider range of movement for Subject 21, with less concentrated movement close to home (upper left) than Subject 5. This variance in the range of mobility also appears in the time prism diagram at the bottom of

Figure 1A and 1B. The vertical lines of the time prism for Subject 5 indicate more time without motion and more time spent closer to home. It also shows shorter trips within a closer distance. This woman's activities are mostly constrained within an areal extent of around 230 × 115 square feet. For the woman with a more diversified livelihood, the area is much larger at 755 × 345 square feet.

Figure 1A implies that women with a less diversified livelihood spend more of their "spatial budget" closer to their homes. Repetitive trips are often conducted for a single activity. One explanation, supported by ethnographic observation and interviews, is that limits on the women's physical carrying capacity of goods or cargo will require multiple daily trips to move goods from home to the point of sale. This diminishes the women's physical energy and increases time poverty by reducing time for more productive work or participation in the rural labor market. Because the women report that they are time poor due to reproductive responsibilities, they often sell their crops and other goods to friends and neighbors who are close by but pay a low price. On a day of the authors' visit, one participant made several trips to a neighbor's home to sell eggs to her buyer. Finding her neighbor away, the woman merely returned later to try again. She walked back and forth for several trips hoping to find her buyer at home. The high frequency of short repetitive trips points to mobility information gaps arising from lack of capacity to gather buyer information.

Transportation factors were analyzed next in the context of mobility. The twenty-seven women were asked about their modes and usage of transportation. All of the women reported walking as the most common mode of transportation, even when they are extremely ill. Approximately half of the women are able to hire a bicycle taxi (BIKE) within the village or to a nearby town approximately once a week or more. One quarter are able to hire a motorcycle taxi (MOTORBIKE) once a week or more. Bus and car use are infrequent, though, with less than 20 percent reporting usage several times a month.

Bicycle taxis are common in the Nakagongo area, but almost one third of the women cannot afford to pay for any form of transportation on a regular basis. Take the case of two participants who are thirty-nine (12) and forty (3) years old, respectively. Both are engaged in subsistence farming, growing sweet potatoes and bananas, but it is barely sufficient to feed their families. Participant 12 also raises pigs and chickens to supplement her meager means but needs to work on other farms to earn extra money. Ability to diversify one's livelihood depends on a more flexible spatial budget:

> I travel to different places by foot because I don't have any bicycle at home. I try to use a bicycle taxi on long distances and I would use the bus but cannot use these because of being poor. (Participant 12)

> I do walk to other people's homes in order to earn some money for my family but it is too far away. I can't go every day. (Participant 3)

Access by bicycle taxi is used within the village area and the motorcycle taxi is used to travel to larger towns farther away. The participants reported cost as the biggest deterrent to the use of paid transport. Restricted mobility arising from lack of access to wheeled transportation adversely affects not just livelihood but also the ability to meet basic human needs such as health. One woman, age twenty-nine, with very young children informed us that

> Sickness has become a routine in our family and at times when someone is totally ill, we have no quick means of transport, which puts someone's life in danger. (Participant 26)

This woman relies on her eldest child, four years of age, to fetch water. She is unable to sell much of her crops (i.e., peanuts and corn), as the demand lies in farther markets. She reported that she has never ridden in a car or bus.

Wheeled transportation reduces physical exertion through decreased walking and ease of load transfer (Calvo 1994). Reduction of fatigue improves work capability and health. Forty percent of the participants reported that they were unable to work for at least five days per month due to illness, and many complained that they are often too sick or weak to travel to the field or market. The motorcycle taxi is relatively widespread but is outside the budget of most of the women for more than occasional use. Although this mode can reduce travel time and decrease physical exertion, it does not increase carrying capacity of goods or cargo that the women badly need to raise the volume of their sales. Access to motorized transport has become a basic need for these women (Lucas 2011).

Approximately 37 percent of women are only able to get a ride in a car once or twice a year to visit relatives or travel to the hospital. Like Participant 26, several reported that they had never ridden in a car before. Those who used car transportation on a monthly basis combined trips to the hospital, visits to relatives, and the trading of goods in the city. The

Table 1. The effect of transport modes on livelihood

Variable	Car	Bicycle	Motorbike	Bus
AGE	−0.086 (0.007)*	−0.129 (0.076)	−0.115 (0.004)*	−0.015 (0.040)
EDUC	−0.004 (0.079)	−0.219 (0.292)	0.002 (0.045)	−0.258 (0.303)
FAM	0.017 (0.052)	0.237 (0.189)	0.005 (0.029)	0.256 ((0.196)
HEALTH	0.029 (0.115)	0.331 (0.424)	0.109 (0.065)	0.350 (0.442)
CAR	3.391 (0.196)*	—	—	—
BIKE	—	−4.347 (3.142)	—	—
MOTOR-BIKE	—	—	9.593 (0.307)*	—
BUS	—	—	—	−0.777 (2.075)
R^2	0.945	0.248	0.982	0.186

Note: Standard errors are in parentheses.
*$p < 0.01$.

majority of the group, though, uses the bus for occasional trips to larger cities more than a couple of hours away or could not afford to make the trip at all.

Interestingly, several women indicated that the cost of transportation and time away from household and reproductive responsibilities would not be worth the potential net income. One woman whose home is located on a main road near the market area is able to sell excess crops in front of her home that fetch up to three times the price of off-road locations. Heavy household and field chores hinder her ability to do so on a regular basis, however. To improve and stabilize their livelihood, rural families must be less reliant on subsistence food production and engage in activities that enhance diverse income sources (Ellis and Bahiigwa 2011).

The mean age of participants is 36.5 years, and they travel an average of 6 miles per day, of which 2.3 miles are in pursuit of livelihood opportunities, with the remainder of their travel dedicated to reproductive work. Many of our subjects are also integrated into family networks that help reduce their reproductive burden. Family capital, however, is a two-edged sword. Although it increases access to resources and opportunities, particularly access to transportation from family members and relatives, some of the women are also the main caretakers of young children and the elderly in the family network. The survey sought data on family capital by asking the subject to rate the level of assistance from children who contribute to reproductive and field chores. The mean of 2.5 out of 4.0 indicates that assistance from children is only slightly above average. This is perhaps surprising given Porter's (2011) and Porter et al.'s (2012) observation that young African girls will often perform chores like fetching water and other load-carrying activities. Participants who gave this question a lower rating felt that their children were too young or sick to be of much assistance, despite their help with household chores. Those whose children were able to help with farming and animal care rated the assistance higher.

To quantify and test the livelihood–transportation relationship more rigorously, we perform regressions on the four modes of transportation. Because the relationship is likely to suffer from reverse causality, we control for endogeneity using two-stage least squares (2SLS) regressions. In the first stage of 2SLS, transport usage of each mode T_i is estimated as a linear regression of livelihood diversification, productive time, and demographics (age, education). Productive time is included to satisfy the identification requirement because the model requires us to include a variable that directly affects T_i. Women who encounter spatiotemporal constraints because of distant fields or markets and greater reproductive activities have a lower productive time so that

$$T_i = L_i\alpha_1 + X_i\alpha_2 + PT_i\alpha_3, \qquad (1)$$

where T_i is the usage per month of transport mode (car, motorbike, bus, and bicycle), L_i is livelihood diversification, X_i is the demographic variable (age, education), and PT_i is productive time.

In the second stage, L_i is estimated as a linear function of predicted T'_i corrected for endogeneity, demographic factors (i.e., age and education), and other relevant determinants (health, family capital). L_i is a count of various livelihood activities. The higher the count, the more diversified and stable the women's livelihood. A low count, on the other hand, implies that the woman has a more uncertain source of income because she is highly dependent on subsistence agriculture. T'_i is estimated for four principal transportation modes: bicycle taxi, bus, motorcycle taxi, and car.

Age and education are in the model because we expect younger and more educated widows' livelihood to be more diversified, as they are physically stronger or they possess a set of skills to pursue nonfarm or waged work. Along with demographic factors, we also expect that better health and a higher family capital (i.e., assistance from children) increase time budget for productive work. In the second stage of the regression, T_i is replaced by predicted T'_I from Equation 1 such that

$$L_i = T_i\beta_1 + X_i\beta_2 + Q_i\beta_3 + \varepsilon_I, \qquad (2)$$

where L_i is livelihood diversification, T'_i is predicted transport mode use (car, motorcycle taxi, bus, and bicycle), X_i represents individual demographic variables (age, education), and Q_i is other determinants of livelihood (health and family capital).

Table 1 reports the results of the 2SLS. Of the demographic variables, only age is significant for both car and motorcycle taxi at the 1 percent level. The estimate is negative, indicating that a more stable livelihood is associated with younger women. This makes sense because many of the older widows we observed lacked the physical strength to pursue a wider range of income-generating activities. Education, health, and family capital do not significantly influence livelihood. Both car and motorcycle taxi use are highly significant at 1 percent, however. The estimates are positive and quite high at 3.391 and 9.593, respectively. The R^2 values of 0.945 for car and 0.982 are also high, indicating a good fit. This is not the case for bicycle taxi ($R^2 = 0.248$) or bus ($R^2 = 0.186$).

Taken together, the 2SLS regressions suggest that access to a motorized form of vehicle, namely, car and motorcycle taxi, increases the women's ability to spend more time in various livelihood activities by increasing the flexibility of their spatial budget. The women are only able to transport small quantities of products to points of sale, as walking or sitting on the back of a bicycle limits the volume of goods ferried. Car travel occurs largely through friends and relatives at little cost, but motorcycle taxis are more widespread and allow the women to travel to nearby towns and villages for market transactions. They provide the participants with much needed mobility, increase their capacity to move goods in larger volumes to and from more distant markets, and offer increased opportunities to pursue waged or nonfarm work located farther away. They reduce time poverty by freeing time up for livelihood activities. They are also not so affordable to the very poor, however; hence, women with lower usage of both cars and motorcycle taxis are vulnerable to less stable livelihoods. Buses are available in the towns but not in the village areas, which might explain their lack of significance.

Overall the analysis in this section highlights two findings. First, women whose livelihood is less diversified also experience less flexible space–time budgets. Second, wheeled transportation strengthens livelihood and enables the women to pursue a wider range of income opportunities but tends to be cost-prohibitive for many of the women.

Conclusion

Pirie (2009) noted that "a person's rank in a spectrum of geographical mobility should not have to be a matter of birthright or location. In Africa of all places, there is reason to devise a new morality of mobility" (30). Integration of GIS data from GPS tracking, ethnographic observation, and participation, as well as survey and interview data, help to provide insight into the realities of gendered mobility disadvantage among widowed women of Nakagongo, Uganda.

Repetitive patterns of movement illustrate spatial restrictions and a less flexible space–time budget among women with less diversified livelihoods. They increase the women's spatial fixity by forcing the women to trade off time budgets between productive work and basic as well as reproductive needs. Participants with less stable livelihoods find little time to tend to their crops and livestock or join the rural labor market when they are confronted with a rudimentary transport infrastructure. Coupled with caring for the elderly and young children, many of them are also time poor. Space–time patterns show multiple daily trips to the market or buyers' homes that verify the reality of life in a walking world. This life is grounded in fragmented economic spaces that offer challenging options for better mobility. Lack of mobility also translates into unmet basic health needs. Regressions confirm that for women who have access to a car or can afford motorcycle taxis, livelihood is more diversified and less uncertain.

Because access to advanced modes might depend on friends, relatives, or partners, this suggests that women who are networked might have better access to transport resources. Given the findings, one future direction for further research lies in exploring the role of social networks not so much as a resource for reducing the burden of reproductive activities but as a mobility resource that directly affects livelihood.

Acknowledgments

We are grateful to the women of Nakagongo who participated in the study. Thank you to the anonymous referees for their insightful and constructive feedback and to Tim Schwanen for his advice as editor.

Funding

This research was made possible through the financial support of the Institute for Research and Education on Women and Gender, the Mark Diamond Research Foundation, and the University at Buffalo Calkins Applied GIS Award.

References

Abdourahman, O. I. 2010. Time poverty: A contributor to women's poverty? *Journal Statistique Africain* 11:16–36.

ArcGIS, Version 10.0, ESRI, Redlands, CA.

Bebbington, A. 2009. Latin America: Contesting extraction, producing geographies. *Singapore Journal of Tropical Geography* 30:7–12.

Calvo, C. M. 1994. *Case study on intermediate means of transport bicycles and rural women in Uganda.* Sub-Saharan Africa Transport Policy Program Working Paper 12, World Bank, Washington, DC.

Corbett, J. 2001. CSISS classics. Torsten Hagerstrand: Time geography. http://www.csiss.org/classics/content/29 (last accessed 7 May 2014).

Couclelis, H. 2009. Rethinking time geography in the information age. *Environment and Planning A* 41:1556–75.

Ellis, F. 2000. *Rural livelihoods and diversity in developing countries.* New York: Oxford University Press.

Ellis, F., and G. Bahiigwa. 2011. Livelihoods and rural poverty reduction in Uganda. *World Development* 31:997–1013.

Evers, B., and B. Walters. 2000. Extra-household factors and women farmers' supply response in sub-Saharan Africa. *World Development* 28:1341–45.

Hagerstrand, T. 1970. What about people in regional science? *Regional Science Association Paper* 24:7–21.

Hanson, S., and G. Pratt. 1995. *Gender, work and space.* London and New York: Routledge.

International Fund for Agricultural Development 2012. Uganda, Ghana and Cote D'Ivoire: The situation of widows. http://wwwifad.org/gender/learning/challenges.widows/55.htm (last accessed 7 May 2014).

Jaramillo, C., A. Grindlay, and C. Lizarraga. 2012. Spatial disparity in transport social needs and public transport provision in Santiago de Cali (Colombia). *Journal of Transport Geography* 24:340–57.

Kwan, M.-P. 2012. How GIS can help address the uncertain geographic context problem in social science research. *Annal of GIS* 18:245–55.

Kwan, M. P., and J. Weber. 2003. Individual accessibility revisited: Implications for geographical analysis in the 21st century. *Geographical Analysis* 35:341–53.

Lucas, K. 2011. Making the connections between transport disadvantage and the social exclusion of low-income populations in the Tshwane Region of South Africa. *Journal of Transport Geography* 19:1320–34.

McQuoid, J., and M. Dijst. 2012. Bringing emotions to time geography: The case of mobilities of poverty. *Journal of Transport Geography* 23:26–34.

Miller, H. 2005. Necessary space–time conditions for human interaction. *Environment and Planning B: Planning and Design* 32:381–401.

Oluoko-Odingo, A. A. 2011. Vulnerability and adaption to food insecurity and poverty in Kenya. *Annals of the Association of American Geographers* 101:1–20.

Pirie, G. 2009. Virtuous mobility: Moralizing versus measuring geographical mobility in Africa. *Afrika Focus* 22:21–35.

Poon, J. P. H. 2005. Quantitative methods III: Not positively positivist. *Progress in Human Geography* 29:766–72.

Porter, G. 2002. Living in a walking world: Rural mobility and social equity issues in sub-Saharan Africa. *World Development* 30:285–300.

———. 2011. I think a woman who travels a lot is befriending other men and that's why she travels: Mobility constraints and their implications for rural women and girls in sub-Saharan Africa. *Gender, Place and Culture* 18:65–81.

Porter, G., K. Hampshire, A. Abane, A. Munthali, E. Robson, M. Mashiri, A. Tanle, G. Maponga, and S. Dube. 2012. Child porterage and Africa's transport gap: Evidence from Ghana, Malawi and South Africa. *World Development* 40:2136–54.

Pred, A. 1977. The choreography of existence: Comments on Hagerstrand's time–geography and its usefulness. *Economic Geography* 53:207–21.

Scheiner, J., and C. Holz-Rau. 2012. Gendered travel mode choice: A focus on car deficient households. *Journal of Transport Geography* 24:250–61.

Schwanen, T., and M. P. Kwan. 2012. Guest editorial: Critical space–time geographies. *Environment and Planning A* 44:2043–48.

Schwanen, T., M. P. Kwan, and F. Ren. 2008. How fixed is fixed? Gendered rigidity of space–time constraints and geographies of everyday activities. *Geoforum* 39:2109–21.

Shoval, N. 2008. Tracking technologies and urban analysis. *Cities* 25:21–28.

Thrift, N. 2005. From born to made: Technology, biology and space. *Transactions of the Institute of British Geographers* 30:463–76.

Uganda Bureau of Statistics. 2010. *Uganda National Household Survey 2009/2010.* http://www.ubos.org/UNHS0910/unhs200910.pdf (last accessed 20 April 2014).

Walker, W., and S. Vajjhala. 2009. Gender and GIS: Mapping the links between spatial exclusion, transport access, and the millennium development goals in Lesotho, Ethiopia, and Ghana. Resources for the Future Discussion Paper 09–27, Resources for the Future, Washington, DC.

Yamano, T., and Y. Kijima. 2011. Market access, soil fertility and income in East Africa. In *Emerging development of agriculture in East Africa*, ed. T. Yamano, K. Otsuka, and F. Place, 187–202. London: Springer.

Livelihoods as Relational Im/mobilities: Exploring the Everyday Practices of Young Female Sex Workers in Ethiopia

Lorraine van Blerk

Geography, School of Social Sciences, University of Dundee

Age is now considered alongside other differentiating categories for exploring mobility experiences, yet little work has emerged conceptualizing the im/mobilities of marginalized young people living in particularly difficult circumstances. This article, therefore, explores the relational im/mobilities of young female sex workers in Ethiopia aged between fourteen- and nineteen-years-old to understand how their livelihoods are shaped by the connections between their relations with others, im/mobilities, and survival in everyday life. The article draws on detailed narratives and participatory mobility mapping with sixty young sex workers in two locations in Ethiopia. Conceptually this article moves beyond sedentary and nomadic conceptions of mobility to what Jensen (2009) termed *critical mobility thinking*, where lives do not just happen in static enclaves or nomadic wanderings but are connected through multiple communities of interest and across time and space. Through these processes, everyday livelihoods are shaped and experienced. Further, drawing on Massey's (2005) relational geographical theory, where sociotemporal practices constitute places in a complex web of flows, the article reveals that young sex workers' critical im/mobilities are relational: Their livelihoods and identities are shaped within and between places based on their ability to move or not. The article reveals that these relational im/mobilities are important for securing work, protection, and accessing services, both within and between places and across a variety of sex work livelihoods. The article concludes by demonstrating that consideration of livelihoods as relational and mobile is central for the development of appropriate interventions.

当前在探讨能动性经验时, 年龄会与其他区别范畴一同进行考量, 但却鲜少研究对于处于特别艰辛环境中的边缘化年轻人的不/能动性进行概念化。于是乎, 本文探讨埃塞俄比亚中, 年龄介于十四至十九岁的年轻女性性工作者的相对不/能动性, 以理解她们的生计如何透过其与他人的关系、不/能动性和每日求生的连结所形塑。本文运用埃塞俄比亚两地中的六十位年轻性工作者的细緻叙事和参与式能动性製图。本文在概念上超越固着和游牧的能动性概念, 转向詹森 (2009) 所谓的批判能动性思考, 其中生命并非仅只是在静止的飞地或游牧的漫游中发生, 而是透过多重利益团体, 在不同的时空中进行连结。透过这些过程, 每日生活受到形塑并经验之。此外, 透过梅西 (2005) 的关係性地理理论, 其中社会时间实践在复杂的流动网中构成了地方, 本文揭露年轻性工作者的批判不/能动性是关係性的 她们的生计和认同, 是在根据她们能否在地方之中与之间移动的能力形塑之。本文揭露这些关係性的不/能动性, 对于在地方之中与之间、以及在不同的性工作生计中确保工作、保护和获得服务而言是重要的。本文于结论中显示, 将生计考量作为关係性且流动的, 是发展适切介入方法的关键。

La edad es ahora tomada en cuenta junto con otras categorías diferenciadoras para explorar experiencias en movilidad, si bien poco ha sido el trabajo emergente para conceptualizar las in/movilidades de gente joven marginada que vive en circunstancias particularmente difíciles. Por eso, este artículo explora las in/movilidades relacionales de jóvenes trabajadoras sexuales en Etiopía, con edades entre catorce y diecinueve años, para entender cómo su sustento está configurado por las conexiones entre sus relaciones con otros, las in/movilidades y supervivencia en la vida cotidiana. El artículo se construye a partir de narrativas detalladas y mapeo participativo de movilidad con sesenta trabajadoras sexuales jóvenes en dos localidades de Etiopía. Desde el punto de vista conceptual, este artículo va más allá de las concepciones sedentarias y nómadas de la movilidad hacia lo que Jensen (2009) llamó pensamiento crítico de la movilidad, donde las vidas simplemente no ocurren en enclaves estáticos o vagabundeo nómada sino que se conectan a través de múltiples comunidades de interés y a través del tiempo y el espacio. A través de estos procesos es como se configura y experimenta la supervivencia cotidiana. Aún más, basándonos en la teoría geográfica relacional de Massey (2005), en la que las prácticas sociotemporales constituyen lugares en una compleja red de flujos, el artículo revela que las in/movilidades críticas de las trabajadoras sexuales jóvenes son relacionales: Su sustento e identidades están configuradas dentro de los

171

lugares y a través de estos con base en su habilidad de moverse o no moverse. El artículo revela que estas in/movilidades relacionales son importantes para conseguir trabajo, protección y tener acceso a los servicios, tanto al interior de los lugares como entre unos y otros, y a través de una variedad de modos de subsistencia basados en trabajo sexual. El artículo concluye demostrando que la consideración de los modos de subsistencia como relacionales y móviles es central para el desarrollo de intervenciones apropiadas.

A critical mass of research that positions young people as social agents has coincided with recent academic interest in mobility as an everyday practice. This has resulted in young people now receiving more attention as mobile subjects, with age considered significant alongside other differentiating categories for exploring life experiences (Holt and Costello 2011). Skelton (2013), however, noted that young people are absent from conceptualizations of mobility. Therefore, this article aims to conceptualize the connection between mobility and livelihoods through a nuanced reading of diverse forms of relational im/mobilities.

Across sub-Saharan Africa, livelihood strategies are an increasingly important topic of analysis for understanding young lives. The increasingly youthful population means that youth poverty is a significant issue, with Gough, Langevang, and Owusu (2013) highlighting that there remains limited understanding regarding how young people create livelihoods in impoverished contexts. Positioning youth poverty as an effect of power and therefore relational (both inter- and intragenerational) helps to contextualize the way in which young people's livelihoods and identities are shaped. Hajdu et al. (2013) demonstrated how relational practices within families, which remove young people from school to help at home or in agriculture, contribute to reducing potential livelihood options and increasing poverty. Others highlight intergenerational contracts, where young people are obliged to support their parents through a logic of debt (Attias-Donfot and Waite 2012; Katz and Lowenstein 2013), as a key relational framework for understanding livelihood strategies.

Although some livelihood research also elucidates connections between mobility and livelihoods, this has mainly focused on mobility for livelihoods, such as migration from rural to urban areas or to large-scale farms or mines for employment (Attias-Donfot and Waite 2012; Thao 2013), rather than seeing im/mobilities as an integral part of everyday livelihood practice. Similarly, a few studies are now linking mobility with intra- and intergenerational relations (Porter et al. 2010; Veale and Dona 2014), with Skelton (2013)

offering a nuanced conceptualization of im/mobilities as relational. Yet, such research is still to consider the implications for livelihood strategies as both mobile and relational. A few studies draw attention to how impoverished young people employ mobile strategies for survival (Langevang and Gough 2009; van Blerk 2013), but the focus is on survival strategies rather than through the framework of mobility.

This literature shows that both relationality and mobility have important contributions to make to understanding livelihoods, but when considered separately these perspectives only partially explain everyday practices. Bringing mobilities and relations together for understanding young lives is important in African contexts where family relationships can be stretched over significant spatial distances (Ansell and van Blerk 2004). In this article, through a critical conceptualization of young Ethiopian sex workers' livelihoods as mobile, I explore the ways in which relational im/mobilities shape their livelihood practices over different spatial and temporal scales.

The Politics of Im/mobility: Critically Conceptualizing the Everyday Practice of Sex Work Livelihoods

Conceptualizing mobility as everyday practice, bringing together a variety of movement at different scales (Sheller and Urry 2006; Cresswell 2010), has resulted in significant developments in our understanding of the centrality of im/mobilities to the everyday. It could be argued that mobility has, more than ever, come to symbolize the contemporary human condition infiltrating all aspects of daily life (Urry 2007; Dalakoglou and Harvey 2012). This theoretical shift highlights that mobilities are situated within wider political, economic, and social contexts, and therefore society cannot be understood as a-mobile but rather as a domain in which daily life transcends spatial boundaries connected through relational networks. Cresswell (2010) employed the term *politics of mobility* to represent how the diverse practices and forms of im/mobilities are both products of social relations and produced by them. This politics highlights mobility as a powerful, albeit unequal, process that intersects with daily life as the social relations that produce and distribute power are differently accessed, creating an unevenness to mobility that does not always equal freedom (Cresswell 2006). This has implications for young sex workers, their livelihoods, and their identities.

Therefore, it is important to move away from binary oppositions of static and mobile and to engage theoretically with a more nuanced relational understanding of mobility that goes beyond sedentary and nomadic conceptions to what Jensen (2009) termed *critical mobility thinking*. This refers to understanding mobility as transcending the sedentary–nomadic dichotomy, where identities (including livelihood identities) are instead constituted via mobile practices as developed through a relational understanding of place. "Accordingly, there is no fixed and nested sense of place but rather ... networks of connectivity that transcend place as an enclave" (Jensen 2009, 143). For Jensen (2009), lives do not just happen in static enclaves or nomadic wanderings but are connected through multiple communities of interest and across time and space. It is through these processes that everyday identities are shaped and experienced. This suggests that mobility is socially produced and full of meaning and power, both reflecting and reinforcing power relations, and therefore not necessarily a resource that everyone has equal access to (Sheller and Urry 2006). As such, mobility theorists are critical of those who link power and mobility only positively as power enabling mobility or mobility as the freedom to evade power (Cresswell 2006). If power and mobility are linked in this way, where power is seen as a circulating force, an effect mediated through a recirculation of practices and procedures (Amin and Thrift 2002), then it is their relationality that is of particular importance. This maps onto Adey's (2006) call for a relational politics of im/mobilities where the connections among people, places, and objects need to be understood through their relationships to each other and suggests that studies of im/mobililties should be framed within a critical relational geography, drawing on Massey's (2005) relational theory, where sociotemporal practices constitute places in a complex web of flows. This will go beyond explaining the everyday as that which happens in particular spaces or territories to also include what happens in the circulation between places.

Following an outline of the research, the remainder of this article draws on this notion that livelihood identities and practices happen through im/mobilities

that are influenced by both positive and negative power relations. The article explores how, through livelihood relations, girls experienced diverse im/mobilities, their own and others. Through examples of becoming involved in sex work, both the relations present in leaving home and arriving in the city and the everyday spatial and temporal relations of sex work among clients, other girls, and bar or room owners, the article demonstrates a complex interconnectedness between mobility and relationality for understanding sex work livelihoods. The article concludes by conceptualizing livelihoods as both relational and mobile, occurring across a range of spatial and temporal scales, within and between actors, and offers the suggestion that livelihoods need to be conceptualized as relationally im/mobile.

The Research

The research took place in two Ethiopian cities: Addis Ababa, the capital city, and Nazareth, the regional capital for Oromia district located on the trade route toward Djibouti. Sex work is a diverse and complex industry, with girls accessing clients in different ways: as bar girls, red-light area workers, and streetwalkers[1] and at different status and payment levels within society. Their lived experience of sex work varies, with red-light area workers[2] having the most clients, commanding the least money of the three groups, and being least mobile (van Blerk 2008). Streetwalkers and bar girls tend to have fewer clients, often only one per night, although occasionally more, depending on the services requested. These latter two groups also have more freedom to engage in mobility. Streetwalkers tend to use main streets around markets and busy centers and move over greater distances and for longer periods of time, whereas bar girls work from small low-end bars but often move out of the bar to other locations with clients.

Sixty girls, between fourteen and nineteen years old, who were engaged in sex work[3] as part of their survival strategies participated in this research. All were marginalized by poverty; they commanded little money from clients and worked in poor neighborhoods, small local bars, and *Araki/Tella/Tej*[4] houses. Approximately 22 percent (thirteen) were drawn from red-light areas, 37 percent (twenty-two) worked as streetwalkers, and 41 percent (twenty-five) worked in bars.[5] The girls were mainly accessed through nongovernmental organization (NGO) drop-in centers, although snowballing alerted others not engaged with NGOs to also participate.

The research followed a street researcher approach using participatory ethnography and rooted in the work of Freire (van Blerk 2012). This approach positions the researcher as a novice willing to learn from participants who are experts in their own lives and encourages researchers to spend time getting to know participants and building trust. Multiple lengthy visits were made to the drop-in centers where the researcher, assisted by a local assistant and translator, observed, chatted, and helped the girls with routine tasks. This enabled the researcher to better understand the complexity of the girls' lives by talking with them and learning from them. Detailed field notes were made and written up each day. In addition, interviews, focus groups, and participatory activities were carried out. This article particularly draws on the analysis and discussion of mobility maps. These maps were drawn by the girls to explain their mobility and were then used as a visual prompt to elicit further explanation. Full notes were made for analysis of the verbal and visual material. Through these maps and discussions, the girls were able to highlight the places they moved between as part of their employment strategies and to discuss the implications of these mobility processes as relational networks in their livelihood strategies.

Research ethics were considered in detail and followed guidelines for working with young people (Alderson and Morrow 2011). As many of the girls had little or no contact with their families, parental consent was not sought. Instead, informed consent was gained from the girls and organizations working with them. Appropriate channels were made available for counseling through the relevant NGOs. At the end of the research, workshops were held with the girls to discuss the findings and to think through policy recommendations.

Livelihoods as Relational Im/mobilities

As noted, critical mobility thinking positions networks of connectivity as central, yet they can take diverse forms for young sex workers. Figure 1 exemplifies the range of mobilities young sex workers engage in. Hanna[6] identified that her networks of connectivity take place at different scales and over different lengths of time. Her map shows several places within Addis where she works but also shows movement between places within Addis as well as between Addis and other towns. It is interesting to note that Hanna's arrows go in both directions, indicating movement between places in a continual process as she develops her own relational im/mobile livelihood strategy.

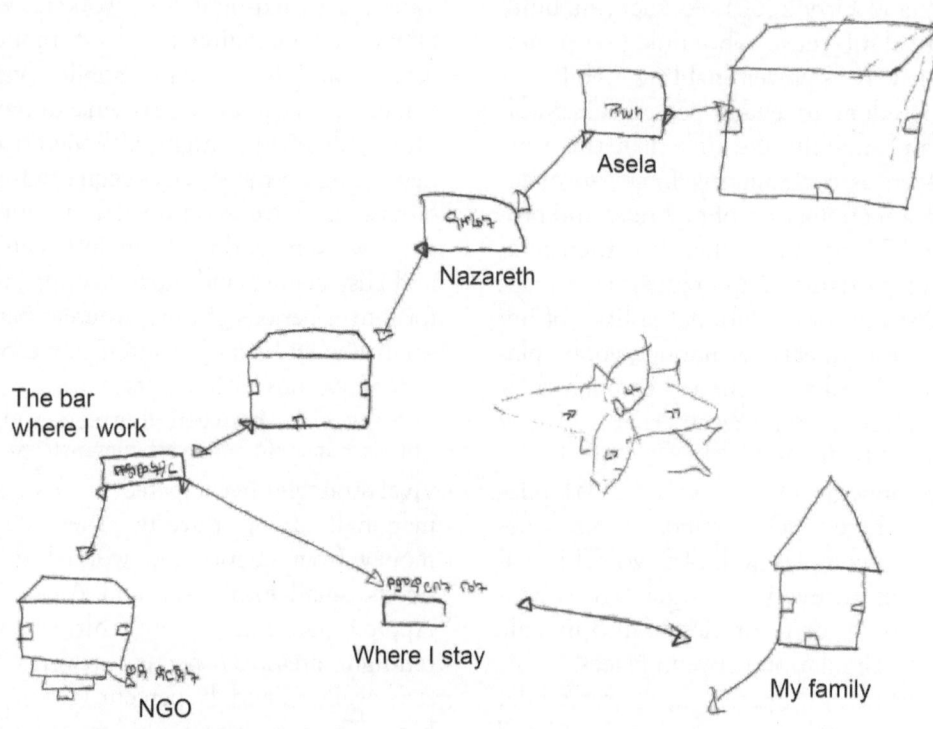

Figure 1. Hanna's map (Hanna, age eighteen, Addis).

174

Hanna's map also acts as an exemplar for highlighting two themes that emerged from the data, which help to conceptualize livelihoods as relational im/mobilities at different scales: the process of becoming involved in sex work and everyday spatial and temporal sex work livelihood strategies. Through these two themes, the article highlights the complexity of livelihoods as shaped by everyday mobilities within cities as well as long-term mobility across greater distances and how they are shaped by powerful positive and negative relations.

Becoming Involved in Sex Work

The process of becoming involved in sex work is highly mobile, but as Jensen (2009) pointed out, it is not simply a matter of nomadic wandering but instead young sex workers become connected through multiple communities of interest and their mobility has a purpose both for leaving and arriving. Many girls engage in mobility, moving from rural areas to larger towns for social and economic reasons. Some girls discussed moving to urban areas because of poverty in their families and as the eldest child they felt somewhat compelled to ease the burden on parents. The significance of intrafamilial relations is important here, with girls' mobilities bound up with their identities as eldest siblings and "good" daughters.

> It was difficult at home and there was not any food to eat. My parents had small children and because I was the eldest I thought it is up to me to look for work. (Kidist, age seventeen, Addis)

For others, such as Atema and Abeba, the relational networks within families were further complicated by social and cultural factors, with girls leaving home and exploring urban opportunities because they did not want to get married (another strategy for overcoming household poverty), to avoid practices related to reaching adulthood such as female circumcision, or because of stigma associated with friendships with young men. In all cases mobility was both purposeful and connected to family and community relations, although often through negative power relations positioning them as "bad" daughters due to their unwillingness to follow cultural practice. This demonstrates a critical edge to girls' mobility practices that positions their leaving with their rural identities.

> I was fifteen when I left home and came to Nazareth. I was not a good girl at home and as my father was not alive it was my brother who forced me out. (Atema, age fifteen, Nazareth)

> My town was Sire which is near to Arsi and is far from here. At my parents' house I used to go to school but they forced me to get married when I was only thirteen. I stayed with my husband only five months, he was twenty-three, then I left him and my aunt came and brought me here to Nazareth. (Abeba, age eighteen, Nazareth)

The importance of place as a complex web of relations (Massey 2005) coupled with the idea of relational identities as stretched over space (Ansell and van Blerk 2004) further adds to the understanding of the relational nature of livelihood mobility. In some cases the process of moving was related more to going somewhere rather than leaving, with girls being enticed to try out working in the city. Here the interplay between relations that connected places separated by distance and the relational identities girls had within their families resulted in some mobility decisions occurring. For example, Kalkidan demonstrated how she was enticed to the city by the glamourous stories of opportunities, working in the coffee board.[7] This, coupled with a precarious home life linked to household poverty, offered Kalkidan an opportunity to create a new livelihood. The journey as a relational process was facilitated by a family friend who shared a connection with Kalkidan through the rural village but whose identity now stretched into the unfamiliar city.

> I am from Wolkete, which is a place far out of Addis. I lived with my parents and we were poor. Because of this my stepfather never treated me like his children, my siblings who were younger than me. My mother used to fight with him to treat me well and I never liked this so I left home with my rich relatives and worked with them as a waitress. They promised me a good life in the city but then I escaped and a broker took me to Gore. I left Addis because my relatives owned a local araki drinking house and I was scared of getting involved in drinking and prostitution. (Kalkidan, age sixteen, Addis)

As Jensen (2009) pointed out, mobility is a complex process of connection across trajectories, and particularly this last example alludes to the diversity of young people's im/mobilities as both real and imagined, as livelihood opportunities are shaped and reconstituted through relations that occur between places. The im/mobilities of beginning sex work (termed *business* by the girls) are complex, however, and not always associated with traveling over distance, from rural to urban places, or from poverty to imagined prosperity. They are sometimes localized within the microgeographies of the city with girls moving only to new districts. This was particularly evident when the reputation of girls was seen to bring shame on the family and particularly if they had become pregnant, as both Yeti and Beti identified. This clearly shows that intergenerational relations feature prominently in understanding sex work livelihoods as mobile, demonstrating the need for livelihood conceptualizations to be considered as both relational and mobile.

> I am from Arat Kilo but I do business in another place that is not near to my parents' home. (Yeti, age seventeen, Addis)

> I left home because when I was in tenth grade I got pregnant and had a child. I was a student and not married so my parents did not allow me to live with them anymore. (Beti, age seventeen, Addis)

Here, the girls mentioned moving within urban centers just to another location to disassociate themselves from their families because of the impact of their relationships. Crucially, the link between mobility and relationality is important where the act of movement enables new identities to be constructed and others forgotten or supressed. These im/mobilities of beginning business are also sometimes practiced because of the stigma attached to sex work. Girls talked of "doing business" in a different area to where they stayed to separate their work from home, moving between these locations on a daily basis as part of their strategies for maintaining a sex work livelihood.

The relational aspect of these critical im/mobilities is immensely important. The girls' movements are not static or nomadic but rather connected to specific relationships that are played out within and between local places (Massey 2005), illustrating that these girls' mobility is socially produced based on particular relations of power between them and their families.

Spatial and Temporal Im/mobilities: Sex Workers' Relational Networks Within and Between Places

In this section, the article explores the everyday im/mobilities that constitute sex work and demonstrates that such livelihoods are mobile processes beyond migration for livelihood opportunities but also small-scale relational im/mobilities within places and between bodies. Considering these everyday practices also provides a greater understanding of how and why mobility is enmeshed in sex work as a livelihood strategy. This section explores im/mobilities within places and between places that vary both spatially and temporally, highlighting a particular geography of sex work as a mobile livelihood strategy.

Sex Work Livelihoods as Small-Scale Relational Im/mobilities

For some girls, most notably those who worked in red-light areas, the everyday practice of doing business might appear to be immobile and their identity imagined as static, tied to that place as fixed enclave. These girls are often confined for long hours to single rooms, usually with little freedom to engage with other places located at a distance to the immediate space of their work. Their livelihood identity is tied to the microspace of the room. Yet, if we consider the notion of place as relational and mobility as networks of connectivity beyond the sedentary–nomad dichotomy (Jensen 2009), then it is possible to move beyond sex work as a static process. By considering the girls as relational subjects (Adey 2006), their immobility can be seen as a process through which they interact with other important actors necessary for sex work to take place. Therefore, if place is constituted by flow (Massey 2005) and relations within place are seen as moments of encounter (Amin and Thrift 2002), the girls' identities as sex workers in the red-light area are intrinsically relationally mobile. It is the movement of clients into and out of that area, creating fleeting moments of encounter, that enables the girls in that place to construct their livelihoods.

> Most of the time my clients are government soldiers that come here to visit. (Zenash, age sixteen, Addis)

For others, their own movement within and between places is part of the connections that enable their livelihoods. Some of the girls mentioned moving their place of business to a new bar or house or to a different part of town, depending on what was taking place at that time and perhaps varying their places of work at weekends or holidays (see Rita and Meberet's stories). Others highlighted that their business-related mobility is often dependent on their relationships with clients who might prefer to go to places such as guesthouses or hidden areas. On occasion, girls chose to persuade the client to go somewhere they were known if they felt uncomfortable or lacked trust that they would remain safe. Here relations with other girls were an important part of these mobility strategies, with girls going to places where their friends could support them if anything went wrong. For some, like Tigist, developing positive relations with other girls had an enabling effect, supporting her to break away from relationships with owners and establish her own business strategies to create a higher earning potential.

> I moved to another bar for two months when I had a fight with the woman owner here because I refused to pay her ten birr. Although we go away, she always accepts us back. (Rita, age sixteen, Nazareth)

> On Saturdays and Sundays we often move the place we stand for business to the markets because we get good business then. (Meberet, age seventeen, Addis)

> I used to work on a house but I had to share the money with the owner. If I earn ten birr I must give five birr to her. After a year a friend told me I can do the same in the street and not have to share my money so now I stand in the street for business. (Tigist, age sixteen, Addis)

It is important to note that not all small-scale mobility flows are positive (Cresswell 2006), and having a critical perspective helps to uncover how these girls' relational im/mobilities within places are often due to power differentials in their encounters with clients and house or bar owners. Some girls talked of moving to a new place because of fights with other girls, negative interactions with security and police, abuse from clients, and even just feeling ashamed when insulted as they walked near their place of work. These negative relational processes all impinge on girls' livelihoods and their feelings of self-worth. Therefore, many talked of moving to other locations, red-light areas or bars, on both a short-term and long-term basis, as means for avoiding further denigration regarding their working practices. Tannesh explained how this mobility can have a detrimental impact on their livelihoods.

> Sometimes we move house if we don't get on with the owner or we want to avoid a client but moving between houses can create a problem because those regular clients who like us may not find us. (Tannesh, age fifteen, Addis)

Through examining the small-scale spatial and temporal im/mobility relations, a greater understanding of young people's livelihoods can be gleaned as distinctly mobile. Relations that occur with other girls, clients, bar owners, the public, and others illustrate how sex work as a livelihood is practiced through connections across space that vary spatially and temporally but without which the girls would not maintain their livelihoods.

Sex Work Livelihoods as Connections Between Places

Given the difficulties of remaining in one place, maintaining a successful sex work livelihood was highly connected to relational im/mobilities that also sometimes resulted in traveling greater distances, between places, sometimes for a long period of time. By drawing on three examples of mobility between places, where girls travel to access work, to visit home, or because of their physical health, the importance of positioning livelihoods as relational and mobile is further evidenced.

Many of the girls identified a well-developed mobility strategy that sought to make best use of their relational networks and maximize their potential for earning through business. As Yetimwerk highlighted, at specific times of the year, they move to new places to create greater opportunities for moments of encounter, as potential clients would also travel to the same place. Genet pointed out that at particular harvest seasons the girls travel to smaller towns where migrant laborers and farmers also go for livelihood opportunities, and Zenash mentioned traveling to different cities at the time of functions or conferences to interact with attendees also traveling there. The social relations that this mobility creates maintains the girls' livelihoods.

> Sometimes I go to Dire Dawa . . . every year there is a big church ceremony and I go there when there are lots of visitors going to look for fun . . . when business is slow in Addis I also go to the Djbouti border because a lot of people are coming there for trade and business is good. (Yetimwerk, age fifteen, Addis)

> Sometimes I go to different places depending on the seasons. If onions are being harvested then I go to the place where they grow onions as that will be good for business. (Genet, age eighteen, Addis)

> Once I went to Debre Zeit because I heard that after the road construction truck drivers from Djbouti were using that road. (Zenash, age seventeen, Nazareth)

Sex Work Livelihoods as Concurrently Mobile and Immobile

Sex work as a mobile livelihood strategy also occurs through relations that position girls as immobile. This is highlighted through two examples: The first relates to health and HIV status and the second to the flow of money. Although the girls were not asked their status, approximately one third disclosed that they were either HIV-positive or suspected that they might be, with the majority too afraid to be tested. Within each moment of encounter, where bodies could be considered immobile, the transmission of bodily fluids always carries the threat of mobility of HIV through this relational encounter. This did result for some in the spread of HIV between the bodies of clients and girls. In turn, this resulted in the girls employing different mobility strategies to ensure the continuation of their livelihood. Mobile practices were used to create new sex worker identities in different places and break associations to their current location that might hinder their ability to continue working. The stigma associated with AIDS meant that as girls fell sick and began to lose weight (a clear sign of HIV-related illness), they were often unable to attract clients. By moving to a new town, they were able to hide their previous identity as "fat and healthy" and break relational networks associated to the previous place.

> We are all HIV positive, and we are very careful that no one knows about our status or we will have to stop doing business only our friends can know. One friend was found out by the client so she moved to Gondar so that she could continue to do business. (Tibilis, age sixteen, Addis)

The relational connections between where girls undertake their livelihoods and home further highlight the im/mobility of livelihood strategies that are both real and imagined. Several girls stated that they often thought of their families and would send money home to provide support, especially if they were the eldest sibling. The flow of thoughts and money maintained a relational connection across space but avoided the shame of direct contact—highlighting livelihood mobility in situations where the girls' bodies are considered immobile. For those who did make the occasional journey home, this was a temporary encounter, requiring a embodied expression of a home identity (wearing long skirts and no makeup) and temporarily giving up the embodied sex worker identity (visible through appearance, behavior, and consuming alcohol or marijuana).

> Sometimes I go back to (rural) Asela to visit my mother. I go every couple of months but I don't stay long. It's not the same now. (Tannesh, age fifteen, Nazareth)

These connections to home complete the picture of livelihoods as mobile practices, as these girls then bring others to engage in sex work, enthralling them with stories of employment and good earning potential

working in the coffee board. This examination of the various spatial and temporal scales provides a nuanced approach to how im/mobility is a crucial aspect of this livelihood strategy.

Conclusion

This article has examined the everyday practices of young sex workers in Ethiopia and explores how their livelihoods are developed through critical relational im/mobilities. Exploring how girls use their relational networks and mobility to become involved in sex work provides an understanding of the need to think through livelihoods beyond any static notion of place. Drawing on Jensen's (2009) critical mobility thinking enabled a more nuanced understanding of these livelihood processes. By conceptualizing mobility as an everyday practice and therefore exploring the everyday relational mobilities girls engaged in, such as moving within the city and between places to maintain and enhance sex work as a livelihood, this article demonstrated that their livelihoods are inherently mobile across a range of spatial and temporal scales. Through employing a critical relational im/mobility lens, this article suggests that their livelihood strategies were mobile both through the movement of their own bodies and also through the connections they had with others including clients, other girls, and families. These examples demonstrate that the mobilities of the young people in this article were not independent singular moves but, rather, complex relational im/mobilities that were both real and imagined.

What emerges from this analysis is that livelihoods need to be considered as more than relational and more than mobile: rather as relationally im/mobile. Mobility is therefore key to understanding how young people create and maintain their livelihoods, a gap in understanding as highlighted by Gough, Langevang, and Owusu (2013). The article highlights that a focus on mobility advances our understanding of the conditions and practices of sex work as a livelihood strategy and positions this work as highly mobile and changing according to different spatial and temporal relations that are also culturally and context specific. The article also advances the way in which we understand engagement in sex work for girls and suggests that interventions need to account for how such livelihoods are relational and mobile. Therefore, the provision of static services (e.g., health, education) does not account for girls' movement or indeed the movement of their clients as key relational actors in their livelihood strategies.

Acknowledgments

I would like to express sincere gratitude to the girls who told me their stories and thank referees whose comments helped to improve the final version.

Funding

I am grateful to the Nuffield Foundation for funding the research this article is based on, although any shortcomings therein are entirely my own.

Notes

1. This research focused on the poorest groups.
2. Red-light area girls work from small informal wooden rooms in poor communities. The rooms are usually owned by someone else who controls the girls' time in the room with clients.
3. It is acknowledged that sex work definitions are complex, particularly in poor communities, where sex might be exchanged for income or other needs at times of hardship. This article focuses on girls who exchange sex for cash with several (unknown) men as their main source of livelihood.
4. These are locally produced traditional drinks that are sold in small bars, often the size of one room.
5. A smaller number of red-light workers participated in the research due to constraints on their mobility.
6. All names used are pseudonyms.
7. The coffee board is a pseudonym for sex work to lure others to the city, unaware of the work they will do.

References

Adey, P. 2006. If mobility is everything then it is nothing: Towards a relational politics of im/mobilities. *Mobilities* 1 (1): 75–94.

Alderson, P., and V. Morrow. 2011. *The ethics of research with children and young people: A practical handbook.* London: Sage.

Amin, A., and N. Thrift. 2002. *Cities: Reimagining the urban.* Cambridge, UK: Polity.

Ansell, N., and L. van Blerk. 2004. Children's migration as a household/family strategy: Coping with AIDS in Lesotho and Malawi. *Journal of Southern African Studies* 30 (3): 674–90.

Attias-Donfot, C., and L. Waite. 2012. From generation to generation: Changing family relations, citizenship and belonging, In *Citizenship, belonging and intergenerational relations in African migration,* ed. C. Attias-Donfot, J. Cook, J. Hoffman, and L. Waite, 40–62. London: Palgrave Macmillan.

Cresswell, T. 2006. *On the move: Mobility in the modern Western world.* London and New York: Routledge.

———. 2010. Towards a politics of mobility. *Environment and Planning D: Society and Space* 28:17–31.

Dalakoglou, D., and P. Harvey. 2012. Roads and anthropology: Ethnographic perspectives on space, time and (im)mobility. *Mobilities* 7 (4): 459–65.

Gough, K., T. Langevang, and G. Owusu. 2013. Youth employment in a globalising world. *International Development Planning Review* 35 (2): 91–102.

Hajdu, F., N. Ansell, E. Robson, and L. van Blerk. 2013. Rural young people's opportunities for employment and entrepreneurship in globalised southern Africa: The limitations of targeting policies, *International Development Planning Review* 35 (2): 155–74.

Holt, L., and L. Costello. 2011. Beyond otherness: Exploring diverse spatialities and mobilities of childhood and youth populations. *Population, Space and Place* 17 (4): 299–303.

Jensen, O. 2009. Flows of meaning, cultures of movement—Urban mobility as meaningful everyday life practice. *Mobilities* 4 (1): 139–58.

Katz, R., and A. Lowenstein. 2013. Theoretical perspectives on intergenerational solidarity, conflict and ambivalence. In *Ageing and intergenerational relations: Family reciprocity from a global perspective,* ed. M. Izuhara, 29–56. Policy Press Online.

Langevang, T., and K. Gough. 2009. Surviving through movement: The mobility of urban youth in Ghana. *Social and Cultural Geography* 10 (7): 741–56.

Massey, D. 2005. *For space.* London: Sage.

Porter, G., K. Hampshire, A. Abane, E. Robson, A. Munthali, M. Mashiri, and A. Tanle. 2010. Moving young lives: Mobility, immobility and inter-generational tensions in urban Africa. *Geoforum* 41:796–804.

Sheller, M., and J. Urry. 2006. The new mobilities paradigm. *Environment and Planning A* 38:207–26.

Skelton, T. 2013. Young people's urban im/mobilities: Relationality and identity formation. *Urban Studies* 50 (3): 467–83.

Thao, V. 2013. Making a living in rural Vietnam from (im)mobile livelihoods: A case of women's migration. *Population, Space and Place* 19:87–102.

Urry, J. 2007. *Mobilities.* Cambridge, UK: Polity.

van Blerk, L. 2008. Poverty, migration and sex work: Youth transitions in Ethiopia. *Area* 40 (2): 245–53.

———. 2012. Berg-en-See street boys: Merging street and family relationships in Cape Town, South Africa. *Children's Geographies* 10 (3): 321–35.

———. 2013. New street geographies: The impact of urban governance on the mobilities of Cape Town street youth. *Urban Studies* 50 (3): 556–73.

Veale, A., and G. Dona. 2014. *Mobility-in-migration in an era of globalisation.* London: Palgrave Macmillan.

Mobilities at Gunpoint: The Geographies of (Im)mobility of Transgender Sex Workers in Colombia

Amy E. Ritterbusch

School of Government, Universidad de los Andes

Drawing from geo-ethnographic data collected during a participatory action research (PAR) project funded by the National Science Foundation and subsequent research conducted in Colombia with marginalized youth populations, this article explores the sociospatial exclusion and (im)mobility of the oppressed, subjugated, and persecuted through the social cartographies, geo-narratives, and auto-photographic images of transgender sex workers that were displaced by paramilitary-led gender-based violence and forced to leave their birth cities and rural communities in Colombia at an early age. As is the case for thousands of victims of the armed conflict in Colombia, displaced transgender populations seek refuge and opportunity in the streets of Bogotá, Colombia. The (im)mobilities of transgender sex workers are explored in two stages—the forced, violent mobilities of their displacement, followed by their experiences of discrimination, sociospatial exclusion, and persecution through hate crimes and social cleansing killings on arrival in Bogotá. This article discusses how research actors constructed their own spaces of cohesion and resistance to the multifaceted discrimination and marginalization from mainstream urban society through PAR. The PAR project presented in this article continues as part of the broader struggle of transgender sex workers to challenge the exclusionary discourses and praxis that limit their mobilities and autonomy in the city. This article concludes with examples of how research actors use the action-driven elements of PAR to negotiate, analyze, and resist the relationships of power and violence embedded within their urban environment and begin to re-present and change the reality of their immobility within the city.

运用一项由国家科学基金会所资助的参与式行动研究 (PAR) 计画中所蒐集的地理—民族志数据，以及随后在哥伦比亚和受到边缘化的青年人口一同进行的研究，本文藉由受到准军事驱动及根据性别的暴力而流离失所，并且被迫在其年幼时离开他们在哥伦比亚的出生城市及农村社区的跨性别性工作者的社会製图、地理叙事和自我摄像照片，探讨受压迫者、从属者和受迫害者的社会空间排除与能 (不) 动性。如同哥伦比亚武装冲突的数千万受害者的情况一般，流离失所的跨性别人口，在哥伦比亚的波哥大街头寻求庇护和契机。跨性别性工作者的能 (不) 动性，在两个阶段中探索之——他们流离失所的被迫且暴力的能动性，随后是他们抵达波哥大后，因仇恨犯罪和社会清洗屠杀而遭受歧视、社会空间排除以及迫害的经验。本文探讨研究行动者如何透过 PAR，建构自身的凝聚和抵抗空间，应对主流城市社会中多面向的歧视和边缘化。本文中所呈现的PAR计画，继续作为跨性别性工作者挑战限制其在城市中的能动性和自主性的排除性论述与实践的广泛斗争的一部分。本文于结论中，以案例显示研究行动者如何运用 PAR 中以行动为导向的元素，协商、分析并抵抗镶嵌在其城市环境中的权力与暴力关系，并着手改变他们在城市中的不动性之现实。

A partir de datos geoetnográficos recabados con ocasión de un proyecto investigativo de acción participativa (PAR) financiado por la Fundación Nacional de Ciencia y la subsiguiente investigación llevada a cabo en Colombia entre poblaciones juveniles marginadas, este artículo explora la exclusión socioespacial y la (in)movilidad de los oprimidos, subyugados y perseguidos por medio de las cartografías, geonarrativas e imágenes autofotográficas de trabajadores sexuales de condición transexual, víctimas de violencia basada en género promovida por paramilitares y obligados a temprana edad a dejar sus ciudades de origen y comunidades rurales en Colombia. Como en el caso de miles de víctimas del conflicto armado en Colombia, la población transexual desplazada busca refugio y oportunidad en las calles de Bogotá, la capital colombiana. Las (in)movilidades de los trabajadores sexuales transexuales se exploraron en dos etapas – las movilidades forzadas y violentas del desplazamiento, seguidas de sus experiencias de discriminación, exclusión socioespacial y persecución a través de la criminalidad del odio y los asesinatos de limpieza social a su llegada a Bogotá. Este artículo discute cómo construyeron los actores de la investigación sus propios espacios de cohesión y resistencia a la discriminación multifacética y marginalización de la corriente principal de la sociedad urbana, a través del PAR. El proyecto PAR que se presenta en este artículo continúa adelantándose como parte de una lucha más amplia de trabajadores sexuales transexuales para desafiar los discursos y prácticas excluyentes que limitan sus movilidades y autonomía en la ciudad. El artículo concluye con ejemplos sobre cómo los actores de la investigación usan los elementos orientados hacia la acción del PAR para negociar, analizar y resistir las relaciones de poder y violencia incrustadas dentro del entorno urbano, y empezar a re-presentar y cambiar la realidad de su inmovilidad dentro de la ciudad.

Alexa watches the rest of the city flash by her on the *Transmilenio*[1] hundreds of times daily as thousands of passersby gaze at her body in the brothel entrance (Figure 1). Whereas millions of Bogotá citizens do move and can move throughout the city en route to work, to school, to wherever they want and need to go, Alexa's life and mobilities are limited to four blocks of the city, as is the case for the majority of transgender sex workers in the city center.

This article explores the lives of transgender sex workers in Bogotá, whose geo-narratives of (im)mobilities are far more complex than the reality imagined by mainstream urban society and passersby in Bogotá. In the Colombian context, the most recent official human rights report on violence against lesbian, gay, bisexual, transgender, and queer (LGBTQ) populations reports 824 homicide victims from 2006 to 2014; of this total, nineteen of the thirty homicide victims during 2013 and 2014 were transgender women and all nineteen of those cases were documented by the government as hate crimes committed due to the victim's gender identity (Colombia Diversa 2015). In terms of police abuse, the report also shows considerably higher incidence among transgender women across multiple regions in Colombia (Colombia Diversa 2015).

In addition to homicide and police abuse, human rights violations are also committed against LGBTQ populations within the context of the armed conflict in Colombia. As is the case for thousands of victims of the armed conflict, displaced transgender populations seek refuge and opportunity in the streets of Bogotá. In this article, the (im)mobilities of transgender sex workers are explored in two stages—the forced, violent mobilities of their displacement, followed by their experiences of discrimination, sociospatial exclusion, and persecution through hate crimes and social cleansing killings in Bogotá.

Conceptualizing Trans (Im)mobilities

Displacement [for me] is having to move from one place to another and not having peace anywhere ... because everywhere I go someone wants to displace me because of my sexual condition or my physical appearance. (Estrella, semistructured interview, 11 September 2014)

Figure 1. Caracas Avenue and the gaze on trans sex work.

As set forth by Cresswell (2010), three interconnected aspects of mobility include "the fact of physical movement—getting from one place to another; the representations of movement that give it shared meaning; and, finally, the experienced and embodied practice of movement" (19; see also Cresswell 1999, 2001). Cresswell (2010) also discussed six facets of a politics of mobility, including "the starting point, speed, rhythm, routing, experience and friction ... that can serve to differentiate people and things into hierarchies of mobility" (26). To examine particular cases of (im)mobility, Cresswell argued that we must take into account how the movement, representation, and practices of particular people are "implicated in the production and reproduction of power relations. In other words, how they are political" (Cresswell 2010, 26).

Within this conceptual framework, in the following sections I untangle the movement, representation, and practice of trans (im)mobility in Colombia, starting with their experiences of forced displacement, what I refer to as mobilities at gunpoint, and contextualizing their subsequent (im) mobility limited to four blocks in the center of Bogotá. Nash and Gorman-Murray's recent work connects Cresswell's constellations of mobility to the geographies of sexualities through their conceptualization of transformations in LGBTQ neighborhoods (Gorman-Murray and Nash 2014; Nash and Gorman-Murray 2014). For them, the politics of mobility and "who or what is 'mobile' ... has historical and geographical specificity and is constituted through relations of power between social groups based on such categories as class, race, age, gender, and ... [as they argue], sexuality" (Gorman-Murray and Nash 2014, 627).

The case of transgender sex workers' forced displacement and their subsequent immobility in Bogotá illustrates how gender shapes movement and how "gendered processes create, reinforce or change patterns of daily mobility" (Hanson 2010, 8; Silvey 2004). (Im)mobilities are forcibly imposed as a means of maintaining traditional, heteronormative relations in society and as a means of eliminating deviant gender identities from society or keeping them at a distance (Adey 2006; Giddings and Hovorka 2010; Hanson 2010). Within this context, Namaste's discussion of "genderbashing" is crucial to the conceptualization of how and why the construction of transgender bodies and the public presentation of transidentities become a factor of exclusion, violence, and forced (im)mobilities. As argued by Namaste, "a perceived transgression of normative sex/gender relations motivates much of the violence against sexual minorities, and ... an assault on these 'transgressive' bodies is fundamentally concerned with policing gender presentation through public and private space" (Namaste 2006, 585; see also Namaste 2000). Multiple embodied practices of violence and exclusion, from explicit threats at gunpoint to looks of disgust and disapproval, indicate where transgender sex workers should and can be in Colombia.

Previous scholarship on the movement of LGBTQ populations has explored the relationship between transgender identity and immigration (Cerezo et al. 2014), gay men's migration within countries (Lewis 2014), the definition of queer migration (Gorman-Murray 2009), and conceptualizations of the overlap between the mobilities and forced migration literature (Gill, Caletrío, and Mason 2011). As argued by Schapendonk and Steel (2014), the incorporation of a mobilities approach within migration research pushes interrogation beyond the descriptive analysis of migratory movement toward a more profound examination of the politics shaping these processes.

To conceptualize the politics of immobility of transgender sex workers in Bogotá's public spaces, it is necessary to explore the factors contributing to violence against this population identified in previous work. Edelman (2011) contextualized the exclusionary practices and policies generated by the implementation of "prostitution-free zones" and the relationship between urban improvement initiatives and the elimination of trans bodies from public spaces. In recent work in Atlanta, Doan (2014) highlighted the relationship between official efforts to clean up public space and the removal of transgender sex workers from these spaces. In terms of the incidence of violence and aggression against LGBTQ populations in U.S. cities, Doan (2007) identified high levels of verbal and physical violence against transgender populations as well as high levels of fear about safety.

Within the Latin American context, previous work has examined the often-conflated social categories surrounding the sexual orientation and gender identity of transgender sex workers in Brazil and Mexico (Kulick 1997; Prieur 1998). These studies raise important considerations for the Colombian context in which transgender sex workers are subject to both physical and discursive violence used by police officers that refuse to recognize their gender identity. Namaste (2006) highlighted the danger of the fusion of sexuality and gender in the examination of violence and suggested that "an attack is justified not in reaction to one's sexual identity, but to one's gender presentation ... women and men who transgress acceptable limits of self-presentation, then, are among those most at risk for assault" (588). As the following empirical sections demonstrate, violence against transgender sex workers in Bogotá is justified in reaction to their gender presentation and their transgression of gender norms, identified principally at the scale of the body.

The growing subfield of trans geographies also emphasizes the importance of prioritizing the lived experiences of transgender people and including their voices in the generation of knowledge about their struggles in different contexts (Hines 2010; Nash 2010).

Methodology

As part of a long-term participatory action research (PAR) initiative, the Mobilities at Gunpoint project was designed collectively with transgender sex workers to document the human rights violations experienced by transgender populations in Colombia and to attain the social justice–oriented goals established with the transgender and activist community in Bogotá. Our PAR work in Colombia is driven by what I have referred to elsewhere as "sustainable structures of care" forged by university–community partnerships in our collective fight for the social justice and rights of transgender sex workers and other historically marginalized communities (Ritterbusch 2012).

To conceptualize trans (im)mobilities in Colombia, I draw principally from semistructured interviews conducted with ten transgender sex workers in 2014. All qualitative data collection, including interviews, auto-photography, and mapping, was completed within a PAR framework and transgender sex workers were involved in decision making and action research design throughout all project phases. The ten project phases were (1) definition of social justice and transformative objectives; (2) collective design of research instruments; (3) mobilization of the research community; (4) immersion and trust-building activities; (5) point of entry focus group and story sharing; (6) semistructured interviews; (7) destabilizing the camera: visual methods workshop; (8) auto-photography exercises in urban space; (9) participatory data analysis: "story ranking"; and (10) reclaiming the city: mobilizing against gender-based violence.

Our approach triangulates photographic, mapping, and ethnographic data collection techniques to visualize the (im)mobilities of transgender sex workers and ensures that these data will be used for social justice purposes by completing the project within a structure of activism already constructed through previous initiatives involving the same PAR team.

Our team used particular inclusion criteria for the interviews, including variation in the year of displacement, region, and the profile of the illegal armed actors causing displacement (guerrilla vs. paramilitary actors). To provide additional empirical evidence supporting the immobility argument, I also draw from interviews conducted with transgender sex workers from a previous project conducted from 2009 to 2012.[2]

The semistructured interview process included a social cartography exercise to visualize the physical trajectory of forced displacement and research actors'[3] mobilities, activities, and perceptions of Bogotá, including the spatial variables listed in Figure 2. Within the instrument, each research actor drew a symbol representing each category and the third column is an explanation designed by peer leaders of each spatial variable. Additionally, research actors were asked to associate an emotion with each trajectory or movement traced on the map as a means of uncovering the representations surrounding their mobilities.

During the auto-photography phase, we conducted a group session of walking, talking, and photographing spaces to capture individual, place-specific visions of violence and exclusion in the city. The auto-photography or self-directed photography approach has been extensively employed in projects led by children geographer's in multiple research contexts and aims to "document ways in which ... [individuals] transact with their environment" (Aitken and Wingate 1993, 66). Auto-photographic exercises seek to give research actors control over the camera, a traditionally exploitative tool used to capture bodies and places in time.

In the following empirical sections, I prioritize the voices and vision of transgender sex workers through the presentation of their cartographies, photographs, and interview excerpts that illustrate their experiences of forced displacement and immobility in Bogotá. The stories shared in this article were chosen during a participatory data analysis exercise we refer to as story ranking (see Figure 3). In this exercise, we read through the interview transcriptions as a group and collectively prioritize which story excerpts represent the most urgent data for action. At the beginning of the exercise, it is

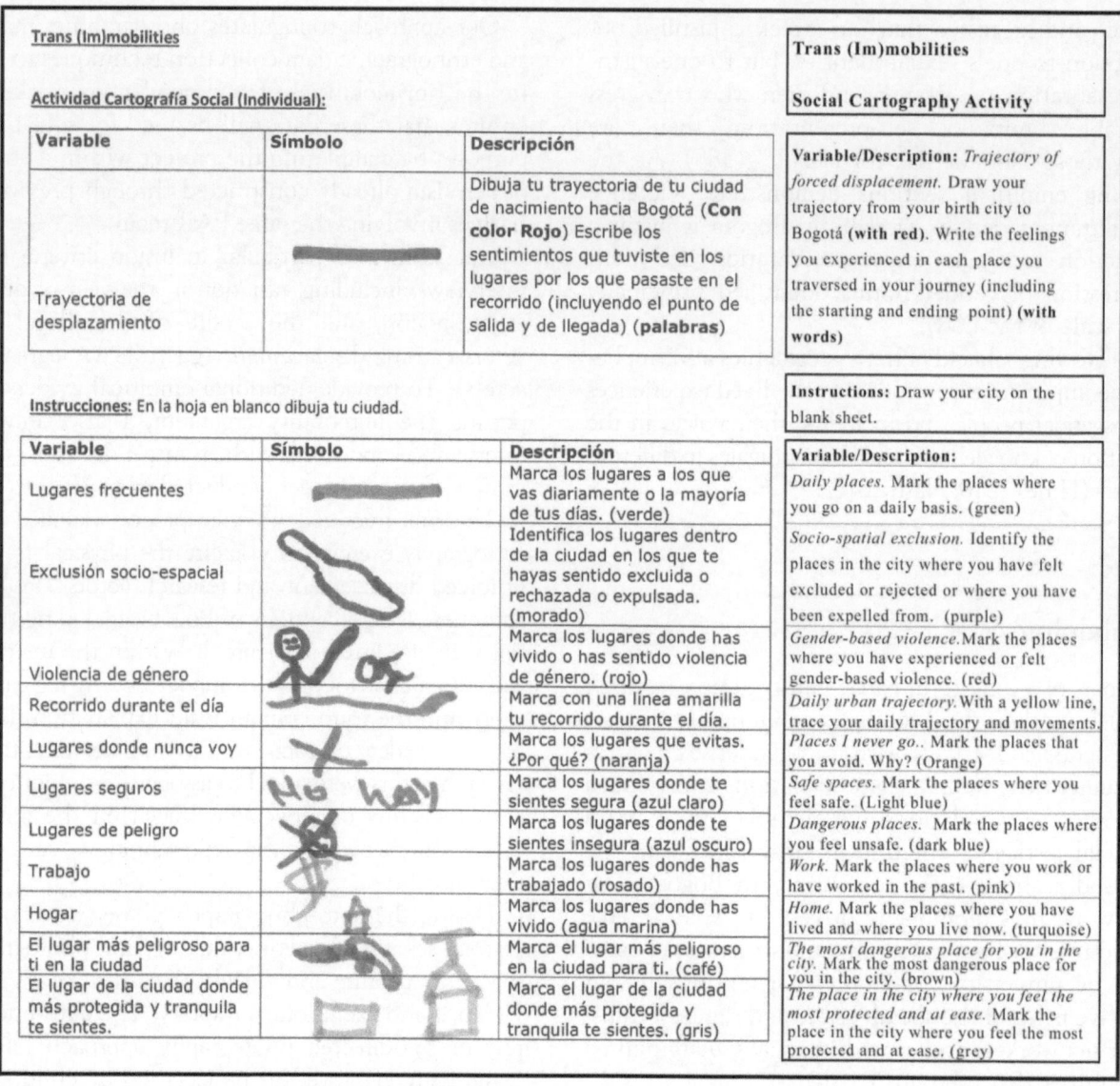

Figure 2. Social cartography research instrument and spatial variables.

crucial to acknowledge the importance of each story and voice and to clarify that the excerpts will be chosen by the group based on their communicative power.

Stories of Forced Displacement: Mobilities at Gunpoint

Good boys ... go to sleep early, bad boys like robbers, gays and junkies are "put to sleep" by them. (Madonna, semistructured interview, 12 September 2014)

In the following empirical section on the forced mobility of transgender sex workers, Cresswell's mobility constellations are grounded through maps that illustrate the physical movement from one town or city to another in Colombia. The geo-narratives drawn from interview excerpts describe in detail the meanings or representations of each movement in the lives of transgender sex workers. The voices of transgender sex workers contextualize how they were forced at gunpoint to leave their homes at a young age (ranging from eight to eighteen years of age) by illegal armed groups (including guerrilla and paramilitary groups).

The forced displacement of the research actors of the Mobilities at Gunpoint project spans twenty municipalities and twelve departments of Colombia (see Table 1). Previous work in Colombia has also

Figure 3. Participatory data analysis exercise: Story ranking.

documented that this violent phenomenon occurs across multiple regions of the country (see Prada et al. 2012).

During the social cartography exercise, Estrella traced her forced movement from Yotoco to Cali to Pasto to Cali to Bogotá, associating multiple emotions representing her experiences of mobility throughout the country. Estrella's emotions range from anger (*rabia*) and disillusion (*desilusión*) in reference to the violent rupture of family ties at the age of eleven years old to calmness (*calma*) and peace or harmony (*paz*) associated with the distance now between herself and the armed actors who threatened her life (see Figure 4). This approach to mapping out emotions over time and as connected to particular forced movements provides a method for empirically grounding Cresswell's discussion of the elements of a politics of mobility, in which he called for an exploration of the feelings connected to particular

Figure 4. Estrella's trajectory of forced displacement.

mobilities (Cresswell 2010). How does it feel, then, to be constantly on the move to flee from death threats and acts of gender-based violence?

The majority of cases include different forms of verbal abuse, discrimination, and physical violence, and all research actors were given a limited amount of time (between twenty-four hours and one week) to leave their hometown and never return. It was made clear that if they remained in their hometown or city they would be killed.

In the case of all research actors, paramilitary or guerilla actors issued violent warnings throughout the community:

> They told us "motherfucking fag I don't want to see you around here or I'm gonna shoot the hell out of you … you know I don't wanna see you around here motherfucking fag gonorrhea [an insult not referring in this context to a sexually transmitted disease, but rather to the repulsion they feel]" … then they fired gunshots into the air. (Francy, semistructured interview, 13 September 2014)

> "We give you eight days to leave or you know what will happen." … they told my Mom I had to leave because if not they will kill her and everyone. (Tulia, semistructured interview, 12 September 2014)

Table 1. Trajectory of forced displacement

Research actor	Movement between cities/towns (state names in parentheses)
Yurleny	Ocaña (Norte de Santander)–Aguachica (Cesar)–Bucaramanga (Santander)
Alexa	Viterbo (Caldas)–Chinchiná (Caldas)
Francy	Cartagena (Bolívar)–Carmen (Bolívar)
Estrella	Yotoco (Valle del Cauca)–Cali (Valle del Cauca)–Pasto (Nariño)
Michel	Pereira (Risaralda)–La Virginia (Risaralda)
Madonna	Chinchiná (Caldas)–Medellín (Antioquia)
Adri	Primavera (Vichada)–Cali (Valle del Cauca)
Tulia	Mesetas (Meta)–Medellín (Antioquia)
Marbel	Calarcá (Quindío)–Quebrada negra (Quindío)–Armenia (Quindío)
Violeta	Magangué (Bolívar)–Valledupar (Cesar)–Venezuela–Medellín (Antioquia)–Bucaramanga (Santander)

Additionally, multiple research actors discussed the violent details of the torture and murder of other LGBTQ peers who did not make it out in time:

> Olimpo was left paralyzed ... we were seven, from those seven, three remain alive, I'm fine thank God, Olimpo is in a wheelchair, he was stabbed eight times for being a fag because someone passed by and he said, "What a hot boy" and pam! ... They stabbed him eight times ... two of them are disabled, the other one ended with a permanent limp and the other one is me ... one month later they killed Ricardo, they chopped him up and they put his penis inside his mouth and he was found in the canal. ... They chopped him up with a machete. (Madonna, semistructured interview, 12 September 2014)

Madonna's narrative contextualizes the violent practices causing the mobilities of transgender women throughout Colombia and the consequences of resisting forced movement, as in the case of Olimpo and peers.

The voices of these research actors demonstrate the multiple embodied experiences of violence that resulted in their forced movement throughout the country, from village to village, from city to city, in pursuit of refuge, protection, and a place where they are accepted and respected. During the story ranking exercise, research actors discussed their feelings associated with these embodied experiences of mobility since childhood, including anger, fear, outrage, and powerlessness as the most frequently stated emotions and also sadness, pain, hate, danger, terror, discrimination, panic, and instability.

Through transgender sex workers' narratives representing their feelings attached to particular movements, it is evident that in this context mobility is not a privilege and becoming a mobile subject is not about going somewhere to fulfill a purpose or desire but rather is about getting any place where their lives are not at risk. Thus, for transgender sex workers, movement is a strategy for survival and their associated embodied practices include whatever means enable them to flee from threat and risk.

Police Abuse, Urban Violence, and Immobility in the City: "My Life in Four Blocks"

In the case of all research actors, gender-based violence and others forms of violence did not end with their experiences of discrimination and forced displacement within their towns or cities of origin. On arrival in Bogotá, all research actors arrived directly to Santa Fe, a zone in the city center spanning from 19th to 24th Street and between Caracas and 18th Avenue.

While in this area of the city, research actors found a new sense of belonging, family, affordable body transformation practices, love, and work. Within this same space, transgender women also experience social cleansing killings, police abuse, and other forms of gender-based violence.

> This is where Wanda was killed. Afterward all of us were frightened. ... Santa Fe was desolate ... all of us went running in panic to other cities. (Tulia, auto-photography 11 November 2014)

One of the first places identified in Santa Fe as a space of violence during the auto-photography exercise is the corner where Wanda Fox, trans community leader and activist, was killed during a drive-by shooting led by paramilitary actors. The community-based organizations and leaders within the community designed a mural in honor of Wanda's life and service to the community. As stated by one of the leaders of Red Comunitaria Trans, one of the nongovernmental organizations leading the initiative, "This mural represents sparks of color that reflect light and energy in the gray, solemn context of these streets" (Daniela, Facebook, 26 June 2014).

During the auto-photography exercise, research actors chose this corner to raise consciousness surrounding social cleansing killings and hate crimes against transgender sex workers and community leaders. In reaction to this process of group reflection among research actors, we participated in the annual protest led in memory of Wanda and other transgender sex workers who have been killed by hate crimes, social cleansing campaigns, and police violence. In addition to the tragedy of Wanda's death for the community, the following interview excerpts describe research actors' multiple experiences of violence and police abuse in Santa Fe, which should technically be the "safe" zone for sex workers as the zone where prostitution is legally permitted:

> He took the nightstick and pram ... he broke my head ... he was always around making my life difficult to the point that I had to stop prostitution for a period of time ... he threatened me and various times he took me, after beating my head in, to the police station that is far from the city ... and beat me with a water hose. (Violeta, semistructured interview, 10 September 2014)

They took us to the 11th station ... and at midnight they said, "Okay, say your last prayers ... because we are going to kill you." ... They were going to kill us [up in the woods above the Circunvalar road]. ... They started yelling. "All of you take your clothes off in the patrol car." ... We had to go running down the hill naked ... they chased us with gunshots ... they forced us to jump into the freezing stream and left us soaked, without makeup, without anything ... barefoot ... without high heels ... to simply bully and laugh at us ... to banish us ... this is torture ... it was their entertainment laughing at us getting wet. (Marbel, semistructured interview, 12 September 2014)

The dynamics of police abuse and violence in Santa Fe include excessive use of force, abuse of power, arbitrary arrest, sexual violence, physical and verbal abuse, use of discriminatory language, and forced displacement within the city to keep them from working and to make their life unlivable. As a means of resistance against police abuse, transgender sex workers use a strategy of cutting their arms to scare away the police and to be able to keep working (Figure 5). Madonna described this mechanism:

I did like this, watch [showing us how she cuts her arm] ... "Touch me and I'll give you AIDS." ... The saying is that we trans are the ones who have AIDS ... so you cut yourself, they see you bleeding and no one touches you, they let you die. (Madonna, semistructured interview, 12 September 2014)

These images and quotes illustrate the manner in which self-inflicted violence has paradoxically become a strategy of resistance to violence, which has been replicated not only in Bogotá but also in other parts of the country in response to police abuse and harassment. The embodied practices of resistance to forced displacement or movement within the city also evidence the HIV-related stigmas surrounding transgender populations and sex workers and the fear of their bodily fluids. Instead of running the risk of being infected with HIV or of a sex worker dying on his or her watch, police officers prefer to leave them alone as soon as blood is shed.

It is not, however, only police who violate the human rights of transgender sex workers. Other actors in the community are also responsible for acts of violence, as demonstrated in the following testimonies.

One time when I was stabbed ... it was here on 19th Street and they were six [men] ... one told the other "Let's give it to those fags." ... They stabbed me here [pointing to her heart] ... they almost killed me. (Yurleny, semistructured interview, 10 September 2014)

24th Street scares me a lot. They kill a lot there ... the paramilitary of the zone kill a lot of trans there. (Alexa, semistructured interview, 13 September 2014)

In light of these experiences of violence within Santa Fe, making the zone a paradoxically unsafe and violent "safe zone," one might assume that transgender women would choose to leave this area of the city; however, there are several reasons why research actors are attached to Santa Fe and avoid the rest of the city. First, their sense of belonging and ability to transform and exercise their gender identity and corporeal image keeps them from leaving. Second, when they leave the imaginary borders of Santa Fe and cross Caracas Avenue, the experiences of rejection,

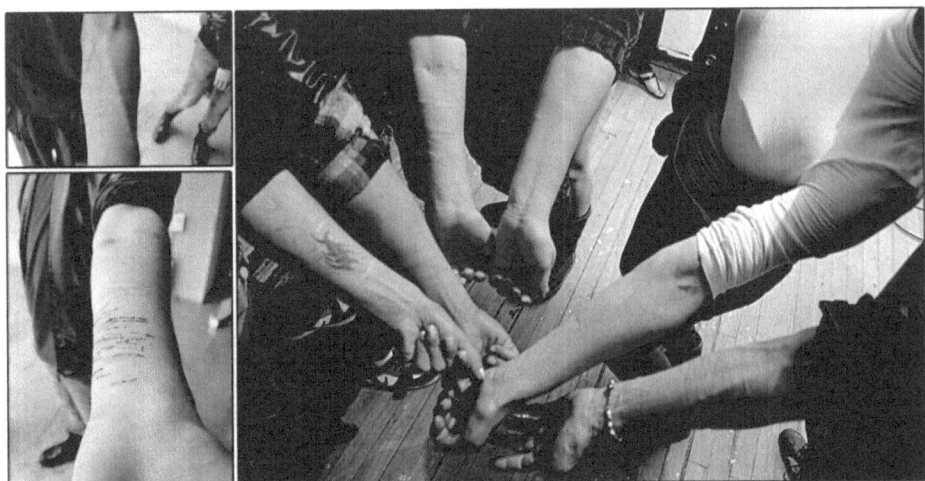

Figure 5. Self-infliction of violence to defer police abuse. (Color figure available online.)

disgust, and social disapproval keep them from leaving the four blocks of their spatial existence in the city. The experience of transgender sex workers is an interesting case for understanding Cresswell's discussion of correct mobilities and friction. What are the correct mobilities for transgender populations in Colombia and when and why does their mobility stop (Cresswell 2010)? All research actors mentioned that during their multiple trajectories of forced movement throughout Colombia they were directed by allies to Bogotá, which is seen by many as a safe haven for LGBTQ populations in comparison to other cities. This perception shaping transmobilities partially contextualizes why the majority of research actors' mobilities stopped in Bogotá.

In the following cases, research actors described this reality.

> The "zone of tolerance" … is my place, it is my territory … I feel safe, I don't feel badly … because … I demand respect [here]. I don't leave Santa Fe. (Michel, semistructured interview, 15 September 2014)

This is evident as well in the maps of Bogotá collected during the social cartography component of the project.

As discussed in the methodology section, the social cartography exercise was guided by multiple questions about research actors' activity spaces and place perceptions associated with violence, rejection, and exclusion in the city. Findings suggest that research actors' activity spaces are limited to four blocks within the city. Twenty-four hours per day, seven days a week, research actors follow the same sociospatial routines to protect themselves from experiences of gender-based violence and rejection outside of the imaginary boundaries separating Santa Fe and the rest of the city.

As illustrated through Carolina's account of the consequences of leaving Santa Fe and attempting to work or socialize in other parts of the city, transgender sex workers cannot move throughout the city without experiencing physical, verbal, or psychological violence:

> [The police officer screams] "You can't be here [referring to the north of the city] … here no … here no. … Get out of here." … So I left and came back later and [as soon as] he saw me … he came to beat me to the ground. (Carolina, semistructured interview, 8 May 2010)

Figure 6. Street occupation campaign: Mobilizing trans bodies in the city. (Color figure available online.)

Carolina's account is one of many acts of aggression against transgender sex workers that occur in multiple spaces outside of the Santa Fe safe zone, which demonstrates that transgender sex workers' immobility in the city is not merely perceived or self-imposed but rather reflects the violent reality of gender bashing in multiple cities in Colombia. These accounts of immobility further contextualize why transgender sex workers' mobilities stop so abruptly in Bogotá and how particular frictions in urban space limit their movement to four blocks of the city.

As part of the action component of the PAR project, research actors decided to plan a protest in both public transportation and privileged places in the city to communicate their message and project findings to society: "We, too, are victims of the armed conflict and we, too, have the right to the city" (see Figure 6).

We collectively designed the activity, got dressed and prepared materials together, and took to the streets to communicate our message to mainstream society. To do so, we chose two strategic locations in the city: the *Transmilenio* route to the northern (affluent) end of the city and one of the designer shopping districts (known as *la zona 'T'*). Throughout our trip to the T, we shouted "We are trans-women ... and we are here to stay!!!"

On arrival, we silently entered in a line and walked to the center of the T to shout, one by one, the messages on our posters: "I, too, have the right to the city," "I am also a victim of the armed conflict," "I deserve a voice in society," "I want to be heard," "I deserve respect," "No to discrimination," "No to gender-based violence," and "I am more than Santa Fe."

The mobility constellations surrounding our protest in the T and transgender sex workers' abilities to move through the city can be explained by the

Figure 7. Tulia's story: "Neither here nor there."

force of collective movement. The meaning and purpose of the mobile practice of taking public transportation to move from point A to B in the city is radically different in the context of the T protest. Moving in a crowded, public space of transportation was a strategic action chosen by the group to communicate our social justice message about trans (im)mobilities to a broad audience while enacting resistance through the occupation of spaces not normally occupied by transgender populations. Additionally, this very movement, from Santa Fe to the T zone in the north, is neither safe nor comfortable for transgender sex workers, and without the collective strength of group movement, it would not have been possible.

Conclusion

In this article I have illustrated how transgender sex workers, such as Tulia (Figure 7), are trapped within a vicious cycle of violence and feel as though they are unable to be anywhere in Colombia: neither here nor there.

This article presents the forced mobility of transgender youth from their towns or cities of origin and their subsequent immobility on arrival in Bogotá illustrated through maps tracing their movement, narratives contextualizing what these movements mean in trans lives, and photographs opening a window on their world of exclusion and immobility.

Additionally, the voices and vision of transgender sex workers support the underpinning argument that the movements, practices, and representations constituting transmobility in Colombia should not be conceptualized as a privilege but rather as a way of disciplining and displacing transgressive bodies from public, heteronormative spaces. Both empirical cases illustrate the importance of examining the movement, practices, and representations of marginalized, persecuted, and oppressed populations at the juncture of geographies of sexualities and mobilities literature as a means of more profoundly interrogating the power dynamics, social differences, gendered relations, and politics implicit in the movements (or lack thereof) of these populations.

Within this framework, a closer examination of transgender sex workers' immobilities reveals that their movement restricted to four blocks of the city is not self-imposed but is a reality shaped by violent reactions to their gender presentation in other parts of Bogotá. Furthermore, the article discusses how these forced immobilities in the city are resisted at the scale of the body, including the mobilization of the fear of blood contamination and through the force of collective movement in public space as a means of resisting exclusionary urban practices of segregation.

Acknowledgments

I would like to express my profound appreciation and gratitude to all research actors involved in and touched by the PAR process and to all those who continue as part of our social justice community in Colombia. I would also like to thank my colleagues of the organization PARCES for their dedication to the movement and Sebastián León Giraldo, Julian(a) Salamanca Cortés, and Andrea Correa for their leadership and vision throughout the action and visual method components of the project.

Funding

This article draws from research funded by the National Science Foundation under Grant No. BCS-0903025.

Notes

1. This is Bogotá's rapid transit bus system.
2. The total research population was thirty-three street-connected youth (including sex workers, drug users, and homeless individuals in Bogotá).
3. I use the term *research actors* in referring to participants to discursively work against the hierarchies traditionally maintained between university and community-based research actors.

References

Adey, P. 2006. If mobility is everything then it is nothing: Towards a relational politics of (im)mobilities. *Mobilities* 1 (1): 75–94.

Aitken, S., and J. Wingate. 1993. A preliminary study of the self-directed photography of middle-class, homeless, and mobility-impaired children. *The Professional Geographer* 45 (1): 65–72.

Cerezo, A., A. Morales, D. Quintero, and S. Rothman. 2014. Trans migrations: Exploring life at the intersection of transgender identity and immigration. *Psychology of Sexual Orientation and Gender Diversity* 1 (2): 170–80.

Colombia Diversa. 2015. *Cuando la guerra se va la vida toma su lugar: Informe de derechos humanos de lesbianas, gay, bisexuales y personas trans en Colombia* [When the war concludes life takes its place: Human rights report on lesbian, gay, bisexual and trans individuals in Colombia]. Bogotá, Colombia: Colombia Diversa.

Cresswell, T. 1999. Embodiment, power and the politics of mobility: The case of female tramps and hobos. *Transactions of the Institute of British Geographers* 24 (2): 175–92.

———. 2001. The production of mobilities. *New Formations* 43:11–25.

———. 2010. Towards a politics of mobility. *Environment and Planning D: Society and Space* 28 (1): 17–31.

Doan, P. 2007. Queers in the American city: Transgendered perceptions of urban space. *Gender, Place and Culture: A Journal of Feminist Geography* 14 (1): 57–74.

———. 2014. Regulating adult business to make spaces safe for heterosexual families in Atlanta. In *(Sub)urban sexscapes: Geographies and regulation of the sex industry*, ed. P. J. Maginn and C. Steinmetz, 97–218. London and New York: Routledge.

Edelman, E. 2011. "This area has been declared a prostitution free zone": Discursive formations of space, the state, and trans "sex worker" bodies. *Journal of Homosexuality* 58 (6): 848–64.

Giddings, C., and A. J. Hovorka. 2010. Place, ideological mobility and youth negotiations of gender identities in urban Botswana. *Gender, Place and Culture: A Journal of Feminist Geography* 17 (2): 211–29.

Gill, N., J. Caletrío, and V. Mason. 2011. Introduction: Mobilities and forced migration. *Mobilities* 6 (3): 301–16.

Gorman-Murray, A. 2009. Intimate mobilities: Emotional embodiment and queer migration. *Social and Cultural Geography* 10 (4): 441–60.

Gorman-Murray, A., and C. Nash. 2014. Mobile places, relational spaces: Conceptualizing change in Sydney's LGBTQ neighborhoods. *Environment and Planning D: Society and Space* 32:622–41.

Hanson, S. 2010. Gender and mobility: New approaches for informing sustainability. *Gender, Place and Culture: A Journal of Feminist Geography* 17 (1): 5–23.

Hines, S. 2010. Queerly situated? Exploring negotiations of trans queer subjectivities at work and within community spaces in the UK. *Gender, Place and Culture: A Journal of Feminist Geography* 17 (5): 597–613.

Kulick, D. 1997. The gender of Brazilian transgendered prostitutes. *American Anthropologist* 99 (3): 574–85.

Lewis, N. 2014. Moving "out," moving on: Gay men's migrations through the life course. *Annals of the Association of American Geographers* 104 (2): 225–33.

Namaste, V. 2000. *Invisible lives: The erasure of transsexual and transgendered people*. Chicago: The University of Chicago Press.

———. 2006. Genderbashing: Sexuality, gender, and the regulation of public space. In *The transgender studies reader*, ed. S. Stryker and S. Whittle, 584–600. London and New York: Routledge.

Nash, C. 2010. Trans geographies, embodiment and experience. *Gender, Place and Culture: A Journal of Feminist Geography* 17 (5): 579–95.

Nash, C., and A. Gorman-Murray. 2014. LGBT neighbourhoods and "new mobilities": Towards understanding transformations in sexual and gendered urban landscapes. *International Journal of Urban and Regional Research* 38 (3): 756–72.

Prada, N., S. Herrea, L. Lozano, and A. M. Ortíz. 2012. *ÀA mi me sacaron volada de allá! Relatos de vida de mujeres trans desplazadas forzosamente hacia Bogotá*

[I was kicked out of there!: Transgender women's life histories of forced displacement]. Bogotá, Colombia: Office of the Mayor of Bogotá and National University of Colombia.

Prieur, A. 1998. *Mema's house, Mexico City: On transvestites, queens and machos*. Chicago: University of Chicago Press.

Ritterbusch, A. 2012. Bridging guidelines and practice: Toward a grounded care ethics in youth participatory action research. *The Professional Geographer* 64 (1): 16–24.

Schapendonk, J., and G. Steel. 2014. Following migrant trajectories: The im/mobility of sub-Saharan Africans en route to the European Union. *Annals of the Association of American Geographers* 104 (2): 262–70.

Silvey, R. 2004. Power, difference, and mobility: Feminist advances in migration studies. *Progress in Human Geography* 28 (4): 490–506.

Mobilities in Rural Africa: New Connections, New Challenges

Gina Porter

Department of Anthropology, Durham University

Fluid interdependencies of mobility—physical and virtual—are growing rapidly in sub-Saharan Africa: The remarkable expansion of mobile phone networks is bringing a tangible new dimension of connectivity into mobility, transport, and access equations on the ground. This article draws on in-depth field research, including co-investigation with two groups often disadvantaged in their physical mobility, youth and older people, to explicate some current African developments and their departure from prevailing Western-based conceptualizations of space–time interactions (regarding the potential for space–time flexibility and microcoordination afforded by mobile phones). Despite the fact that face-to-face interaction is often of great significance in Africa, when the value attached to personalized relationships is balanced against factors of widespread poverty and irregular, sometimes very dangerous transport, the potential for phone substitution appears greater than in many Western contexts. Better distance management through phone use could be particularly closely associated with populations with very low disposable incomes or those whose physical mobility is limited; for instance, by disability, infirmity, age, or gender.

在撒哈拉沙漠以南的非洲, 能动性的流动相互依赖性——实体与虚拟——正快速地成长: 行动电话网络非比寻常的扩张, 正将连结性的实际新方向, 带进日常生活中的能动性、运输与可及性的均衡。本文运用深度田野研究, 包含与身体能动性上经常处于不利位置的两大群体——年轻人与老年人——进行共同调查, 以阐明当前非洲的部分发展, 及其偏离以西方为基础的盛行时空互动概念 (有关行动电话所提供的时空弹性及微观协调之潜能)。儘管面对面的互动在非洲经常具有相当大的重要性, 但当附加于个人化关係的价值受到广泛的贫穷和不规律且有时是非常危险的交通因素所抵消时, 电话的替代性, 便较诸多西方的脉络而言更显得重要。透过电话使用进行更佳的远距管理, 特别能与具有微少可支配所得、或是身体能动性受到诸如残疾、体弱、年龄或性别限制的人口紧密相关。

Fluidas interdependencias de movilidad—físicas y virtuales—están creciendo rápidamente en el África subsahariana. La notable expansión de las redes de teléfonos celulares está aportando una nueva dimensión tangible de conectividad dentro de las ecuaciones de movilidad, transporte y acceso en el terreno. Este artículo se basa en investigación de campo a profundidad, incluyendo la coinvestigación de dos grupos a menudo en desventaja en lo que a movilidad física se refiere, la juventud y la gente más vieja, para explicar algunos de los actuales desarrollos y su alejamiento de las conceptualizaciones dominantes de origen occidental sobre las interacciones espacio-tiempo (en relación con el potencial de la flexibilidad espacio-tiempo y la micro-coordinación que pueden ofrecer los teléfonos móviles). A pesar del hecho de que la interacción cara a cara a menudo reviste gran significación en África, cuando el valor asignado a las relaciones personalizadas se compara con factores de pobreza generalizada y transporte irregular, a veces muy peligroso, el potencial de la sustitución por teléfono aparece más grande de lo que ocurre en muchos contextos occidentales. Un mejor manejo de la distancia mediante el uso del teléfono podría estar estrechamente asociado en particular con poblaciones con muy bajos ingresos a su disposición, o con aquellas cuya movilidad física es limitada; por ejemplo, por discapacidades, enfermedad, edad, o género.

The potential for some substitution of physical by virtual mobility has long been present in the West, through the widespread availability of fixed landlines. For Africa's poor, by contrast, this has only become feasible in the last decade, with the advent of cheap mobile phones (Aker and Mbiti 2010; Porter 2012; Chavula 2013). Commercial network coverage, albeit as yet incomplete, is reaching ever-remoter areas: In off-road villages, airtime offers are displayed at local kiosks, and a brisk trade in SIM cards, airtime, and phone charging (using solar panels) is commonplace. Sharing of phones among family, friends, and neighbors reduces capital expenditure in very poor households, but for many—especially youth—the phone and its associated running costs are a priority, sometimes even above food. Phone contact in Africa is now commonly perceived not merely as a significant conduit for business interactions but, above all, as key to the everyday maintenance of the social networks so essential for protecting and supporting individuals and families in times of stress. The implications of phone usage for daily mobility practices in Africa are still emerging, but data presented in this article suggest intriguing variation from Western practices: Precarity brings to the fore dimensions of the mobilities time–space nexus little considered to date.

Phone usage associated with the coordination of mobilities in everyday life in the West is contrasted in this article with empirical evidence for two rural districts in Africa, where phones are now changing the mobility landscapes of commonly disadvantaged groups. One case study focuses on older people in Tanzania, and the second centers on young people in Malawi. Both are extremely poor countries, close to the bottom of the Human Development Index in 2014 (Tanzania 159th, Malawi 174th, out of 187 countries). The final section of the article considers how extensive distance management through phone use, and associated reductions in face-to-face contact, derives from conditions of deprivation and constraint, in populations with very low disposable incomes and those for whom physical mobility is restricted, whether by infirmity, age, or gender. This encourages wider reflection on the significance of precarity for virtual and physical mobility interactions.

Mobile Phones and Daily Mobility Practices: Global Perspectives

Phone connectivity can transform experiences of space and time, either by substituting for or reconfiguring physical mobility. Kwan (2006) pointed to the need for investigation of the phone's role in alleviating mobility-related social exclusion, but the majority of research continues to focus on resource-rich societies. Thus, a growing literature attests to how the reconfiguration of time–space geographies through mobile phones is offering new possibilities for microcoordination and rescheduling on the move in cities in the West. Here, mobile phones and Internet access can

help accommodate uncertainties and travel challenges, especially for women, by augmenting rather than directly replacing corporeal mobility, although the potential to soften space–time fixity constraints will depend, in part, on the household sociospatial context, lifestyle, and personalities of those involved (Schwanen and Kwan 2008). Nevertheless, more space–time flexibility is now feasible for many, because information and communication technology (ICT) has loosened the associations among activity, time, and space: "Punctual time" transforms into "negotiated time," and travel time itself can be productively occupied (Jain and Lyons 2008; Elliott and Urry 2010; Ben-Elia et al. 2014). Coordination of mobility occurs both before and during actual travel, but without any evident impact on modal choice (Peters, Kloppenburg, and Wyatt 2010; Line, Jain, and Lyons 2011).

Urry (2012) emphasized, however, that the need for occasional copresence with significant faces remains: Sufficient physical travel is required to satisfy particular social obligations and to observe "the rituals and sustained quality time often at particular moments and within specific kinds of ambient place, places appropriate for a certain affective quality" (26). Given the fundamental differences between face-to-face interaction and electronically mediated exchange, this importance attached to face-to-face interaction is hardly surprising. As Nohria and Eccles (1992) observed, copresence in time and place allows a cycle of interruption, feedback, and repair that is virtually instantaneous. This has implications for negotiating identity, uncertainty, and ambiguity; reducing duplicity; and establishing and maintaining multidimensional, robust relationships: It might be essential for mobilizing collective action in situations of uncertainty and risk. The phone, by contrast, might filter out not only social context clues such as location but the full range of psychoemotional reactions, such as discomfort or attraction.

Rural Africa offers a very different context within which to observe new connectivities and their impacts. As Schwanen and Kwan (2008) emphasized, as the social, physical, and technological realms are mutually constitutive of one another, we cannot assume that the sociospatial implications of new technologies found in the West will be replicated in other sociophysical contexts and networks elsewhere. Africa is characterized by both widespread poverty and irregular, sometimes very dangerous transport but also an oft-observed significance of face-to-face interaction (including in business, where personalized relationships are commonly crucial). This raises interesting questions: Are mobile phones encouraging new patterns of microcoordination and rescheduling on the move, as in the West? Or is further space–time flexibility simply unnecessary, given that activity times have, of necessity, commonly tended toward flexibility, because of travel uncertainties? Is ICT having any impact on transport modal choice in this very different context? Because available resources are so sparse, will the tipping point at which copresence is deemed essential be delayed? An examination of two African rural contexts facilitates reflection on these questions.

Accessibility Challenges in Rural Africa: Where Phones Can Make a Difference

The challenges faced by Africa's rural people in accessing distant services and markets are enormous. Poor provision and maintenance of road infrastructure plus poor transport service availability, unreliability, high fares, and safety issues are widespread constraints on rural travel, especially because ownership of motorized vehicles and intermediate means of transport is often restricted to rural elites. Certain disadvantaged groups—the old, the young, infirm, and women—face particular mobility difficulties; in the case of women and girls, this could include cultural constraints on their movement (Porter 2011). In rural contexts, new phone-enabled connectivities appear to have the potential to ameliorate poor access conditions, whether by reducing the need to travel to services, or by enabling more efficient travel when physical mobility is essential. M-health, in particular, is enabling some substitution of virtual for physical mobility (Deglise, Suggs, and Odermatt 2012) and, for emergency health care, mobile phones now widely facilitate access to essential transport. In the agricultural sector, the potential of mobile phones to facilitate rural produce trade, especially when allied to mobile money transfers, has been evident for some years (Overa 2006).

The two case studies presented here illustrate the growing interdependencies between physical and virtual mobility and the implications of the new connectivities for everyday mobility practices in rural lives. In each case study, consideration is given both to questions of direct phone substitution for travel and to how phones may facilitate more efficient travel organization as transport operators and their customers connect. These studies employed a similar methodology, in which co-investigation with non-academic community members, recruited as researchers, played a key part in establishing vital questions for subsequent qualitative and survey research (Porter and Abane 2008; Porter et al. 2010; Porter 2014; Porter et al. 2014). This is particularly helpful when researching relatively disadvantaged groups.

Older People's Mobility and the New Connectivities in Kibaha District, Tanzania

In 2012, a ten-village study (one settlement on the paved road, the remainder off-road) aimed at understanding older people's mobility (in particular, their access to health services and livelihoods), a little-explored issue. Interviews conducted by twelve older people with their peers ($n = 74$), academic-led, in-depth checklist interviews with older people and key informants ($n = 194$), and a small questionnaire survey administered to older people ($n = 339$) pointed to the emergence of significant new connectivities associated with the recent introduction of motorcycle taxis and mobile phones. The impact on older people's lives, especially in off-road villages, has been substantial.

Motorcycle taxi services (*boda-boda*) emerged only between about 2007 and 2009 in Kibaha district, associated with the availability of cheap Chinese imported motorcycles. They are now the principal transport mode, except along the paved main road: Previously, residents had to walk or cycle. It is no exaggeration to state that boda-boda has transformed rural lives: In the week prior to the survey, 18 percent of older women and 31 percent of older men had used their services, and there is widespread attestation to their significance in accessing health and other facilities, especially in emergencies. Despite the discomfort of pillion travel and expressed concerns about the speed at which boda-boda are driven, the only real off-road alternative is walking. Even in the roadside study settlement where buses are available (and cheaper), boda-boda are valued because they enable door-to-door vehicle access (an important attribute in infirmity or when carrying heavy loads), and they ply their business through the night when other transport has stopped (Porter et al. 2013). For young men, meanwhile, boda-boda offer a significant new livelihood option that provides a year-round income, unlike farming. In-depth interviews were conducted with thirty-five drivers between fourteen and thirty-eight years old; most drove motorcycles owned by their father, rural elites, or ex-charcoal producers living nearby. Those driving for nonfamily members normally pay a standing daily rate and then keep the balance, which encourages long hours and high speeds, thus contributing to high accident rates. Between eight and thirty boda-boda operators were based at each village station.

Meanwhile, the massive expansion of mobile phone ownership in Kibaha has brought an important complementary connector into the rural access equation. This is evident from both transport user and operator perspectives. For infirm older people, even a short walk to the village boda-boda station can present a massive hurdle: The potential to call transport operators to their home is a substantial benefit: "I have a phone and in my phone contact I have one number of a boda-boda operator who I usually call in case I need [him]" (Widow, sixty-seven years old). Remarkably, 41 percent of older men (sixty and older) surveyed owned phones and 15 percent of older women ($n = 339$) did, often a gift from their children in town; 58 percent of men and 49 percent of women reported the presence of a phone in their household that is available for them to use. Beyond the immediate household, phones are also widely available through relatives and friends—sharing is the norm. Meanwhile, all but one of the boda-boda drivers interviewed owned a mobile phone and reported that up to twenty clients had their number stored. The one nineteen-year-old driver who did not possess a phone observed that this was making his business difficult.

This is not to suggest that everyone can afford regular use of boda-bodas: As one twenty-four-year-old driver observed, "(Older people) have boda-boda drivers' numbers. Whenever there is a need they call the boda-boda instead of walking to where the boda-bodas park, but airtime and charging are still a (cost) problem." Many older people still walk long distances to the clinic, markets, and other key destinations but, in emergencies, communities often come together to help pay for calls and boda-boda travel.

Although the mobile phone can enable older people to access essential transport, in many cases they reported that phone use has reduced their overall travel: "I don't have to travel so much nowadays—maybe when there is a funeral or a crucial thing for me to travel, but for minor things I use my brother's phone and we talk" (Woman, sixty-six years old). This substitution of virtual for physical mobility is often welcomed by older, less mobile people, especially those whose family members live in distant places. Moreover, city dwellers now frequently send remittances by phone to the villages using phone-enabled mobile money services (e.g., where elderly parents are looking after grandchildren):

I use M-PESA; my children usually send money through my chip (Vodacom-number), then they call my friend through his phone telling how much they have sent through my Vodacom-line, so I just go with my chip to the Vodacom shop to take money. (Man, sixty-six years old)

This brings reported savings in time, cost, and potential travel accidents or theft on the journey. There was, however, an observed downside, for some interviewees:

Most older people have phones now. They call their children who are far away. If you don't remind the children they forget you and your needs. (Man, seventy-one years old, caring for five young orphaned grandchildren)

Phone has changed travel patterns—in the past my children and other relatives used to come to greet me but now they just call. (Widow, eighty years old)

The reduction of face-to-face interaction that is enabled by substitution of physical by virtual mobility ironically leaves some older people feeling more isolated than before, an issue to which I return later.

Young People's Mobility and the New Connectivities in Rural Lilongwe, Malawi

This section draws on ongoing research into young people's phone use and associated mobility practices in a different (three-country) study in sub-Saharan Africa. It refers specifically to evidence from two off-road rural settlements in Lilongwe District, Malawi (one relatively large village with services including a secondary school and health center, the second more remote and with no services at all), although the broader conclusions resonate with emerging project findings for rural locations elsewhere. As with Kibaha, although the villages are located within about sixty miles of the country's capital city, rurality prevails. The methodology employed mirrors that of the previous case study, although here the focus was on youth aged nine to twenty-five years old: qualitative research with young people and other key informants ($n = 138$, including peer-researcher interviews) and a questionnaire survey to young people ($n = 378$; i.e., for these two Lilongwe settlements alone).

Throughout Malawi, unlike in Tanzania's Kibaha District, motorcycle taxis are still relatively rare. Transport services in the study settlements are extremely poor, with bicycle taxis and a few ox carts the only transport modes regularly available. The substantial trade in charcoal is mostly evacuated by externally based pickups organized by private dealers: These occasionally carry a few passengers from the villages. Livings are precarious, so most young people lack resources to pay for the limited transport available, few have bicycles, and almost all walk to school, local markets, and so on.

Mobile phone networks have only recently become accessible in these villages and phone ownership rates are still extremely low, including among youth, although—as elsewhere—they are keen to embrace this new technology. Youth have few resources to facilitate phone purchase, and although some reported being given phones by family members working in South Africa, only 7 percent of young people in the larger village owned a working cell phone, and only 6 percent had phones in the smaller, remoter village (predominantly males in each case). In the larger village, however, 39 percent of respondents and 30 percent in the remoter village said that other members of their household owned a phone. Sixty-eight percent of those surveyed in the larger village had used a cell phone at some point, as had 55 percent in the remoter village: Sharing of phones among household members, neighbors, and friends is evidently widespread (as in Kibaha). A few young people in both settlements reported having received funds through mobile money services.

In the context of limited transport in the locality, the potential opportunities to benefit from connectivities between phones and commercial transport operations are fewer than in Kibaha. In the major roadside settlement to which these villages are connected by earth road, bicycle taxi operators report significant benefits from phone ownership in building a customer base, which extends into the surrounding rural settlements:

> When a customer comes, we all fight, compete ... unless the customer has a specific preferred bicycle taxi driver. Therefore, those of us with phones have a fair advantage over our friends. Our customers call us to pick them from various places, so we compete less with others. ... I now have many customers. (Bicycle taxi operator, twenty-six years old)

In the study villages, however, none of the youth interviewed referred to coordination of transport arrangements by phone.

By contrast, there are many interview reports of opportunities being taken to substitute phone contact for travel, especially to more distant locations, whether for business or social reasons:

> (Foster child) calls her father who stays in Chilobwe, once in a while ... just to chat ... she doesn't have to travel all the way to see and talk. (Grandfather, seventy years old)

> I sometimes call the wholesaler ... to find out if baking flour is in stock (before walking there). (Male tearoom owner, twenty-three years old)

Comparisons between transport costs and a phone call are often drawn

> (Before) I was being forced to travel (to see how relatives are). With a cell phone it's cheap since I just call them. (Male, twenty-seven years old)

> I only used K50 for the airtel units (to inform a relative about his grandmother's funeral, rather) than ... K100 for a bicycle taxi, so I feel it was cost-cutting. (Male, twenty-eight years old)

For long journeys, the cost advantages are particularly clear. As one unemployed sixteen-year-old girl, unable to continue school because of lack of examination fees, observed, why save up 6,000 Kwacha for a return ticket to visit her sister, when a call costs only 100 Kwacha. Even for journeys where no direct monetary cost would be incurred, time saved in avoiding needless long walks is appreciated:

> Before I started using the phone in my business, sometimes I used to travel (to town) only to come back without anything because ... people have not started selling their produce or the prices are too high. (Woman farmer/ground nut dealer, twenty-two years old)

In the survey, young people who had used a mobile phone in the last twelve months were asked how this had affected their travel: Although approximately 60 percent overall said it had made no impact, 30 percent in both settlements reported that use of mobile phones had led to a reduction in their longer distance (irregular) journeys, compared to under 10 percent, who said it had led to an increase. Data disaggregation by gender indicates that impact of phone use on journey reduction was particularly great among males (who are likely to have made more long journeys than females prior to phone adoption, given commonly greater access to resources). Forty-six percent of males in the small, remoter settlement and 35 percent in the large settlement reported taking fewer long journeys as a result of phone usage (as opposed to 10 percent in both settlements making more long journeys. For small, local, day-to-day journeys, mobile phones are less clearly associated with travel reduction among either gender, because regular household phone usage currently imposes regular additional local trips for charging batteries and buying airtime: These will probably reduce as small phone-service businesses emerge within the settlement.)

Taking the qualitative and survey data together, it appears that, for some youth—especially males—significant gains are being made through more efficient use of transport facilitated by mobile phones for longer distance journeys. Advance calls are highly advantageous for checking on the prior availability of people or goods and associated journey planning, before committing funds to travel. As with older people, however, the importance of at least occasional face-to-face meetings for maintaining personal relationships was sometimes raised; too much reliance on the phone was seen to encourage what one respondent called "taking away the beauty of people meeting face-to-face" (Life history, male, thirty-one years old). A thirty-five-year-old mother, whose phone was broken, observed, "Although the phone made the difference [fewer visits to her three children living with grandmother in a village two hours' walk away], I could feel the distance between us, so I would organize a trip to go and see them." Copresence is discussed further later.

Discussion and Conclusion: Distance Management in Contexts of Remoteness and Deep Poverty

Fluid interdependencies of mobility—physical and virtual—are growing rapidly in Africa: The remarkable expansion of mobile phone networks is bringing a tangible new dimension of connectivity into mobility, transport, and access equations on the ground. For rural populations with very low disposable incomes, the potential for better distance management offers considerable benefits. Such advantages are likely to be compounded among people whose physical mobility is also constrained. The case studies illustrated how this is working out for two different age groups, in two different countries, yet in fairly similar types of rural place.

In Kibaha, Tanzania, motorcycle taxis have already considerably improved older people's access to health, other services, and overall well-being. This strongly echoes earlier findings from my research on Nigeria's Jos Plateau (1991–2001), where perceptions of improved well-being among off-road populations were directly linked to the arrival of motorcycle taxis. Although costly to hire, they could negotiate rough tracks with relative ease. Despite fares double or triple those by bus (where available) on the same route, motorcycle taxis brought not only greater security in the event of emergencies but also a new sense of connectedness to the wider world (Porter 2002). Although many African governments are concerned to regulate (or ban) motorcycle taxis, in view of high

accident rates and a perceived association with unruly youth, their spread to new areas continues, seemingly inexorably: They fulfill hitherto unmet needs across the continent.

The recent rapid spread of mobile phones across rural Africa has compounded the benefits now afforded by motorcycle taxis in many locations, as transport can be called up when required, rather than having to search for a vehicle (often involving a long walk to a distant paved road). Despite relatively high fares, for the less mobile—including many older people—this integrated connectivity brings a very significant sense of security in the absence of alternative transport, especially in emergencies.

In both case studies, phones not only help organize access to transport but also enable transport substitution. In Lilongwe District, where there is little locally available transport of any type, phone calls reduce the number of required long journeys, many of which are occasioned by social obligations or the need to obtain material or financial resources like school fees. If physical mobility is essential, whether to see someone in person or to obtain or sell goods, travel now usually takes place only once the caller is assured of availability. In both case studies, respondents contrasted the cost of specific journeys with the cost of a phone call (rarely with texting or messaging) to illustrate the benefits they gained by substituting virtual for physical mobility. In contexts of deep poverty, the phone now presents a much valued tool in survival strategy kits, especially for sourcing external resources through family contacts. Essential journeys continue to be made, but they can be more efficiently planned and executed; this is especially valuable where transport services are sparse and incomes are low. Evidence of increased space–time flexibility, of the type that has emerged in Western cities through microcoordination and rescheduling on the move, is absent and would be difficult, given (as yet) often limited rural phone network connectivity. In any case, as result of persistent rural travel uncertainties, time flexibility is already deeply embedded in rural lives. Some reworking of current space–time flexibilities to encompass the opportunities that the new connectivities present is evident. Messages now flow freely to inform about imminent key events such as funerals, when in the past such information often arrived too late for participation. Modal choice has been affected (by contrast with Western contexts) because, wherever motorcycle taxi services emerge, their phone-enabled drivers offer convenience that has hitherto been unavailable.

In both case studies, though, some respondents observed the limitations of the phone as a travel substitute and the importance of copresence, particularly for emotional well-being. This was often expressed simply as a desire to meet with close family or lovers resident at a distance. The centrality of mobility beyond basic survival, for people's social and emotional lives, appears as strong in Africa as in the West: however, respondent narratives suggest that, in practice, face-to-face meetings with distant others are now frequently rationed. This is certainly a factor of poverty among both youth and elders but also relates to prevailing mobility constraints. Poor transport availability and potential breakdowns, accidents, and harassment en route all encourage careful assessment of a journey's value. In the West, those with strong network capital have "the capacity to engender and sustain social relations with those people who are not necessarily proximate" and so generate emotional, financial, and practical benefit (Urry 2012, 197). Many respondents in these African cases are now busily engaged in efforts to build their network capital through the phone; however, the tipping point at which copresence is deemed essential is evidently delayed in conditions of precarity.

On a related theme, little reference was made in these rural spaces to any benefits of travel time (Jain and Lyons 2008), apart from occasional comments by young people about chatting while walking with friends. Lilongwe youth reported frequent journeys occasioned by social obligation linked to their lowly position in local hierarchies of power; these they willingly substituted with a short phone conversation (although from the perspective of an older relative, the failure to visit might be perceived as a significant loss). For older people in Kibaha, long walks are commonly perceived as a harsh imposition, to be endured, not enjoyed: Among the poorest, motorcycle taxi journeys are limited to health emergencies. There seems little resonance, as yet, with Western scenarios where the phone is a travel companion and support, used to amuse or improvise and reschedule on the move by activating a network of connections telephonically for maximum flexibility (Licoppe 2004). Variable mobile network provision in remoter rural areas and consequent interrupted connectivity is currently a significant constraint: The mobile phone is, as yet, principally a home-based technology (albeit of enormous value as such, in the absence of landlines). Network coverage is expanding exponentially in Africa, however: The potential for novel on-journey amusements and improvisations is on the horizon.

To conclude, reflections on the association between phones and travel behavior in this article extend a debate that has continued for at least thirty years in Western contexts, where the overall assessment indicates no concrete evidence of major decline in distance traveled (Aguilera, Guillot, and Rallet 2012; Ben-Elia et al. 2014). These case studies suggest a different scenario. Although social networks are densely threaded through the lives of respondents, the friction of distance is evidently stronger than in Western contexts, especially in rural locations where both transport and financial resources are extremely scarce and other mobility constraints, such as those associated with infirmity or age, could come into play. Urry's (2012) argument that relationships that are maintained at a distance often involve substantial personal, emotional, and relationship costs, clearly holds up in Africa as elsewhere, but against this must be set the reality of costs, financial and otherwise, of travel in what remains largely a walking world. Variation across the globe in patterns and practices of ICT use, mobility, and associated connectivities are to be expected: The ways in which perceived needs for copresence intersect with and mediate physical travel are complex and contingent. The mobility conceptualizations developed within late capitalist urban societies have limited application in rural Africa because extreme precarity interposes different constraints: As Schwanen, Kwan, and Ren (2008, 2120) observed, "Geography is certainly not dead in the Information Age"!

Acknowledgments

This article draws on research from various projects I have led. I am most grateful to the very many people who have participated and collaborated in these studies and to the helpful comments from the editor and three anonymous reviewers. I wish to make particular acknowledgment of the contribution of James Milner of the University of Malawi, who sadly died following a car accident in Malawi in September 2014.

Funding

The research in Tanzania was funded by the UK Department for International Development under its Africa Community Access Programme (AFCAP) through HelpAge International. The research in Malawi is funded by the UK Economic and Social Research Council (ESRC) and the Department for International Development (Grant ES/J018082/1, for which data will be archived in accordance with ESRC policy).

References

Aguilera, A., C. Guillot, and A, Rallet. 2012. Mobile ICTs and physical mobility: Review and research agenda. *Transportation Research Part A* 46:664–72.

Aker, J. C., and M. Mbiti. 2010. Mobile phones and economic development in Africa. Working Paper 211, Center for Global Development, Washington, DC.

Ben-Elia, E., B. Alexander, C. Hubers, and D. Ettema. 2014. Activity fragmentation, ICT and travel: An exploratory path analysis of spatiotemporal interrelationships. *Transportation Research Part A* 68C:56–74.

Chavula, H. K. 2013. Telecommunications development and economic growth in Africa. *Information Technology for Development* 19 (1): 5–23.

Deglise, C., L. S. Suggs, and P. Odermatt. 2012. SMS for disease control in developing countries: A systematic review of mobile health applications. *Journal of Telemedicine and Telecare* 18 (5): 273–81.

Elliott, A., and J. Urry. 2010. *Mobile lives.* London and New York: Routledge.

Jain, J., and G. Lyons. 2008. The gift of travel time. *Journal of Transport Geography* 16 (2): 81–89.

Kwan, M.-P. 2006. Transport geography in the age of mobile communications. *Journal of Transport Geography* 14:384–85.

Licoppe, C. 2004. "Connected" presence: The emergence of a new repertoire for managing social relationships in a changing communication technoscape. *Environment and Planning D: Society and Space* 22:135–56.

Line, T., J. Jain, and G. Lyons. 2011. The role of ICTs in everyday mobile lives. *Journal of Transport Geography* 19 (6): 1490–99.

Nohria, N., and R. Eccles. 1992. Face-to-face: Making network organizations work. In *Networks and organizations,* ed. N. Nohria and R. Eccles, 288–308. Boston: Harvard Business School Press.

Overa, R. 2006. Networks, distance and trust: Telecommunications development and changing trading practices in Ghana. *World Development* 34:1301–15.

Peters, P., S. Kloppenburg, and S. Wyatt. 2010. Coordinating passages: Understanding the resources needed for everyday mobility. *Mobilities* 5 (3): 349–68.

Porter, G. 2002. Improving mobility and access for the off-road rural poor through intermediate means of transport. *World Transport Policy and Practice* 8 (4): 6–19.

———. 2011. "I think a woman who travels a lot is befriending other men and that's why she travels": Mobility constraints and their implications for rural women and girl children in sub-Saharan Africa. *Gender, Place and Culture* 18 (1): 65–81.

———. 2012. Mobile phones, livelihoods and the poor in sub-Saharan Africa: Review and prospect. *Geography Compass* 6:241–59.

———. 2014. Exploring collaborative research methodologies in the pursuit of sustainable futures. In *Sustainable development: An appraisal from the Gulf Region,* ed. P. Sillitoe, 419–35. Oxford, UK: Berghahn.

Porter, G., and A. Abane. 2008. Increasing children's participation in transport planning: Reflections on methodology in a child-centered research project. *Children's Geographies* 6 (2): 151–67.

Porter, G., K. Hampshire, M. Bourdillon, E. Robson, A. Munthali, A. Abane, and M. Mashiri. 2010. Children as research collaborators: Issues and reflections from a mobility study in sub-Saharan Africa. *American Journal of Community Psychology* 46 (1): 215–27.

Porter, G., A. Heslop, F. Bifandimu, E. Sibale, A. Tewodros, and M. Gorman. 2014. Exploring intergenerationality and ageing in rural Kibaha Tanzania: Methodological innovation through co-investigation with older people. In *Intergenerational space,* ed. R. Vanderbeck and N. Worth, 259–72. London and New York: Routledge.

Porter, G., A. Tewodros, F. Bifandimu, M. Gorman, A. Heslop, E. Sibale, A. Awadh, and L. Kiswaga. 2013. Transport and mobility constraints in an aging population: Health and livelihood implications in rural Tanzania. *Journal of Transport Geography* 30:161–69.

Schwanen, T., and M.-P. Kwan. 2008. The internet, mobile phone and space–time constraints. *Geoforum* 39:1362–77.

Schwanen, T., M.-P. Kwan, and F. Ren 2008. How fixed is fixed? Gendered rigidity of space–time constraints and geographies of everyday activities. *Geoforum* 39 (6): 2109–21.

Urry, J. 2012. Social networks, mobile lives and social inequalities. *Journal of Transport Geography* 21:24–30.

The Way They Blow the Horn: Caribbean Dollar Cabs and Subaltern Mobilities

Asha Best

Department of American Studies, Rutgers University

In this article, I map subaltern mobilities: practices of movement that I define as flexible, vernacular, and specific to postcolonial subjects. I do so through a six-month ethnography of "dollar cabs" used by Caribbean immigrants in Brooklyn, New York—taxis recognized not by exterior color or medallion but by the way they blow their horns, the familiarity between driver and passengers, and other diacritics this article critically attends to. These discursive geographies and practices allow Caribbean immigrants to navigate the U.S. urban landscape and to interact with each other in unique ways. Because dollar cabs often operate outside of dominant structures of licensure, they have been studied primarily as informal paratransit systems. This article offers a critique of the framework of informality as it relates to mobilities of subaltern subjects and argues that, given their focus on systems rather than practices, scholars have foreclosed on the analytical possibilities of fully understanding the social within these geographies of mobility. Through this ethnography I make a significant theoretical and methodological intervention by showing how both international and local subaltern movements and flows have disrupted, produced, and been affected by the global city.

我在本文中绘製从属者的能动性：我所定义为弹性、本土且特别对后殖民主体而言的移动实践。我透过对纽约布鲁克林区的加勒比海移民所使用的"一元出租车"进行为期六个月的民族志来完成上述工作——该出租车并不以外在的颜色或徽章加以识别，而是透过它们的鸣笛方式，驾驶与乘客之间的熟识，以及本文批判性地关注的其他区分进行识别。这些论述地理与实践，使得加勒比海移民能够以特别的方式，在美国城市地景中游走并进行互动。由于一元出租车经常是在核发执照的统治结构之外运作，因此它们主要被视为非正式的辅助运输系统而进行研究。本文提出一个非正式性架构的批判，而它关乎从属主体的能动性，并主张，有鉴于学者们聚焦系统而非实践，因而失去了全面理解这些能动性地理中的社会性的分析可能。我透过此般民族志，藉由显示国际与在地从属者的移动与流动如何扰乱及生产全球城市，并受到全球城市所影响，以此创造显着的理论与方法论介入。

En este articulo, mapeo movilidades subalternas: prácticas de movimiento que defino como flexibles, vernáculas y específicas para sujetos poscoloniales. Hago esto con base en una etnografía de seis meses a los "taxis de dólar" usados por inmigrantes caribeños en Brooklyn, Nueva York—taxis que se reconocen no por su color exterior, o por el medallón, sino por la manera como hacen sonar la bocina, la familiaridad entre el conductor y los pasajeros y otras diacríticas de las cuales se ocupa críticamente el artículo. Estas geografías discursivas y prácticas permiten a los inmigrantes caribeños navegar en el paisaje urbano de los EE.UU. e interactuar entre sí de maneras únicas. Debido a que los taxis de dólar a menudo operan fuera de las estructuras dominantes de licencia, se han estudiado primariamente como sistemas de paratránsito informal. Este artículo presenta una crítica del marco de la informalidad en cuanto se relaciona con las movilidades de sujetos subalternos y arguye que, dado su enfoque sobre sistemas más que sobre prácticas, los eruditos han centrado su atención en las posibilidades analíticas de comprender completamente lo social dentro de estas geografías de la movilidad. A través de esta etnografía hago una intervención teórica y metodológica significativa mostrando cómo los movimientos y flujos subalternos internacionales y locales han afectado, producido y han sido afectados por la ciudad global.

In Haiti there are *tap-taps*, in Manila *jeepneys*, in Ghana *tro-tros*, in Kenya *matatus*, in South Africa *combis*, in Barbados and Jamaica minibuses or minivans, and in Trinidad and Tobago maxi taxis. In Brooklyn, New York, there are dollar vans.[1] You can find them in predominantly Caribbean neighborhoods like East Flatbush, picking up along Avenue D, Utica, Church, Nostrand, Glenwood, and Flatbush. Like the tro-tro, the dollar van, or dollar cab, gets its name from the original fare. When dollar cabs came about they cost a dollar per ride; otherwise, passengers could pay with subway tokens.[2] The driver picks up and drops off almost anywhere along a fixed route, blowing the horn to hustle for fares. Like the tap-taps and jeepneys, this form of quasi-public transportation is a critical site of sociality, particularly for Caribbean immigrant communities living in the Brooklyn borough. A material response to urban divestment in the 1970s and 1980s, the dollar cab allowed black immigrant communities to live and work, to move between sites of leisure and labor, and to make the strange familiar by making available in the United States a mode of travel popular in the Caribbean. It allowed black immigrants to inhabit the space through the everyday practice of getting around it.

By mapping the racial and spatial landscape of postindustrial Brooklyn, documenting cycles of urban blight and gentrification, the loss of manufacturing jobs in the postwar era and subsequent fiscal crises, divestment in and increased privatization of public transportation, urban historians have well attended to "complex relations of conflict" (Barnard College 2012) and spatial management that script post-1960s New York (Mollenkopf 1983; Rose 1994; Wilder 2000; Osman 2011). As a practice and as a form of urban life that is firmly situated in the context of these complex relations and shifting terrains, the dollar cab should be located within these broader histories. And yet, because dollar cabs in the United States often operate outside of dominant structures of licensure, they have been studied primarily as informal paratransit systems without a historical or social life (Cervero 1997). In approaching the study of dollar cabs and the dollar cab market in this way, urban planning literature in particular but also urban histories have often foreclosed on the analytical possibilities of fully understanding the social within these geographies of mobility rather than situating dollar cabs within broader frameworks of transnational life, race, gender, class, citizenship, labor, space, and speculation.

In what follows, based on six months of ethnographic field work in East Flatbush, Brooklyn, I attend to both the cultural logics of these mobilities and the intellectual strictures that have produced either descriptive or prescriptive understandings of dollar cabs.[3] I do so by putting mobilities studies in conversation with subaltern studies and black geographical analysis to map what I refer to throughout this discussion as subaltern mobilities, or practices of movement defined as flexible, vernacular, and specific to postcolonial subjects.[4]

I argue that the dollar cab market can be understood as a form of "subaltern urbanism" (Roy 2011b). Subaltern urbanisms here are defined as practices of "embodiment and habitation" that "encompass complex relations of conflict and transaction between dominant protocols for organizing life, and practices of survival by those targeted for relocation and elimination from the city, and who are subject to various zoning as well as coding practices designed to regulate their presence in the city" (Barnard College 2012). As an analytical framework, this allows us to attend to those seemingly banal urban practices that make life possible in the face of exhausting spatial management—practices that are often studied under the rubric of informality. Specific types of life and labor, including transit systems like dollar cabs or dollar vans, fall under such a rubric as they emerge from the interstices created by shifts in capital and allow people to survive when differential forms of mobility are imposed on them. I use archival data, ethnographic interviews, and observation not to delve into the everyday lives of dollar van drivers or passengers (which I read as descriptive) or to make inferences about how to legitimize this form of transit through legalism (which I read as a prescriptive) but rather to look at how the movement of black people, of capital, of these particular types of vehicles, through urban space redistributes categories of life and labor in city spaces like Brooklyn. I posit that analytical attention to everyday, banal movements and flows offers a productive rubric for understanding black, immigrant, popular, postcolonial life.

Origins, Global Circuits, and "Ungeographic" Terrain

> Wherever car culture reshapes society ... that triumph places technology and power squarely in the middle of ordinary life which is transformed by the ways that cars have redefined movement and extended sensory experience. We need to be able to examine the workings of this ordinary potency within power's wider circuitry: governmental, metropolitan, ecological and, of course, economic. The history of black communities and automotivity can help accomplish these tasks. (Gilroy 2001, 89)

The dollar cab has its own sort of origin story. It is more teleological than genealogical or historical, composed of a set of incomplete narratives around its nascency. The first has to do with the 1980 transit union workers' strike, which greatly restricted movement around the city of New York. The emergence of the dollar cab is sometimes connected to this twelve-day event (King and Goldwyn 2014). The second set of narratives is more elaborate and has to do with a major court case involving the right to legally operate a dollar cab. In 1997, after a seven-year struggle, a van cooperative called Brooklyn Van Lines won a lawsuit against the City and State of New York. The suit challenged a statute passed in 1993 that required the city council to approve licenses for dollar van drivers and granted the city council the authority to void licenses approved by the New York City Taxi and Limousine Commission (TLC; Best 1997; Broome 1997; Mellor 1997). It was a resolution that would stunt the growth of the dollar cab market and criminalize vehicles in operation.

The print archive occludes grassroots organizing efforts and the large-scale protests by dollar van drivers who blocked the Brooklyn Bridge in support of the suit. Instead, most of the print media coverage of the legal battle focuses on either one of two men, Vincent Cummins of Brooklyn Van Lines or Hector Ricketts, who operated dollar vans in Queens, two drivers framed as exceptional, nonradical immigrants with both a desire and a propensity for hard work. A letter to Speaker Peter Vallone from one Caribbean civic group in support of the lawsuit described Caribbeans as "hard working people" who "believe in the 'American Dream' and subscribe to our 'Free-Enterprise System.'" Further, the letter states, "The 'Dollar Van System' encourages self-sufficiency, entrepreneurship and reduces reliance on public social support systems." In the wake of the court battle, neoliberal discourse around entrepreneurial success and disdain for social services were operationalized and tagged to the dollar cab, effectively linking masculinity, individualism, legalism, automobility, and the free market—its own antiblack spatial imaginary.

The scholarly literature on dollar cabs leaves this schema largely untouched. In suggesting that this court case opened up the dollar cab market, the literature, in part, reproduces neoliberal narratives around entrepreneurship. As Kasinitz (1992) noted, the "image of West Indians as ghetto entrepreneurs characterizes much of the academic literature" (90). My goal is not to challenge this idea by arguing that dollar cabs represent some part of a Caribbean "community economy," located outside of, or in opposition to, itineraries of contemporary capitalism (Roy 2011b, 229). I argue the opposite: Dollar cabs and dollar vans—this

market of informal mobilities—is very much invested in and directed by capitalist itineraries. Mathew's (2005) study on New York's yellow taxi industry mapped a clear distinction and class struggle between working-class unionized drivers and medallion owners who also own corporatized taxi companies. This binary does not quite translate in the case of dollar cabs, though. Those who are part of the dollar cab market occupy various class positions, but across the segments a deep attachment to neoliberalism resides.

Instead I want to focus on the fact that there were a number of geographies linked by intersecting movements and flows that prompted the growth of this market and that allow us to historicize the emergence of the dollar cab with greater complexity. Mapping this landscape moves us past the opportunistic immigrant entrepreneur narrative and suggests that the dollar cab can be understood as an aesthetic object through which we can link immigrant communities and technologies of race to global circuits of capital and processes of globalization.

As a point of departure, we can start by considering that many parts of Brooklyn were regarded as marginal or "ungeographic" spaces (McKittrick 2006, 58), rendering the bodies that inhabit those places illegible and their proximity to global circuits of capital incalculable. Salient forms of urban deprivation (e.g., a lack of transportation) meant to manage these populations affected the mobility of working-class black immigrants living in the boroughs. Post-1965 immigrant flows from the Caribbean to New York, specifically the boroughs, brought a new mode of getting around, a mode of living that was already part of popular life in the Caribbean. Later, with the solidification of the North American Free Trade Agreement (NAFTA) in the 1990s, auto parts, alongside immigrant workers from the Caribbean, South Asia, and other regions in the Global South, flowed across international borders and traveled to the emergent global cities, ultimately reassembling themselves on the streets of places like Queens and East Flatbush.

The transit strike of 1980 should not be regarded as phenomenal; it can be understood as a result of ongoing labor struggles happening across the city of New York in the wake of divestment in public services. The Grand Marquis, Crown "Vics," and Econoline vans, referred to in the vernacular as "dollar cabs," running along the streets of Brooklyn became a viable way to navigate these shifting terrains. Dollar cabs forged a sort of "priority lane" for black working people traveling throughout parts of Brooklyn, connecting to

Manhattan via train. There was a strong opposition to and consequent informalization of these urbanisms by the city and political engagement by many van drivers. As Roy (2011b; 2012) reminded us, though, modes of work and life generally understood to be unregulated, or to fall outside of dominant understandings of life and labor, are central to the dominant mode of producing space. There has since been intense police surveillance and overregulation of this type of mobility even while it is generally understood to be an unregulated form of transportation. It is through the very organization and regulation of space and the regulation of movement throughout urban space that understandings of bodily difference are produced.

Rather than engendering this form of subaltern urbanism, then, the 1997 Brooklyn Van Lines court case marks the limits of subalternity for a segment of the dollar cab market. In creating and conferring "institutional agency" (Spivak 2013) to the figure of the Caribbean immigrant entrepreneur, the court victory of 1997 did not spark but rather changed the nature of the dollar cab industry and prompted the rise of the owner-operator in the dollar cab market. Ultimately this resulted in a segmented market giving way to tensions between "plastics" (drivers who operate without TLC licensure) and those who have TLC authorization and can therefore form van collectives and lease their vehicles to other drivers. In my research, I interviewed van drivers from both categories. It is the plastics who still inhabit the position of the subaltern and whose caricature has become the controlling image for the dollar van market. Although both plastics and TLC licensed drivers are subject to heightened police surveillance and contact, it is the former who are the figure of racial difference and foreignness. The plastic engages in a form of automobility that, although it allows him to earn a living wage, excludes him from the neoliberal definitions of freedom and autonomy promised by automobility.

Technologies of Automobility[5]

As we approach the Botanical Gardens, the driver speaks into the mic on his CB radio "De Southside of the library—blue and white." He warns the other drivers about the heightened police presence. "Flatbush hot today." The drivers on the CB continue to chatter about police, and traffic, sometimes joking. They laugh, retelling the story of someone who "defecated himself like a foul" at a police stop.

"You haffe drive 5 miles per hour . . . cause if you drive 35 dem a say you drive 50," the driver says to no one in particular.

In popular media and even in some academic discourse, the dollar cab market is often framed through the discourse of criminality, the response to which is heightened contact with and surveillance by the police justified by concerns about speed. As a fundamental trope through which the city of New York is imagined, realized, and "worlded" (Sassen 2001; Roy 2011a), as a technology of mobility fetishized in popular culture, and as a practice that has real and devastating bodily consequences in the world of driving, speed is a diacritic that calls for a bit of exploration. For dollar van drivers, the speed at which they travel does not signal efficiency and freedom but "recklessness," a word that was used by almost each passenger that I interviewed and two interviewees from state agencies. Speed in this context is a sign of irrationality, difference, and cause for oversight. As the drivers that I interviewed explained, however, speed is also a result of having to hustle for passengers in a growing street economy.

Here is an observational note: I was in a dollar van among a group of passengers headed downtown Brooklyn via Flatbush Avenue when another driver skirted in front of us to pick up a passenger in a bus zone, quickly ducking back into traffic with smoothness and ease. Leaning forward on the wheel, he swerved in and out of traffic. Chuckling to himself and shaking his head, the driver of the van in which I was a passenger managed to pull alongside the swerving van. They were familiar with one another. Through his window one driver warned the other that TLC was out ticketing that day. "Fuck the TLC," the swerving driver replied. It was then that I wondered about the relationship between recklessness and refusals.

As Hannam, Sheller, and Urry (2006) observed, "Mobility is always . . . caught up in power geometries of everyday life" (3; see also Cresswell 2006; Urry and Sheller 2006; Cresswell and Merriman 2011). These power geometries are brought into sharp relief along the corridor between Lefferts and East Flatbush, a heavily policed area that creates a boundary between gentrified Brooklyn and other sites under speculation, suggesting that the informalization of this market and the attendant police contact acts in the service of Brooklyn's most recent cycle of gentrification. Initiatives like Vision Zero, a campaign aimed at ending traffic fatalities and creating "safe" streets, and the installation of bike lanes indicate an imagined New York that is more walkable, less congested, and structured to accommodate alternative mobilities. Although this is a hopeful spatial imaginary, this new New York, this new Brooklyn, this space of unfettered mobility relies on regimes of spatial management, police contact, and surveillance.

As Morley and others have reminded us, "In all these discussions of 'speed,' the often overlooked question concerns its obverse: waiting—which is of course, often the fate of the poor, or those who lack the qualifications which give access to the relevant 'fast-track' or priority lane" (Morley 2011, 753; see also Massey 1991; Hutchinson 2003). In the dollar cab market, waiting serves a different function: It has become a useful mode of counterveillance. An analysis of waiting in this context gives us the opportunity to rethink the potential equation between speed, time, and labor. I am thinking here of Tadiar's (2012) explanation of "life-times as an attempt to account for the productivity of social practices of life and experience, which appear to lie outside the formal sites of labor exploitation" (11). It was explained to me by several drivers that their downtime overlaps with the time of day that the police regularly change shifts and are apt to distribute more tickets and increase traffic stops. Sometimes downtime is used to attend to other business (many drivers have other sources of work). Sometimes downtime becomes life time. Because there are few authorized dollar van stops (they exist but are sparsely located in residential neighborhoods and not well labeled), dollar cabs and vans make improvised hubs at local businesses with available parking lots and within close range to Wi-Fi access. I observed one van disassembled to make something akin to a bus shelter or bench: A row of seating was pulled out of the van and placed on the curb to make outdoor seating for the driver and his entourage as they share beverages and bump music through a sizable sound system. The seat was reinstalled at the end of this downtime.

The CB radio, another mode of counterveillance and recognition, has become a fixture in dollar cabs and vans, a material response to police surveillance and hyperregulation. This "lo-fi" technology outfits most dollar vans and some dollar cabs, representing a unique technoscape, challenging the discourse that locates black people, immigrants, outside of the realm of the digital. As Morley (2011) argued, it is critical to attend to "the factors governing the differential mobility of the people who constitute media audiences, in terms of available transport technologies, techniques for the regulation of both on- and offline territories ... and the regulation of flows of both messages and people" (744). Discreet vernaculars of the Caribbean, communicated through the digital, have materialized in response to these power asymmetries. Police are often identified through the word "Babylon," Jamaican patois signifying a system of policing, corruption, and government repression. Here we can see the dollar cab, or taxi, as what Sharma (2008) described as a medium, "a communicative space between driver and fare ... the sharing of knowledge about the local and global" (457).

Life in the Rearview: On Aesthetics and Affective Labor

> Marcia is running late this morning so instead of walking to Utica, we walk over to Ave D. We cross the street and hop in a sedan. Hanging on the rearview mirror there's a laminated card that says "Haiti Labor Day" and the music coming from the radio is patois.

Vernacular mobilities produce unique material cultures and moments of communication in postcolonial street life (Suzuki 1985; Chattopadhyay 2009). In this article, it is my hope to highlight this and other ways of knowing, other modes of recognition of these subaltern movements and flows. Here, I focus briefly focus on van and cab interiors and decorative hangings placed on rearview mirrors. In doing so, I follow Chattopadhyay's explanation of the importance of visual text in studying subaltern social life:

> If the ethnographer's habit of citing voices from interviews ... insinuates a claims to authenticity ... it is important to remember that the problem of studying the culture of marginalized groups resides not in the absent voice of the subaltern. Voices of the marginalized have always been present in one form or another; if not in oral and written traces, then in material culture and performance. (114)

Although the political conditions under which "the automobile as a vehicle of modernity has been adopted" into places like the Haiti and Ghana are different than those under which the automobile was integrated into Brooklyn, New York, there is utility in locating dollar vans in Brooklyn within a broader network of postcolonial geographies of mobility. I asked participants who were passengers in dollar cabs and

Figure 1. Jamaican flag on the rearview mirror of dollar van, Brooklyn, New York, 2014.

vans how they knew whether or not the driver was Caribbean. Several participants noted that there were often flags hanging on the rearview.

I often noted flags or other symbols of national affiliation (see Figure 1). In the preceding vignette, a laminated card from Haiti's Labor Day celebration hangs from the driver's rearview, and in the image a Jamaican flag is tethered to the mirror. I posit that this card, like the flag, is a form of affective labor, part of the aesthetics of nationalism; the driver is continually engaged in a discursive sort of nationalism by displaying a flag or other national memento. This invites the uncomfortable possibility, then, that nationalism offers a sense of place to those whose bodies are treated as out of place through the informalization of their labor and, thus, their everyday lives.

Conclusion

By treating the dollar cab prescriptively at best (as a thing that needs revision or solution) or at worst as a form of piracy (gypsies that steal passengers from legitimate modes of public transport), it is difficult if not impossible to recognize this market as integral to the production of space or to think of dollar cabs as inhabiting their own cultural location. One of the goals of this article is to illustrate how both international and more localized mundane movements and flows have

disrupted, produced, and been affected by the global city. To do so I have organized this article into three main sections.

First, I map the historical terrain in which the dollar cab emerged, attending to the multiplicity of flows that engendered and reshaped Brooklyn's street economies and the dollar cab as a form of transport. I give some attention to questions of regulation, formality, and labor in relation to mobility. I argue that if it is as Roy (2012) proposed—that informality is "the fickle, arbitrary, unstable relationship between the legal and the illegal, legitimate and illegitimate, authorized and unauthorized" (689)—dollar cabs inhabit a space of contingency. At times the dollar cab is necessary to the city when other forms of mobility have been interrupted (e.g., during Hurricane Sandy when transit came to a halt, dollar cabs were still operable). At others times, dollar cabs are treated as unsafe and parasitic, stealing passengers from the Metropolitan Transit Authority and recklessly endangering the welfare of pedestrians and bike riders. I draw on the theoretical impetus of studies of subaltern urbanism as an entrée to critique of neoliberal logics of entrepreneurship that often underwrite how particular practices of labor are discussed in urban studies. I then suggest how definitions of automobility function in the context of subaltern mobilities and how understandings of automobility might be revised in the context of subaltern geographies of mobility. Finally, I highlight the expressive and performative cultures through which the subaltern "speaks," gesturing toward the production of an aesthetic material culture that allows communication between drivers and passengers, as well as drivers and other spectators.

I want to conclude by proposing that to recognize the dollar van in a more substantive and capacious way requires a critical race perspective. One of the disconcerting overlaps between academic and popular discourse around dollar cabs has been how technologies of automobility are not only framed but how the uses of these technologies become perverse modes of racialization. When downtime becomes dead time, becomes "lurking," and the very presence of a van in the traffic gridlock so common to other parts of New York is described as the cause of "pollution" and "congestion" it is necessary to think about how these terms have been used historically to code racial and ontological difference. It is critical to expand research on subaltern urbanism through an analysis of the consequences of these signs and values for the bodies that move throughout the dollar cab market, namely, its black immigrant passengers and drivers.

Notes

1. Although a real count of the number of dollar cabs would be difficult to ascertain, it has been estimated that there are more than 200 dollar cabs operating in Brooklyn alone.
2. Although dollar cabs and dollar vans are different types of vehicles, I sometimes use the term *dollar cab* to refer to either the car or the van because, for passengers, they perform the same function and the terms are often used interchangeably.
3. This project is based on a mobile ethnography conducted over the course of six months. I interviewed twenty research participants, including dollar cab drivers, passengers, those from government agencies, and those from community organizations. When permitted, I also observed drivers and passengers on their daily travel routes.
4. There is a growing body of work that analyzes subalternity within the context of mobilities studies (see, e.g., Samanta and Roy 2013). I attempt to build on this work by thinking about subaltern mobilities in the U.S. context but find it useful to define how I am using the term *subaltern* and why I find it instructive. Therefore, in keeping with what Spivak (1994) proposed in "Can the Subaltern Speak?" as well as her reflections in the postscript "The Subaltern and the Popular" (Spivak 2012), I employ the term subaltern not as a synonym for other geopolitical identity markers like immigrant, third-world, or postcolonial, although it is related to all of these things. Nor do I use subaltern to indicate a class position. I use the term subaltern as a referent for popular practices of mobility enacted by postcolonial subjects that presently lack viable intellectual recognition precisely because they are "imbricated with the idea of non-recognition of agency" (Spivak 2012, 432).
5. For a robust discussion of automobilities, see Sheller and Urry 2000; Edensor 2004; Thrift 2004; Urry 2004, 2007; Seiler 2008.

References

Barnard College. 2012. Anupama Rao, Saidiya Hartman and Neferti Tadiar: "Subaltern Urbanisms." Year-Long Heyman Center Faculty Seminar. https://womensstudies.barnard.edu/node/20791 (last accessed 26 January 2016).

Best, T. 1997. Attack on van plan: City Council committee votes to impose moratorium on license. *The New York Carib News* 23 September 1997:5.

Broome, R. 1997. Give dollar vans a square deal. *Caribbean Life*.

Cervero, R. 1997. *Paratransit in America: Redefining mass transportation.* Westport, CT: Praeger.

Chattopadhyay, S. 2009. The art of auto-mobility. *Journal of Material Culture* 14 (1): 107–39.

Cresswell, T. 2006. *On the move: Mobility in the modern western world.* London and New York: Routledge.

Cresswell, T., and P. Merriman. 2011. *Geographies of mobilities: Practices, spaces, subjects.* Farnham, UK: Ashgate.

Edensor, T. 2004. Automobility and national identity: Representation, geography and driving practice. *Theory, Culture & Society* 21 (4–5): 101–20.

Gilroy, P. 2001. Driving while black. In *Car cultures*, ed. D. Miller, 81–104. Oxford, UK: Berg.

Hannam, K., M. Sheller, and J. Urry. 2006. Mobilities, immobilities and moorings. *Mobilities* 1(1): 1–22.

Hutchinson, S. 2003. *Imagining transit: Race, gender, and transportation politics in Los Angeles.* New York: Peter Lang.

Kasinitz, P. 1992. *Caribbean New York: Black immigrants and the politics of ace.* Ithaca, NY: Cornell University Press.

King, D., and E. Goldwyn. 2014. Why do regulated jitney services often fail? Evidence from the New York City Group Ride Vehicle Project. *Transport Policy* 35:186–92.

Massey, D. 1991. A global sense of place. *Marxism Today* 38:24–29.

Mathew, B. 2005. *Taxi! Cabs and capitalism in New York City.* New York: The New Press.

McKittrick, K. 2006. *Demonic grounds: Black women and the cartographies of struggle.* Minneapolis: University of Minnesota Press.

Mellor, W. 1997. Let the vans roll: An open letter to City Council Speaker Peter Vallone. *The New York Carib News* 5 August 1997.

Mollenkopf, J. H. 1983. *The contested city.* Princeton, NJ: Princeton University Press.

Morley, D. 2011. Communications and transport: The mobility of information, people and commodities. *Media, Culture & Society* 33 (5): 743–59.

Osman, S. 2011. *The invention of Brownstone Brooklyn: Gentrification and the search for authenticity in postwar New York.* New York: Oxford University Press.

Rose, T. 1994. *Black noise: Rap music and black culture in contemporary America.* Hanover, NH: University Press of New England.

Roy, A. 2011a. Postcolonial urbanism: Speed, hysteria and mass dreams. In *Worlding cities: Asian experiments and the art of being global*, ed. A. Roy and A. Ong. Oxford, UK: Blackwell.

———. 2011b. Slumdog cities: Rethinking subaltern urbanism. *International Journal of Urban and Regional Research* 35 (2): 223–38.

———. 2012. Urban informality: The production of space and practice of planning. In *Oxford handbook of urban planning*, ed. R. Weber and R. Crane, 691–705. New York: Oxford University Press.

Samanta, G., and S. Roy. 2013. Mobility in the margins: Hand-pulled rickshaws in Kolkata. *Transfers* 3 (3): 62–78.

Sassen, S. 2001. *The global city: New York, London, Tokyo.* 2nd ed. Princeton, NJ: Princeton University Press.

Seiler, C. 2008. *Republic of drivers: A cultural history of automobility in America.* Chicago: University of Chicago Press.

Sharma, S. 2008. Taxi as media: A temporal materialist reading of the taxi cab. *Social Identities* 14 (4): 457–64.

Sheller, M., and J. Urry. 2000. The city and the car. *The International Journal of Urban and Regional Research* 24 (4): 737–57.

Spivak, G. 1994. Can the subaltern speak? In *Colonial discourse and post-colonial theory: A reader*, ed. P. Williams and L. Chrisman, 66–111. New York: Columbia University Press.

———. 2013. *An aesthetic education in the era of globalization.* Cambridge, MA: Harvard University Press.

Suzuki, P. T. 1985. Vernacular cabs: Jitneys and gypsies in five cities. *Transportation Research Part A: Policy and Practice* 19 (4): 337–47.

Tadiar, N. X. M. 2012. Life-times of becoming human. *Occasion: Interdisciplinary Studies in the Humanities* 3:1–17. http://occasion.stanford.edu/node/75 (last accessed 8 October 2015).

Thrift, N. 2004. Driving in the city. *Theory, Culture & Society* 21 (4–5): 41–59.

Urry, J. 2004. The "system" of automobility. *Theory Culture and Society* 21:25–39.

———. 2007. *Mobilities.* Cambridge, UK: Polity.

Urry, J., and M. Sheller. 2006. A new mobilities paradigm. *Environment and Planning A* 38:207–26.

Wilder, C. S. 2000. *A covenant with color: Race and social power in Brooklyn.* New York: Columbia University Press.

Fixing Mobility in the Neoliberal City: Cycling Policy and Practice in London as a Mode of Political–Economic and Biopolitical Governance

Justin Spinney

School of Planning and Geography, Cardiff University

Academic interest in utility cycling has burgeoned in recent years with significant literature relating to the health and environmental benefits of cycling, the efficacy of cycle-specific infrastructure, and the embodied experiences of cycling. Yet with few exceptions, none of these accounts conceptualizes cycling as a mode of neoliberal governance through which circulation and quality of labor are improved. This article seeks to address this absence, positioning cycling in relation to broader biopolitical and political–economic governance in two ways: Focusing on the recent experiences of London (UK) it argues first that cycle promotion is a principally biopolitical "mobility" fix that seeks to operate through shaping individuals as entrepreneurs of the self who will move more efficiently. Second, and in relation to this biopolitics of cycling, it suggests that the minimal spatial fixing embodied in cycle-specific infrastructure represents a metaphorical fix resulting from tensions created between the enterprise of cycling, the realities of practicing it in hostile urban environments, and the temporary networks of actors that produce it as such. In doing so, the article contends that what is being materialized is a narrow productivist framing of cycling both materially and discursively in the shape of commuter-focused infrastructure and promotion that effectively marginalizes subaltern and alternative performances. The article concludes by arguing that theorizing the promotion and practice of cycling as part of broader processes of neoliberalization should help to direct future research agendas for cycling and critical mobilities scholarship more broadly.

近年来, 学术界兴起了对实用自行车的兴趣, 并伴随着有关骑自行车的健康及环境益处、自行车专属的基础建设之效力, 以及骑车的身体化经验之大量文献。但除了几个特例之外, 这些说法没有一个将骑车概念化为藉此促进劳动循环和质量的新自由主义治理模式。本文旨在应对此一阙如, 以两种方式将骑自行车置放于广泛的生命政治与政治经济治理之中: 本文首先聚焦伦敦 (英国) 的晚近经验并主张, 自行车提倡主要是生命政治的 "能动性" 修复, 并企图透过将个人形塑为移动更有效率的自身企业家来进行操作。再者, 与上述骑车生命政治相关的是, 本文主张, 自行车专属的基础建设所体现的最小化空间修补, 呈现出由自行车骑乘事业、在不友善的城市环境中骑车的现实, 以及生产该环境的暂时性行动者网络之间创造出的紧张关系的象徵性修补。本文藉着这麽做, 主张被实现的是骑自行车在物质与论述上的狭义生产性框架, 并以有效边缘化从属者和另类展演的以通勤者为核心的基础建设及推广为形式。本文于结论中主张, 将骑自行车的推广与实践理论化作为广泛的新自由主义化过程的一部分, 能够协助指引骑自行车的未来研究议程, 以及更为广泛的批判能动性学术研究。

El interés académico en el ciclismo utilitario ha florecido en años recientes, generando una importante literatura relacionada con los beneficios sanitarios y ambientales del ciclismo, la eficacia de la infraestructura específicamente asociada con la bicicleta y las experiencias que el ciclismo encarna. Sin embargo, con pocas excepciones, ninguno de estos recuentos conceptualiza el ciclismo como un modo de gobernanza neoliberal a través del cual se mejore la circulación y la calidad del trabajo. Este artículo pretende llenar este vacío, posicionando de dos maneras al ciclismo en relación con una más amplia gobernanza biopolítica y político–económica: Concentrándose en experiencias recientes de Londres (RU), el artículo argumenta primero que la promoción de la bicicleta es un artificio principalmente de "movilidad" biopolítica, que busca operar configurando los individuos como empresarios del ego que se desplazarán más eficientemente. Sugiere, en segundo término, y en relación con esta biopolítica del ciclismo, que el mínimo de adaptación espacial inmersa en la infraestructura específicamente ligada con la bicicleta, representa un ajuste metafórico que resulta de las tensiones que crearon la empresa del ciclismo, las realidades de practicarlo en entornos urbanos hostiles y las redes temporales de actores que lo producen como tal. Al hacer esto, el artículo sostiene que lo que se está materializando es una enmarcación productivista estrecha de ciclismo, material y discursivamente, en forma de una infraestructura y promoción centradas en los migrantes pendulares, que marginaliza con efectividad actuaciones subalternos y alternativas. El artículo concluye con el argumento de que teorizar la promoción y práctica del ciclismo como parte de procesos más amplios de neoliberalización, debe ayudar a orientar con mayor amplitud las agendas de futura investigación sobre ciclismo y la erudición sobre movilidades críticas.

Cycling—as a former PhD student recently noted—is the new "apple pie": Everyone likes cycling, right? It is certainly tempting to see cycling as a panacea for a variety of urban afflictions, the promotion of which can only lead to positive outcomes. As Stehlin (2014) has commented, many advocates point to "its importance in contributing to urban vitality, freedom, invigorated commercial districts and sociability" (22). Certainly whether situated in a mobilities or transport ontology, the vast majority of cycling research is conducted by academics and activists who are often cyclists themselves and wish to see more cycling. As Stehlin (2014, 22) goes on to say contemporary cycling advocacy and practice must be critically situated within broader processes of capitalist urbanism.

Taking London (UK) as its focus, this article seeks to do just that by connecting recent developments in cycling policy to biopolitical and political–economic governance. The first section provides a brief review of the undertheorized relationship between transportation and political economy in mobilities scholarship. The second section explores the links between mobility and productivity and the inability of successive administrations to provide a spatial fix to decrease the circulation time of capital and labor. The third section brings together insights on governmentality and, bio–politics with political–economic theorizations of, productivity "fixing" to argue for seeing cycle promotion as a principally biopolitical project that seeks to operate through shaping individuals as entrepreneurs of the self. In doing so it argues that cycling policy can be viewed as a mobility fix (Minn 2013) because cycling represents a way of governing and fixing that works primarily through producing alternative ways of moving rather than producing space (Bærenholdt 2013).

Finally, and in relation to this biopolitics of cycling, the article demonstrates that what little spatial fixing is evident in cycle-specific infrastructure can be understood as an improvised and contingent metaphorical fix (Jessop 2004) resulting from tensions between attempts to mobilize the enterprise of cycling on the part of individuals and the realities of practicing it in hostile urban environments.

The article concludes by discussing some of the consequences of the neoliberalization of urban cycling, particularly a narrow productivist framing of cycling in the shape of commuter-focused infrastructure and promotion that effectively marginalizes subaltern and alternative performances (Hoffman and Lugo 2014;

Stehlin 2014). The concluding hope of the article is that theorizing the promotion and practice of cycling as part of broader processes of neoliberalization (Tickell and Peck 2003) should help to direct future research agendas for cycling (and critical mobilities scholarship more broadly) that focus on issues of inclusivity, representation, and inequality.

Mobilities and Political Economy

As Bærenholdt (2013) has stated regarding mobilities scholarship, despite attention to everyday mobile social practices, "When dealing with issues of power, hegemony and social order, mobility studies are rather vague" (21). In relation to cycling, for example, we have seen many papers that have looked at what cycling as a practice means to those who do it as a sensory engagement (Spinney 2006, 2007; Jones 2012; Nixon 2012; Jungnickel and Aldred 2014), as a subcultural practice (Furness 2005; Fincham 2006), as a gendered practice (Cresswell and Uteng 2008), and as a response to health and environment concerns (Newman, Kenworthy, and Vintila 1995; Button and Nijkamp 1997; Bohm et al. 2006; Bonham 2006). Although insightful, it is fair to say that these accounts have failed to critically engage with the relationship between cycling and the neoliberal state.

Cresswell (2010), for example, excavated a politics of mobility as practiced, usefully illuminating elements of mobile practices that are produced by and productive of social relations (and hence power). Nixon (2012) and Fincham (2006) both produced accounts that tie cycling into the politics of urban mobility and energy use. While insightful, these accounts support the idea that if mobilities scholarship has been viewed as apolitical, it is because its examinations of micropolitics have largely stalled when it comes to relating these to wider political–economic processes.

In one of very few papers linking increases in cycling with broader urban processes, Danyluk and Ley (2007) found an overrepresentation of cycle commuting in gentrifying areas of Canadian cities. Similarly, Pucher and Buehler (2011) confirmed that the majority of cycling in U.S. cities occurs in gentrifying areas.

More recently, papers by Lugo and Hoffman (2014) and Stehlin (2014) have begun to make links between cycling and gentrification in a more overtly political way. Stehlin (2014), for example, highlighted the fact that neighboring poorer districts (and those who cycle in them) have none of the cycle-specific facilities evident in gentrifying areas. Similarly, Hoffman and Lugo (2014) discussed the relationship between the creative class and bicycle infrastructure, arguing that despite good intentions, the emerging geography privileges access for some while also being implicated in urban redevelopment and competitiveness strategies.

One of the few accounts that attempts to link UK cycling policy with a broader neoliberalization is Aldred (2012). Aldred (2012, 96) argued that although concepts such as "hollow state" and the "responsible individual" have gained traction within social policy, their use in transport policy is less evident. She went on to argue that by positioning cycling in relation to health and environment debates (rather than just as transport), the individual becomes "urged" to act as part of an agenda of responsibilization (Aldred 2012, 99). Although such a reading begins to link cycling policy to neoliberal and biopolitical modes of governmentality, it sees UK cycle policy as affected by neoliberal doctrine rather than conceptualizing it as a principle mode of neoliberal government.

With few exceptions, it is evident that transport geography and mobility studies have failed to link cycling (as a particular manifestation of mobility) with broader political–economic processes. Although city form has long been considered as a physical articulation of social relations and ideologies, with changes to their structure providing insight into broader political change (Hackworth 2007, 79), cycling and its infrastructures seem to have somehow been regarded as "value neutral" rather than as an agent of capital accumulation.

Productivity, Urban Mobility, and Biopolitics

Harvey (2003b) has long argued that mobility is central to productivity, acknowledging that many of what he has termed spatial fixes are more accurately spatiotemporal fixes because they are about mobility as providing solutions to contradictions in capital accumulation. Put simply, Harvey theorized mobility as enhancing capital accumulation in two key ways: First, as far as physical goods are concerned, mobility is required to realize the surplus value in goods by bringing them to market. Technological advances that speed up mobility serve to annihilate space through time. The benefit of this to the capitalist is that the time capital spends in circulation (as unproductive) is reduced. As a consequence, Harvey (2003a) pointed

out, "massive amounts of capital and labor have been invested in the sorts of immobile fixed capital we see in airports, commercial centers, office complexes, highways, suburbs, container terminals" (28). Second, these same technological advances in mobility can serve to bring labor together with capital to enhance productivity: "The more mobile the labourer, the more easily capital can adopt new labour processes and take advantage of superior locations" (Harvey 1982, 381).

A key problem highlighted by Harvey (1982), however, is that the physical infrastructures on which mobility relies are "immobile in space and highly vulnerable to place-specific devaluation" (380). As Harvey goes on, "Capitalism increasingly relies upon fixed capital (including that embedded in a specific landscape of production) to revolutionise the value productivity of labour, only to find that its fixity (the specific geographical distribution) becomes the barrier to be overcome" (394). In many cities across the world, the success of particular forms of mobility (notably the car system) has resulted in the circulation of capital and labor slowing down to the point where these systems of mobility reduce the rate of capital accumulation (Hickman and Bannister 2015).

Harvey's argument is exemplified in cities like London where, according to a recent report, "companies located in Central London are highly dependent on the transport of staff" and to a lesser extent that of goods for their productivity (Greater London Authority [GLA] Economics 2005, 53). As the same report states, though, average traffic speeds in the city have decreased consistently since the mid-1970s (GLA Economics 2005) as business and residents have been attracted back to the center and congestion has worsened.

The minimum quantifiable cost of these transport delays to London commuters and businesses has been calculated at £1,190 million per annum (GLA Economics 2005). Including costs for nonwork purposes, the GLA Economics report concluded that the overall cost of congestion in London is £6.7 million per business day, equivalent to 0.9 percent of London's gross domestic product (GLA Economics 2005). For London, the revalorization of inner-city areas with a corresponding increase in the volume of goods and people to be moved within aging and underfunded mobility spaces has evidently come at a price.

Accordingly, successive governments have grappled with the kinds of fixes required to improve the circulation time of capital and labor to increase productivity. In the present there is evidence of a major spatial fix occurring in London in the form of Cross Rail—a new east–west underground rail line. Although this is costly (£14.8 billion), its spatial impact on the surface is relatively limited—the vast majority of this spatial fix is subterranean. In contrast to the UK road trunking program of the 1960s,[1] what has become abundantly apparent is that municipal administrations in cities such as London are unwilling to institute spatial fixes in the form of urban road building.[2]

This reluctance is exemplified in the latest Transport for London (2014) *Roads Modernisation Plan*, which outlines £4 billion worth of road spending between 2014 and 2021. The ambition of its spatial fixes is modest, with no new roads planned; the £4 billion is to be spent on alterations to the existing network including junction treatments, surface treatments, bridge treatments, and signal upgrades. Apart from £1 billion to be spent on cycle infrastructure (discussed later), the £4 billion is little more than a maintenance budget for the existing network. Indeed, as the Transportation for London Roads Task Force (2013) report states, "London is approaching the limit of what can be achieved through road space reallocation alone" (8). In other words, London—like many other cities—has little appetite for terrestrial spatial fixes to the transport system. There are three principal reasons for this: First, and certainly in the current post-crash economic cycle, neither the state nor the capitalist (if we can even still distinguish) is willing or able to take the "massive devaluation" that wholesale rebuilding of the roads system would require to function significantly more efficiently[3]; second, such direct interventions are not favored in neoliberal modes of governance; and third, because the notion that falling productivity is solely due to deteriorating circulation time of capital and labor is far from the whole story.

Foucault (2010) argued that economic growth cannot be explained through the speeding up of labor or capital alone. Rather, explanation must also turn to the ways in which human capital has been augmented and how this contributes to increases in productivity (232). Although often criticized for downplaying extra-economic factors such as human capital, Harvey is not blind to this issue. Writing in 1989, Harvey suggested that new entrepreneurial forms of city governance essentially manifest in four key ways, the most salient being increasing competitiveness through the creation of new forms of labor power (Harvey 1989).

In the modern city this means, for example, not only a more skilled labor force but a healthier one. It is evident that although sedentary forms of mobility such as the car, bus,[4] and train have historically reduced circulation times of labor—thus improving

productivity, they have lowered its productivity in other ways by producing more sedentary and less healthy laborers. A recent report quantifying the economic costs of sedentary transport stated that "the average employee has 6.8 sick days per year in the UK. A conservative estimate is that increased exercise can lower this sick leave by 6%. The value of lowering sick leave by this 6% is equal to £47.68 per working age adult across England based on an annual GVA per employee of £37,000" (SQW Consulting 2008, 19). Accordingly, a key goal of UK transport policy in recent years has been to encourage more active modes of travel such as cycling and walking. Certainly a spatial fix to increase car use would not satisfy the health economic argument for changes in travel behavior.

The issue for municipal governments, then, is not just one of decreasing circulation time through fixing mobility but also enhancing human capital through mobility in other ways. Harvey (1982) implied as much when he stated, "In so far as many aspects of physical and social infrastructure are fixed in space, the problem of geographical mobility is converted into one of transformation of the social and physical environment within which other forms of capital circulate" (406). The role of particular fixes in urban governance "has thus become much more oriented to the provision of a good business climate" and to the construction of "all sorts of lures to bring capital into town" (Harvey 1989, 11). As Cerny (2006, 690) went on to point out, the competitive state[5] is not only one with economic policies that focus on promoting competitiveness but one that attempts to incorporate competitiveness into everyday social reproduction. Accordingly, Foucault (2010) argued that modern policies of growth are "focused precisely on one of the things that the West can modify so easily, and that is the form of investment in human capital" (232). Indeed, as Guthman (2009) argued, the material contradictions of capitalism are increasingly resolved through bodies. Accordingly, in the next section the work of Foucault is used to theorize cycling as a form of self-government and "entrepreneurship of the self" situated within a biopolitics of obesity and sedentarism.

Urban Cycle Promotion as Biopolitical Mobility Fix

It is here that I want to introduce cycling in London—once championed only by "mad radicals" such as the London Cycling Campaign and Friends of the

Earth (Spinney 2008)—as a mobility practice increasingly promoted to enhance urban productivity and competitiveness.

A report by SQW (2007), "Valuing the Benefits of Cycling," set out the economic argument to invest in cycling, stating that overall value accrues from a unique combination of improvement in general health and fitness, reduced pollution and the emission of CO_2,[6] and help in tackling congestion. The study calculated that a return to 1995 levels of cycling (an increase of 50 percent) would provide health, pollution, and congestion savings of more than £1.3 billion (SQW Consulting 2008, 2). The Department of Health (DoH) concluded that the SQW analysis—in line with other similar reports—confirmed that the health benefits of cycling are the largest single reason to promote it (DoH 2010).

Galvanized by such valuations, a loose and shifting group of state (Transport for London, Greater London Authority [GLA], London Boroughs) and nonstate actors (London Cycling Campaign, bloggers, academics, news media) has coalesced[7] in the last ten to fifteen years as a kind of Foucauldian "dispositif" (Braun 2014) to promote cycling as a key way of reducing commute times, creating healthier workers, and reducing pollution (for an excellent overview, see Aldred 2012). The GLA's current stated goal is to increase cycling levels to 5 percent mode share by 2025, which would mean an overall 400 percent increase (GLA 2014).[8]

Particularly in the case of cycling, promotion has tended to be in the shape of "soft" measures as Pooley et al. (2013) confirmed: "Recent policies have been reluctant to adopt more interventionist approaches relying instead on persuasion and promotion of active travel mainly on health grounds" (67; see also Golbuff and Aldred 2011). Congruently it should be emphasized that the official consensus in the United Kingdom has until very recently been that encouraging cycling should not require separate infrastructural fixes as expressed in the "hierarchy of solutions" (Institution of Highways and Transportation, Cyclists' Touring Club, Bicycle Association, and Department of Transport 1996; DoH 2010).

In *The Birth of Biopolitics*, Foucault (2010) argued that as neoliberalism cannot intervene in the economic, it must intervene "on society ... in its fabric and depth" (145). In so doing it will become a government of enterprise society rather than economy where the individual becomes the locus of "enterprise and production" (146–47). Using the example of migration (although equally applicable to the kind of intraurban

mobility that cycling represents), Foucault went on to discuss mobility directly in relation to enterprise, theorizing mobility as an investment on the part of the individual because it leads to an increase in status and remuneration (230). Consequently, the individual becomes an investor, "an entrepreneur of himself who incurs expenses by investing to obtain some kind of improvement" (230). So for Foucault the ways in which an individual moves are conceived as a form of enterprise because of the financial and bodily rewards of moving in particular ways.

For individuals to invest in themselves, however, requires the construction of a social infrastructure serving to inculcate self-management and construct "reflexive health entrepreneurs, willing and able to manage their own wellbeing under the guidance of 'distant' experts" (Cramshaw 2012, 200). This is exemplified in strategy documents such as the report *Cycling Revolution London* (Transport for London 2010). Despite reference to some "hard" measures, cycle promotion in the report is mainly concerned with equipping individuals with the correct incentives and skills to cycle, including cycle training, workplace travel planning, wayfinding, and general awareness and communication. Cycle promotion policies are primarily concerned with shaping social (as opposed to spatial) infrastructures to influence entrepreneurs to invest in themselves: "The individual is positioned as entrepreneur and the self as an enterprise to be formed in direct response to the imperatives of neo-liberal choice, itself provided through a system of markets and incentives to achieve self-actualisation through judicious navigation of a series of regulated freedoms" (Cramshaw 2014, 173).

It is evident that situated within a wider constructed politics of obesity and sedentarism (Guthman 2009), cycling becomes a key way in which individuals can invest in themselves and (hopefully) obtain improvement in the form of vitality and increased life expectancy. At the same time, the state and capitalist reap the reward of this "improved" labor in the form of a reduction in sick days, decreased health care costs, reduced circulation time of labor, and reduced environmental costs. As Cramshaw (2012) pointed out, recent public health strategies reflect a shift from the social to the individual as the target for regimens of self-discipline relating to diet, exercise, and the consumption of "risky" substances (202).

Crucially, this is city government acting not only on mobility through a set of desired outcomes but government through mobility: Cycling becomes a terrain through which the individual recognizes and becomes

212

equipped to deal with obesity and sedentarism as public health problems. Hence, as a type of fix, cycling is best theorized as a mobility fix because it represents a way of governing that works primarily through the production of new modes of mobile comportment rather than the production of space.

Urban Cycling as Metaphorical Fix

Although cycle promotion is clearly tied up with a broader biopolitics, Foucauldian analyses of governmental projects should not make us blind to the uncertainties in whether such projects manage to meet their ends. As Bærenholdt (2013) noted, discursive intent does not make the world, and the political landscape is littered with the failures of biopolitical projects (Cramshaw 2012; Nettleton and Green 2014).

On the face of it, the promotion of cycling in London would appear to have been successful, with a rapid increase over the past fifteen years. Census data show a 144 percent increase in cycling in London over the period from 2001 to 2011 (Office for National Statistics 2014) with a daily average of 131,000 cycle journeys in 2014 (GLA 2014).

Leaving aside those who do not respond to exhortations to cycle however, even for those who do, cyclist deaths (particularly under the wheels of heavy goods vehicles) and safety more generally have become an increasing cause for concern. A prominent theme in recent media reports of cyclist fatalities has been a preponderance of blame falling on the abstract systems of road and junction design. Pressured by an extremely bad two weeks in November 2013 when six cyclists died, Transport for London responded with a number of infrastructural initiatives. In addition to its ongoing commitment to a Central London Grid of cycle infrastructure, "combining 20 miles of arterial Superhighways on main roads with a broad 60-mile network of capillary Quietways on calmer backstreets" (Transport for London 2014, 52), in March 2014 the group announced the new "Mini-Holland programme." According to Transport for London, this program aims to emulate the cycle-friendly, low-traffic neighborhoods in Amsterdam. The winning bids included "redesigns of town centres, new suburban cycle superhighways, [and] Dutch-style roundabouts" (Transport for London 2014, 59).

This raft of measures is evidence that the loose alliance of Transport for London and its partners is increasing its efforts to encourage cycling by producing space for it, mainly in the form of redesigning and reassigning existing transit space rather than the production of new transit spaces. The rolling out of the superhighways network and Mini-Holland program must be seen as a direct response to tensions arising between the mobility fix of cycle promotion and the inhospitable environments within which it is practiced (cf. Spinney, Kullman, and Golbuff 2015). Indeed, it is increasingly apparent that as awareness of the risks of cycling grows as more people do it, these risks serve to push back on (Latour 2008) and produce new spatial configurations.

A primary conclusion to be deduced from this is that any spatial fix associated with cycling is a by-product and adjunct to its failures as a biopolitical mobility fix. Evidently what emerges as a mobility fix operating on the individual becomes spatial to some degree because the individual entrepreneur faces numerous spatial barriers in getting an enterprise off the ground (see, e.g., Cramshaw 2012). In responding to these, policy actors are forced to institute a limited suite of spatial fixes to help individual enterprises succeed. This affirms Hackworth's (2007) point that under neoliberalism, "stasis, not volatility, is more common in city landscapes" (84). Cerny (2006) similarly questioned whether urban areas even require spatiotemporal fixes any more, suggesting rather that "perhaps what we are seeing is not merely a reinvention of geography but rather a downgrading of the relative importance of geography itself—although not, of course 'the end of geography'" (692).

A second conclusion is that the mobility fix of cycle promotion and cycle infrastructure also represents what Jessop (2004) termed a metaphorical fix: a "second best" solution characterized by its improvised and temporary nature that at best partially and provisionally seeks to overcome the contradictions and dilemmas inherent in capitalism and regularize accumulation in the long term (Jessop 2004, 13). In their piecemeal and fragmented nature, cycling's spatial fixes appear avowedly metaphorical. Moreover, metaphorical fixes, as Jessop goes on to say, are an outcome of the institutionalization of disparate and compromised agendas and actors coming together (as is the case here) in the kind of ad hoc policy assemblage described by Braun (2014).

Conclusion

The key contribution of this article is to theorize cycling policy and promotion as a mode of neoliberal governance rather than a practice of resistance (see, e.g., Furness 2005) or as being affected by neoliberal policy, as has been previously theorized (Aldred 2012). This article has demonstrated that promotion of cycling has become a key way in which responsibilities are shifted from public to private spheres in relation to a biopolitics that constructs obesity and sedentarism as abnormal and problematic. In doing so, it contends that cycle promotion is essentially a biopolitical mobility fix that attempts to construct preferred bodily modes of comportment and social infrastructures to support them. At the same time, the article has shown that in order for individual entrepreneurs to perform successfully within the existing (hostile) spatial infrastructures necessitates an (albeit minimal and geographically unequal and contingent) spatial and metaphorical fix in the form of cycle-specific infrastructure.

Moving beyond this, there are (at least) three avenues opened up by the critical leverage of the framing of cycle policy and practice as a mobility and metaphorical fix: First, even if we were to agree that cycling is a public good, in theorizing cycle promotion as biopolitical we have to question the inequalities inherent in the differing abilities of societal groups to respond to cycle promotion. Steinbach et al. (2011), for example, demonstrated that cycling in London is a practice deemed acceptable largely only to young white men. Which other groups are excluded through the particular social, spatial, and artefactual infrastructures being produced?

Second, it is evident that the provision of cycle infrastructure is producing uneven urban geographies. For Jessop (2004), spatiotemporal fixes always involve "an internal as well as external differentiation of winners and losers from ... a given fix and to its associated uneven development" (13). At the street scale, new segregated cycle lanes represent a segmentation of road space that serves some and not others, linking up only particular neighborhoods and excluding many potential users (Hoffman and Lugo 2014; Stehlin 2014; see also Minn 2013). Such selective development should be recorded and challenged through critical mobilities scholarship.

Third—and taking both of these preceding points together—we also have to question the marginalization of subaltern and alternative versions of cycling where disparate actors coalesce around an agenda that presents the purpose of cycling as primarily replacing car journeys. A prominent outcome of the linking of cycling with improving labor productivity or enhancing health in the form of promoting commuting cycling is that certain types of cycling are promoted as more productive and therefore more important than others. Stehlin (2014), for example, commented that the focus on cycling commuting undercounts subaltern cycling practices; those of low-wage laborers navigating shift work and social services or underemployed youth riding within estates and locales are often discounted. It is notable that forms of sport and leisure cycling are also exorcised from urban policy framings, particularly those deemed to be "inefficient" or "for kids," like BMX and mountain biking (Spinney 2008, 144). The explicit linking of cycle policy with productivist agendas should prompt us to acknowledge and explore ways in which the ludic nature of cycling can be reclaimed.

Accordingly, a key hope of this article is that linking current cycle policy to broader political–economic and biopolitical processes will accelerate a move away from a first or second wave of cycling research concerned with increasing cycling levels toward a third wave also asking fundamentally political questions about inclusivity, representation, access, class, and race. Cycling's time might indeed have come, but for whom?

Acknowledgments

Thanks are due to Tim Schwanen and Mei-Po Kwan for organizing the original set of 2014 American Association of Geographers sessions in which this article started life and for continuing support and feedback on the article since then. I would also like to thank the four anonymous reviewers whose insightful comments have helped to sharpen significantly the theoretical underpinnings, focus, and narrative of this article. As ever, any errors or misinterpretations are my own.

Notes

1. After a period of unabated road building in the United Kingdom throughout the 1970s and 1980s, enthusiasm began to wane as the "new realism" set in and it became apparent that building roads just generated more traffic, congestion, and associated environmental and health issues (Goodwin et al. 1991, in Docherty and Shaw 2008).
2. In late 2014, the UK government announced a funding package of £15 million for new road schemes in England. However, all of the proposed schemes are for inter-urban rather than intra-urban areas.
3. It is also highly questionable whether any road building program would significantly speed up circulation times, in part because increasing city densities mitigate, against car use in ways that suburban living did and does not.

4. Although a discourse dominates that positions public transport as sedentary, we might want to question the extent to which this is true. Webb (2014), for example, has demonstrated that those who use public transport are 25 percent less likely to be obese than those who do not.

5. It should be noted that the state in such a reading is increasingly the municipal government.

6. Despite the strong (some might say overly so) associations of cycling with environmental issues, this article does not discuss them as a component of the mobility fix. This is partly due to word limits but more that this article views contemporary cycling practice as a feature of neoliberal urbanization that is relatively independent of the sustainability imperative.

7. Rather than suggest a simplistic notion of "the state," it is evident that multiple actors, networks, and flows are producing the resurgence of cycling. Bulkeley (2010) noted how a new wave of municipal networks mobilizes private and nongovernmental actors alongside the state. The multiple actors involved in the resurgence of cycling as an urban phenomena exemplify these kinds of messy, mobile, and multilevel forms of governance comprising various local (Spinney 2010) and international nongovernmental and individual actors and actants (Golbuff 2014; Stehlin 2014).

8. This is against a backdrop of continual decline in UK cycling levels, which peaked in 1949 at 37 percent of all traffic and have been on the slide ever since, particularly between 1950 and 1970, and flat-lining throughout the 1980s (Golbuff and Aldred 2011). By 1989, a European Union survey placed Britain alongside Belgium as the worst cycling nation in Europe (Golbuff and Aldred 2011) with official interest in cycling returning only slowly in the early 1990s.

References

Aldred, R. 2012. Governing transport from welfare state to hollow state: The case of cycling in the UK. *Transport Policy* 23:95–102.

Bærenholdt, J. O. 2013. Governmobility: The powers of mobility. *Mobilities* 8 (1): 20–34.

Bohm, S., C. Jones, C. Land, and M. Paterson. 2006. Introduction: Impossibilities of automobility. *Sociological Review* 54 (S1): 2–16.

Bonham, J. 2006. Transport: Disciplining the body that travels. *Sociological Review* 54 (S1): 57–74.

Braun, B. 2014. A new urban dispositif? Governing life in an age of climate change. *Environment and Planning D: Society and Space* 32:49–64.

Bulkeley, H. 2010. Cities and the governing of climate change. *Annual Review of Environment and Resources* 35:229–53.

Button, K., and P. Nijkamp. 1997. Social change and sustainable transport. *Journal of Transport Geography* 5 (3): 215–18.

Cerny, P. G. 2006. Restructuring the state in a globalizing world: Capital accumulation, tangled hierarchies and the search for a new spatio-temporal fix. *Review of International Political Economy* 13 (4): 679–95.

Cramshaw, P. 2012. Governing at a distance: Social marketing and the bio-politics of responsibility. *Social Science and Medicine* 75:200–207.

———. 2014. Institutionalising commercialism? The case of social marketing for health in the United Kingdom. In *Organising neoliberalism:Markets, privatisation and justice*, ed. P. Whitehead and P. Cramshaw, 155–78. London: Anthem Press.

Cresswell, T. 2010. Towards a politics of mobility. *Environment and Planning D: Society and Space* 28:17–31.

Cresswell, T., and T. Uteng. 2008. *Gendered mobilities*. Farnham, UK: Ashgate.

Danyluk, M., and D. Ley. 2007. Modalities of the new middle class: Ideology and behaviour in the journey to work from gentrified neighbourhoods in Canada. *Urban Studies* 44 (11): 2195–2210.

Department of Health (DoH). 2010. *Value for money: An economic assessment of investment in walking and cycling.* http://www.apho.org.uk/resource/view.aspx?RID=91553 (last accessed 1 January 2014).

Docherty, I., and J. Shaw. 2008. *Traffic jam: Ten years of sustainable transport in the UK*. Bristol, UK: Policy Press.

Fincham, B. 2006. Bicycle messengers and the road to freedom. *Sociological Review* 54 (S1): 208–22.

Foucault, M. 2010. *The birth of bio-politics: Lectures at the college de France 1978–79.* Basingstoke, UK: Palgrave Macmillan.

Furness, Z. 2005. Put the fun between your legs!: The politics and counterculture of the bicycle. Unpublished PhD thesis, University of Pittsburgh, Pittsburgh, PA.

Golbuff, L. 2014. Moving beyond physical mobility: Blogging about urban cycling and transport policy. Unpublished PhD thesis, Cardiff University, Cardiff, UK.

Golbuff, L., and R. Aldred. 2011. *Cycling policy in the UK: A historical and thematic overview.* London: University of East London, UEL Sustainable Mobilities Group.

Greater London Authority (GLA). 2014. Cycle flows on the TFL road network. http://data.london.gov.uk/dataset/cycle-flows-tfl-road-network (last accessed 26 November 2014).

Greater London Authority (GLA) Economics. 2005. Time is money: The economic effects of transport delays in central London. http://www.london.gov.uk/mayor/economic_unit/docs/time_is_money.pdf (last accessed 1 November 2014).

Guthman, J. 2009. Teaching the politics of obesity: Insights into neoliberal embodiment and contemporary biopolitics. *Antipode* 41 (5): 1110–33.

Hackworth, J. 2007. *The neoliberal city: Governance, ideology and development in American urbanism.* Ithaca, NY: Cornell University Press.

Harvey, D. 1982. *The limits to capital.* 2nd ed. London: Verso.

———. 1989. From managerialism to entrepreneurialism: The transformation in urban governance in late capitalism. *Geografiska Annaler* 71B (1): 3–17.

———. 2003a. The fetish of technology: Causes and consequences. *Macalester International* 13:3–30. http://digitalcommons.macalester.edu/cgi/viewcontent.cgi?article=1411&context=macintl (last accessed 30 November 2014).

———. 2003b. *The new imperialism*. Oxford, UK: Oxford University Press.

Hickman, R., and D. Bannister. 2015. *Transport, climate change and the city*. London and New York: Routledge.

Hoffman, M., and A. Lugo. 2014. Who is "world class"? Transportation justice and bicycle policy. *Urbanities* 1:45–61.

Jessop, B. 2004. Spatial fixes, temporal fixes, and spatio-temporal fixes. http://bobjessop.org/2014/01/16/spatial-fixes-temporal-fixes-and-spatio-temporal-fixes/ (last accessed 1 November 2014).

Jones, P. 2012. Sensory indiscipline and affect: A study of commuter cycling. *Social & Cultural Geography* 13 (6): 645–58.

Jungnickel, K., and R. Aldred. 2014. Cycling's sensory strategies: How cyclists mediate their exposure to the urban environment. *Mobilities* 9 (2): 238–55.

Latour, B. 2008. "Where are the missing masses?" The sociology of a few mundane artifacts. In *Shaping technology/building society: Studies in sociotechnical change*, ed. W. E. Bijker and J. Law, 225–58. Cambridge, MA: MIT Press.

Lugo, A., and M. Hoffman. 2014. Who is world class? Transportation justice and bicycle policy. *Urbanities* 4 (1): 45–61.

Minn, M. 2013. The political economy of high speed rail in the United States. *Mobilities* 8 (2): 185–200.

Nettleton, S., and J. Green. 2014. Thinking about changing mobility practices: How a social practice approach can help. *Sociology of Health and Illness* 36 (2): 239–51.

Newman, P., J. Kenworthy, and P. Vintila. 1995. Can we overcome automobile dependence? Physical planning in an age of cynicism. *Cities* 12 (1): 53–65.

Nixon, D. 2012. A sense of momentum: Mobility practices and dis/embodied landscapes of energy use. *Environment and Planning A* 44 (7): 1661–78.

Office for National Statistics. 2014. London residents cycling to work doubles in 10 years. http://www.ons.gov.uk/ons/rel/census/2011-census-analysis/cycling-to-work/sty-cycling-to-work.html (last accessed 30 November 2014).

Pooley, C., D. Horton, G. Scheldeman, C. Mullen, T. Jones, M. Tight, A. Jopson, and A. Chisholm. 2013. Policies for promoting walking and cycling in England: A view from the street. *Transport Policy* 27:66–72.

Pucher, J., and R. Buehler. 2011. Analysis of bicycling trends and policies in large North American cities: Lessons for New York. Final report, University Transportation and Research Center, New York, NY. http://www.utrc2.org/publications/analysis-bicycling-trends-and-policies-large-north-american-cities-lessons-new-york (last accessed 20 November 2014).

Rose, N. 1999. *Powers of freedom: Reframing political thought*. Cambridge, UK: Cambridge University Press.

Spinney, J. 2006. A place of sense: A kinaesthetic ethnography of cyclists on Mt Ventoux. *Environment and Planning D: Society & Space* 24 (5): 709–32.

———. 2007. Cycling the city: Non-place and the sensory construction of meaning in a mobile practice. In *Cycling & society*, ed. D. Horton, P. Rosen, and P. Cox, 25–46. Aldershot, UK: Ashgate.

———. 2008. Cycling the city: Movement, meaning and practice. Unpublished PhD thesis, Royal Holloway University of London, London.

———. 2010. Mobilising sustainability: Partnership working between a pro-cycling NGO and local government in London (UK). In *Low carbon communities: Imaginative approaches to combating climate change*, ed. S. Fudge, M. Peters, and T. Jackson, 89–107. London: Edward Elgar.

Spinney, J., K. Kullman, and L. Golbuff. 2015. Driving the "Starship Enterprise" through London: Constructing the im/moral driver-citizen through HGV safety technology. *Geoforum* 64:333–41.

SQW Consulting. 2007. *Valuing the benefits of cycling: Report to Cycling England*. http://webarchive.nationalarchives.gov.uk/20110407094607/http:/www.dft.gov.uk/cyclingengland/site/wp-content/uploads/2008/08/valuing-the-benefits-of-cycling-full.pdf (last accessed 1 January 2014).

——— 2008. *Planning for cycling: Report to Cycling England*. http://webarchive.nationalarchives.gov.uk/20110407094607/http:/www.dft.gov.uk/cyclingengland/site/wp-content/uploads/2009/03/planning-for-cycling-report-10-3-09.pdf (last accessed 1 January 2014).

Stehlin, J. 2014. Regulating inclusion: Spatial form, social process, and the normalization of cycling practice in the USA. *Mobilities* 9 (1): 21–41.

Steinbach, R., J. Green, J. Datta, and P. Edwards. 2011. Cycling and the city: A case study of how gendered, ethnic and class identities can shape healthy transport choices. *Social Science & Medicine* 72 (7): 1123–30.

Tickell, A., and J. Peck. 2003. Making global rules: Globalization or neoliberalization? In *Remaking the global economy: Economic-geographical perspectives*, ed. J. Peck and H. Yeung, 163–81. London: Sage.

Transport for London. 2010. *Cycling revolution London*. London: Transport for London.

———. 2014. *London's road modernization plan*. London: Transport for London.

Transport for London Roads Task Force. 2013. *The vision and direction for London's streets and roads*. London: Transport for London.

Webb, E. 2014. Bus passes, active travel and health. ICLS Occasional Papers 11(3), University College London, London.

Policies on the Move: The Transatlantic Travels of Tax Increment Financing

Tom Baker,* Ian R. Cook,† Eugene McCann,* Cristina Temenos,‡ and Kevin Ward§

*School of Environment, University of Auckland
†Department of Social Sciences and Languages, Northumbria University
‡Humanities Center, Northeastern University
§School of Environment, Education and Development, University of Manchester

Growing influence of the new mobilities paradigm among human geographers has combined with a long and rich disciplinary tradition of studying the movement of things and people. Yet how policy ideas and knowledge are mobilized remains a notably underdeveloped area of inquiry. In this article, we discuss the mobilization of policy ideas and policy models as a particularly powerful type of mobile knowledge. The article examines the burgeoning academic work on policy mobilities and points toward a growing policy mobilities approach in the literature, noting the multidisciplinary conversations behind the approach as well as the key commitments of many of its advocates. This approach is illustrated using the travels of tax increment financing (TIF) with the role of learning and market-making within efforts to introduce TIF in more cities highlighted. In conclusion, we discuss some of the political and practical limits that often confront efforts to mobilize policy ideas.

人文地理学者的崭新能动性范例逐渐增加的影响，与研究事物和人的移动的长久及丰富的学门传统相互结合。但政策概念与知识如何被动员，显然仍是发展不足的研究领域。我们于本文中，探讨政策概念和政策模型的动员，作为特别强大的移动知识之类型。本文检视迅速兴起的政策能动性之学术研究，并指向文献中逐渐成长的政策能动性取径，关注该取径背后的多重领域对话，及其诸多倡议中的关键承诺。此般取径，运用税收增值信贷 (TIF) 的移动进行描绘，并凸显将 TIF 引进更多城市的努力中，学习和市场创造所扮演的角色。我们于结论中，探讨经常与动员政策概念的努力相互冲突的部分政治及实际限制。

La creciente influencia del nuevo paradigma de las movilidades entre los geógrafos humanos se combina con una larga y rica tradición disciplinaria en el estudio del movimiento de cosas y personas. No obstante, la manera como se movilizan ideas políticas y conocimiento sigue siendo un área de indagación notoriamente subdesarrollada. En este artículo discutimos sobre la movilización de ideas políticas y modelos de política como un tipo particularmente poderoso de conocimiento móvil. El artículo examina el pujante trabajo académico sobre movilidades políticas y destaca la notoria aplicación del enfoque de movilidades políticas en la literatura, notando las conversaciones multidisciplinarias que se desarrollan detrás de ese enfoque lo mismo que los compromisos claves de muchos de sus defensores. Este enfoque se ilustra usando los viajes del financiamiento del incremento tributario (TIF, por su acrónimo inglés), destacando el papel de aprender y crear mercado dentro de los esfuerzos por introducir el TIF en más ciudades. En conclusión, discutimos algunos de los límites políticos y prácticos que a veces tienen que confrontar los esfuerzos emprendidos para movilizar las ideas sobre política.

Tax increment financing (TIF) is an idea that's time has come. This, at least, is the conclusion one might draw from its expanding geography within and beyond the United States. There are two central features to TIF. The first involves establishing a TIF district by drawing a line around part of a city. Within this area, taxes on the value of properties continue to be collected and paid out to tax-receiving agencies, which in many U.S. states include local government, the police, and schools. Establishing the TIF district (for periods ranging from twenty-three to thirty years), however, means that any future increase in the assessed values within in it no longer accrues to these tax-receiving agencies. Instead, the extra increment is paid to the agency overseeing the TIF district. In some cases this agency is a city government, whereas in others it is a specially established redevelopment agency. The second feature of TIF is the creation of debt—often through the issuing of bonds. These debts are accrued against the potential "increment," so that the various stakeholders can finance changes to infrastructure and land use within the district in the hope that these changes lead to increased assessed values.

Currently there are TIF programs in every U.S. state except Arizona. In Illinois, a state with one of the longest standing TIF statutes, Chicago refers to itself and is referred to by many others in the U.S. economic development industry as the "poster child" of the U.S. TIF program. Others are less generous, arguing that the program has caused mass displacement, because the increment is often used to fund gentrification (Wilson and Sternberg 2012). Just over 30 percent of Chicago's land area falls within one of its 163 TIF districts, each of which, once approved, lasts for twenty-three years. These districts collected a total of $454 million in property taxes in 2011. The Chicago City Council has used TIF to finance a range of economic development projects, from the gentrification of the downtown to providing incentives to firms willing to relocate to its declining industrial districts (Weber 2010).

The emergence of TIF across the United States has occurred through a myriad of channels and networks, many of which involve the Council of Development Finance Agencies (CDFA). Established in 1982 as "the conduit linking development finance professionals together,"[1] it operates as a loose assemblage of actors, documents, events, materials, and technologies gathered, some purposively and some by chance, to promote and sell the TIF program to interested city officials globally. It does this through its annual conferences, educational programs, presentations, reports, and webinars.

TIF, then, is a policy that seems to be very much on the move. It has been rendered mobile both inside the United States and beyond its borders. Officials from Australia, Canada, and the United Kingdom have attended conferences, participated in training courses, and spoken to CDFA officers, for example. Yet, as we discuss later, TIF, like all policy ideas, has an uneven geography of implementation, speaking to the continued importance of local institutional context and place-specific politics in the circulation of policy models. Even when a policy finds its time, for ideological, institutional, and political reasons, it must still find its place.

We argue that the study of mobilities benefits from, and is enhanced by, the geographical study of ideas and knowledge. Most contemporary literature on mobilities focuses on air and automobile travel, migration, pilgrimage, and tourism. This focus is reflected in the other articles in this special issue. Although scholars have broadened their remit to the study of everything from water and waste mobilities, the movement of energy and resources, and the ethical and political implications of these mobilities (Adey et al. 2013; Sheller 2014), there is scope for a deeper analysis of the ways in which people move ideas and the sociospatial implications of ideas on the move. Central elements of the geographical literature on policy mobilities have drawn explicitly on the new mobilities paradigm (McCann 2011). Certainly, the recent proliferation of work on policies in motion (e.g., Peck and Theodore 2010, 2015; McCann and Ward 2011; Cochrane and Ward 2012; Temenos and McCann 2013) provides an opportunity to specify and deepen the geographical engagement with mobilities by focusing on how elements of policy—ideas, calculations, expertise, models—and methods of policy implementation circulate in and through institutions and places.

The paradoxical case of TIF—a traveling policy that promotes state-led revenue collection yet has been adopted and advocated by governments that explicitly advance neoliberalization—allows us to demonstrate how policy mobilities are social productions of specific, path-dependent, territorialized, and also global-relational policy landscapes. In the following section, we outline the multidisciplinary conversations that have generated the policy mobilities literature, before discussing what have become key commitments of policy mobilities studies. The article then returns to TIF as a way of illustrating how policy ideas are mobilized through practices of learning and market-making. Throughout this section, we use TIF to exemplify the policy mobilities approach, while also using our discussion of that approach to improve our understanding of TIF. We conclude by discussing some of the ways in which barriers and constraints are important features in the geographies and mobilities of policy.

Multidisciplinary Conversations About Policy and Mobilities

There are seemingly few policy ideas more grounded and fixed than TIF. It is a policy with a clearly defined territorial extent, intent on maintaining and developing local physical infrastructures. Certainly, the geographical study of urban governance, policy, development, and politics has tended, over the years, to be localist and territorialist (McCann and Ward 2010). Indeed, Cresswell and Merriman (2011) argued that geographers of all stripes often assume "a stable point of view, a world of places and boundaries and territories rooted in time and bounded in space" (1). Developing a new approach or paradigm for studying mobilities, they and others

> problematize … both "sedentarist" approaches in the social sciences that treat place, stability, and dwelling as a natural steady-state, and "deterritorialized" approaches that posit a new "grand narrative" of mobility, fluidity or liquidity as a pervasive condition of postmodernity or globalization. (Hannam, Sheller, and Urry 2006, 5)

Although not without its critics (Faist 2013), this renewed emphasis on studying mobility valuably conceptualizes it as a process infused with meaning and power. It sets the terms of analysis to encompass more than the movement of people and objects from A to B. Rather than focus simply on this "desocialised movement" (Cresswell 2001, 14), mobilities scholars turn their attention to the practices and power relations involved in movement. Yet, although "people move, things move, ideas move," as Cresswell (2010, 19) argued, far less attention has been paid to how, where, and with what consequences ideas move and to the people and resources who move them. Ideas are understood in this context to be socially produced. They emerge from individuals and their relations with others.

We argue that the study of policy provides an ideal lens through which to study powerful ideas on the move, like TIF, and to conceptualize the power of those mobilized ideas on social groups and places. *Policy* from this perspective has a specific connotation, succinctly defined by Kuus (2014) as

> the fundamental organizing and productive principle of modern societies. … [P]ublic policies … [are] technologies of power that do not simply serve public interests but also produce these very interests. Policies do not merely regulate existing relationships; they create new relationships, objects of analysis, and frameworks of meaning. (39)

The mobilization and mutation of policy produces policy markets and landscapes through the work of diverse policy actors, themselves operating within wider ideological and structural contexts. Central questions in this approach include the following: Who mobilizes and who is mobilized in policymaking processes? How are policies rendered mobile? What sites and spaces shape and are shaped by mobilization? What are the politics of this global-relational policy and knowledge making?

A series of commitments that motivate many policy mobilities studies, to one extent or another, have emerged around these questions (Table 1). These studies draw on the notion of mobility as peopled and power-laden. They are informed by a conceptualization of policy similar to that described by Kuus (2014), earlier, and that informs Peck's (2011) critique of rational formalism in traditional policy studies. Examples of this work are numerous and include analyses of creativity (Peck 2005; Prince 2010, 2012), design

Table 1. Commitments that motivate policy mobilities studies

Conceptual commitments	Methodological commitments
• To political–economic and social constructivist approaches to policy mobilization that take poststructuralist and postcolonial critiques seriously	• To primarily qualitative investigations of the practice, process, and meaning of policymaking through interviews, observation, site visits, and documentary analysis
• To conceptualizations of policymaking's role in wider geographies of ideas and knowledge	• To empirically tracing the pathways taken by policy through communities, institutions, places, and situations
• To analyses of policies as powerful and productive technologies	• To "extended" or multisited case study analysis
• To analyses of interlocal, rather than necessarily international, mobilizations	• To detailed description, informed by theory and directed toward theory building
• To analyses of assembling, emergence, hybridity, mutation, relationality, and translation	
• To analyses of the immobilities, inertia, barriers, and "differential mobilities" that also constitute policy	

219

(Faulconbridge 2013; MacLeod 2013; Rapoport 2015), education (Geddie 2014), economic development (Ward 2006, 2007; Cook 2008), homelessness (Baker 2014), public health (McCann and Temenos 2015), drug policy (McCann 2008, 2011), sustainability (Temenos and McCann 2012; Fisher 2014; Müller 2015), and transport (Wood 2014).

Unlike some of those working on mobilities more generally, there appears to be no sense yet among policy mobilities scholars that their approach constitutes a coherent paradigm or "canon" (McCann and Ward 2015). According to Peck (2011, 774) work on policy mobilities more closely resembles a "rolling conversation" or, perhaps more appropriately, a series of conversations. Here we focus on just two.

First, drawing on a well-established tradition of scholarship in urban planning (Clarke 2011), the policy mobilities conversation has involved planning historians and geographers, among others (Healey and Upton 2010; Jacobs 2012; Jacobs and Lees 2013; Quark 2013; Cook, Ward, and Ward 2014, 2015). This urban planning work is typically empirically rich, providing insights into the longer-than-often-assumed histories of policy mobilities, particularly in the field of architecture, engineering, and planning where the literature has paid particular attention to work done in moving policy by certain professions' ideas and expertise across particular institutional contexts. A second, still

burgeoning, engagement around policy mobilities is also multidisciplinary in nature. It involves anthropologists and others working on the notion of *policy worlds*—"domains of meaning" that policies both reflect and create (Shore, Wright, and Però 2011, 1; see also Shore and Wright 1997; Wedel et al. 2005). This literature has recently come into conversation with those developing critical geographies of policy (Peck 2011; Robinson 2011, 2013; Roy and Ong 2011; Jacobs 2012; McCann and Ward 2012a, 2012b, 2013; Söderström 2014). This is a conversation both about how to conceptualize policy and policymaking and one focused on questions of methodology (Cochrane and Ward 2012; Jacobs and Lees 2013).

Engaging in what Shore and Wright (1997, 14) termed "studying through" and by "tracing" the travels of policies, anthropologists uncover the ways in which specific arrangements of actors and institutions shape the development of policy landscapes (Wedel et al. 2005; Kingfisher 2013). For those geographers working on policy mobilities, these insights have spurred analysis of the various ephemeral situations, as well as more established tendencies and path dependencies, implicated in policymaking, and have encouraged more detailed understandings of how policy actors, from professionals to activists, assemble local policies through engagements with more extensive circuits of policy knowledge (McCann and Ward 2012b). Thus, actors

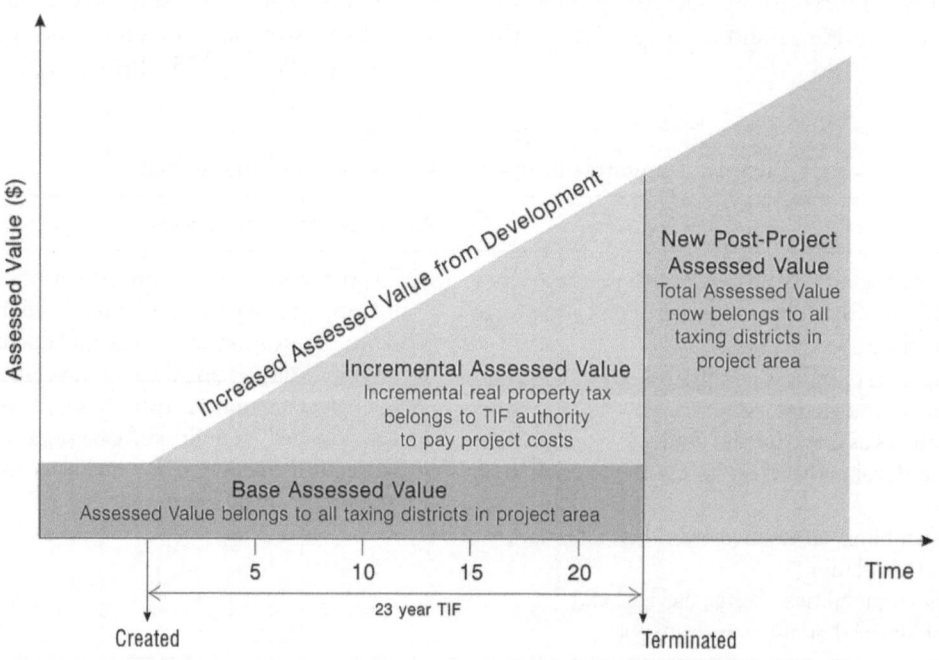

Figure 1. The tax increment financing (TIF) model.

who make and who mobilize policy become important objects of analysis in uncovering how policies and their attendant elements move.

Studying Policy Mobilities Through TIF: Learning and Market-Making

The multidisciplinary nature of the contemporary policy mobilities approach is marked by significant internal heterogeneity and the ongoing emergence of new critiques and (re)orientations. This diversity is paralleled by ongoing conceptual and methodological debates in other disciplines on how policy is transferred and translated (see McCann [2011] for a summary and Mukhtarov [2014] for a recent intervention). More empirical research will strengthen these conceptualizations, but a central tenet of the policy mobilities approach remains: Policies are not generated abstractly in "deterritorialized" networks of experts; rather, they emerge in and through concrete "local" situations that constitute wider networks. Two emerging foci merit discussion in this regard: learning and market making. Here we use TIF to operationalize and explore these orientations. We begin by defining and contextualizing TIF as a policy model.

TIF

As set out in the article's Introduction, TIF is a mechanism for borrowing against predicted revenue streams.[2] At the formation of a TIF district, the established tax receiving agencies, such as local government and schools, have their revenues capped for its duration. A debt is established through the issuing of a bond, which is then used to cover a number of prescribed infrastructure and land use costs. The logic underpinning TIF is that investment in the TIF district will lead to a rise in assessed property values and, thus, tax receipts. If this is the case, the "increment" accrues to the agency overseeing the TIF district: the city government or a specially established redevelopment agency (see Figure 1). If assessed values for the TIF district stagnate or drop, then local government might have to use its general fund to pay down the debts incurred in making the initial investment.

Originating in California in the early 1950s, soon after the Community Redevelopment Act (1945), TIF emerged amid concern over post–World War II urban "blight." Yet, the use of TIF in California was minimal until the late 1970s, when the introduction of Proposition 13 curtailed the capacity of city governments to raise taxes without a popular vote. This made TIF an attractive option. As Klacik and Kriz (2001) noted, "TIF is one of the few locally controlled funding options available to local economic development practitioners that can be used for investment in infrastructure improvements they deem necessary for economic growth" (16). In the context of having limited ability to increase taxation, TIF provided a potential mechanism for generating revenues, albeit one that first involved the creation of debt. This advantage, and the role of transfer agents and infrastructures like the CDFA in promoting the model, has led to its proliferation across the United States since the 1980s. Of course, TIF has also been argued to circumvent the right of electorates to vote on the future development of their cities, to direct revenues away from standard tax receiving agencies, and to subsidize the redevelopment industry through forms of "corporate welfare" (Man 2001).[3] With the mobilization of TIF across states and countries, many of its original features have been transformed, responding to the demands of differing financial, governmental, and legal frameworks.

Learning TIF

Academic work on policy mobilities includes a growing emphasis on practices of policy learning and the role of particular sites and situations in which learning takes place (Cook and Ward 2012; Temenos and McCann 2012, 2013). Learning is understood as more than an additive process whereby an individual simply acquires knowledge. Learning is a growth in perception associated with "specific processes, practices and interactions through which knowledge is created, contested and transformed" (McFarlane 2011, 3). This nuanced notion of learning is particularly appropriate in the context of policymaking. Policy actors often learn at a distance, through e-mail, Web sites, and best practice manuals, for example. These forms of learning mobilize policy ideas. Yet, policy actors cannot only learn at a distance. The ability to gain knowledge of new policy ideas also depends on their periodic gathering with other members of their professional and epistemic communities in specific locations at delimited events such as conferences (Cook and Ward 2012; McCann and Ward 2012a; Temenos and McCann 2013). Furthermore, the increasingly common practice of study tours and policy tourism, where individuals or delegations visit model places or initiatives to experience them firsthand, is also central to how and, importantly, what policy actors learn (Cook and Ward 2011; González 2011; Cook, Ward, and Ward 2014, 2015; Wood 2014).

In the influential report, *Towards an Urban Renaissance*, the then–UK government's Urban Task Force (1999) reflected on a study tour to Chicago: "We were . . . impressed on our visit to the United States . . . [particularly with] the Tax Increment Financing (TIF) scheme. . . . [We] believe this approach has much to commend it" (285). One of the Task Force's policy tourists elaborated:

> Chicago was probably the most influential in terms of the lessons. Because the first day the planners showed us kind of, some of the inner, very badly decayed, hollow core . . . but also some of the bits they were trying to redevelop. And . . . then the following day there was this breakfast think tank, which was extremely good, and I think that's where we picked up a lot of the ideas . . . . that was when it clicked into place, the idea that actually we're not going to draw any lessons about physical redevelopment . . . the interesting stuff is the role of business in regeneration and leadership and so on. (Member 1, Urban Task Force, 20 March 2012)

Learning, in this context, was very much tied to a sense of authenticity and legitimacy springing from the direct (if only fleeting) experience of daily practice for Chicago's economic development professionals, rather than a less tacit, more codified version of TIF expressed in reports and other documents. This was explained by another Task Force member:

> We were taken to an area and simply it was explained to us, you know, this is how the property taxation system works in Chicago. "This is the mechanism that we're using, TIF, here to get the place regenerated." It . . . was probably going for a few years by then. For them it wasn't an experiment it was just the way they did things. (Member 2, 30 March 2012)

TIF, then, was learned and mobilized in part through face-to-face engagement and interaction among peers who shared a common focus on urban regeneration. As two members of the Urban Task Force reported:

> Everyone was taking different things out of the trip, depending on their particular expertise and area of interest. So I suspect my excitement about TIF wasn't actually created from anybody else. It was a kind of nerdy finance reaction. I just kind of got it straight away, because it made sense to me because of my background. So I just thought—I could see all sorts of translation difficulties into the UK but as a way of thinking differently about

the problem it just seemed to me to be a very interesting one. (Member 1, 20 March 2012)

> Certainly there was on-going conversations during the course of the visit and . . . the whole process was a conversation . . . based on the iterative exchange of ideas and building hypotheses and then testing hypotheses and refining them. It was a bombardment really of qualitative and quantitative data and that you were sort of constantly synthesising and part of the synthesis was about conversation and reflecting on what you'd seen and what could be derived from it. (Member 3, 11 May 2012)

Learning and translation continued to happen on the move, or "along the way" (McCann 2011), as members traveled back from Chicago to the United Kingdom and reported on their experiences.

Making Markets for TIF

The mobilization of policy ideas and models among cities and other localities is also defined by the development of variegated, yet structured, policy markets. As Roy (2012, 33) argued, "It is useful to think of policy as commodity." From this perspective, policy markets, like the communities of practitioners through which they operate, are politicized contexts that inform both the supply and demand sides of the policy process. Policy mobilities research seeks to understand the ideological, institutional, and professional parameters that govern the *making* of policy. Policy markets, as part of this process, are conceptualized as

> structured by relatively enduring policy paradigms. . .and, perhaps above all, saturated by power relations. These intensely contested and deeply constitutive contexts, which have their own histories and geographies, shape what is seen, and what *counts*. (Peck 2011, 791, italics in original)

Policy mobilities scholars seek to understand the role that systematic, structuring forces play in the selection of certain policy models and in advancing certain interests over others. Most notably, theories and practices of neoliberalization—referring to national projects of market-oriented state restructuring and urban projects of entrepreneurial governance—have offered a useful lens through which to understand the asymmetric marketplace for policy ideas, particularly in the Global North (Peck and Theodore 2001; Ward 2006). TIF is an example of a policy that emerged in the context of socially progressive state intervention but through its travels has emerged as an example of neoliberal statecraft (Peck

2002) because of how "cities front huge sums for land acquisition and development based on tenuous promises of future value generation" (Weber 2002, 537). More recently, as the geographical ambit of the policy mobilities literature has expanded beyond the Global North to places such as Singapore (Bok 2015; Bunnell 2015), China (Zhang 2012; Barber 2013), and Indonesia (Phelps et al. 2014; Cohen 2015), accounts have identified the power of other political projects, particularly those with developmental and progressive characteristics.

This highly political market making is again evident in the case of TIF's travel to the United Kingdom. In its follow-up report to *Towards an Urban Renaissance*, the Urban Task Force (2005) argued for the introduction of "TIF pilots" and a flurry of events and publications followed in the late 2000s. As a British demand-side market was created, comparisons and references to the U.S. experience were plentiful. As someone involved reflected:

> We looked at the pros and cons, we looked at different forms of TIF at that time. The credit crunch was on us and was emergent at that time. ... But we did use the American experience very closely ... both in London and in Edinburgh we set about writing to ministers, local authorities, going to meet them, pushing the case for TIF late 2007. (Senior Figure, UK Trade Organization, 11 January 2012)

By April 2010 the then–Labour Government had committed £120 million over 2011 and 2012 to pilot some TIF program schemes.

The May 2010 formation of the UK's Conservative-Liberal Democrat Coalition Government involved the introduction of a particular form of "localism," in contrast to a perceived centralization of political power under the previous Labour government. This resonates strongly with the ideological and practical underpinnings of the TIF approach to financing urban infrastructure investment. Since its formation at the end of 2011, this program was given meaning and shape by Parliamentary acts, bills, white papers, green papers, and statements. Referring to TIF, the Deputy Prime Minister Nick Clegg MP, at his party's annual conference in September 2010, outlined publicly for the first time the Coalition Government's position:

> We are different; we are liberal. Because we will put local government back in charge of the money it raises and spends. ... That's why we will end central capping of Council Tax. That's why we will allow councils to keep some of the extra business rates and council tax they raise when they enable new developments to go ahead.

... I assure you it is the first step to breathing life back into our greatest cities.[4]

Picking up on and emphasizing a link between financial decentralization and the establishment in the United Kingdom of TIF, Clegg, together with Conservative politicians, such as Eric Pickles, the Minister for Communities and Local Government, and a range of other actors have been policy mobilizers and market makers for TIF in the United Kingdom. Although their motives and rationalities might not have been the same, they have developed an ideological institutional project operating in tandem with networks of professional expertise that delineate what is possible and desirable from what is not. These shifting fields of practice (Peck and Theodore 2010) thus structure the policy marketplace, anointing certain actors with the power of expert authority and positioning certain policy ideas as worthy of replication by virtue of their congruence with expert opinion.

As this mobilizing and market-making progressed, the UK Coalition Government published the Local Government Finance Act in 2012, which contained details of its approach to TIF and the new ways in which local business rates (taxes) or nondomestic rates (NDR) might be distributed between central and local government. Simultaneously, the government has begun to introduce a range of TIF-like reforms to allow English local governments to borrow against potential future revenue streams.

Whether this will end with the introduction of something called TIF remains unclear. What it does suggest, however, is that TIF, as a mobile idea and policy model, is not naturally best or most appropriate for cities outside the United States. Rather, its growing influence in the United Kingdom at least is the result of sustained political work done by a range of UK actors, from members of the Urban Task Force, to members of national and local business coalitions, to politicians from three national political parties, as well as by transfer agents based on the other side of the Atlantic. TIF, then, is not so much an idea that's time and place has come, as it is an idea that has moved beyond its early sites of experimentation because new places for it have been painstakingly created in professional and ideological landscapes elsewhere.

Conclusion

We have argued that the mobilization of policy models is enacted by coalitions of powerful actors, including politicians, government bureaucrats, economic development professionals, activists, and consultants. A diverse cast of transfer agents makes mobilization possible. People who are responsible for or invested in the mobility of particular policy models, such as TIF, demonstrate that particular constellations of ideas are mobilized in the service of making markets and addressing needs. In this example, there are a number of logics of the market at work in TIF. It encourages tax creation at a municipal level, which acts as a counternarrative to the minimal state mentality of a neoliberal market economy. Yet, the model manages to capture another aspect of market making by placing cities in the role of consumers. It encourages the municipality to enter into a debtor's economy to finance state-led infrastructural projects without having to consult citizens on its borrowing practices.

Calibrating these sorts of paradoxical ideologies so that new policy solutions can be realized in certain places is done, we have argued, through the productive work of mobilizing policy ideas. "Home-grown" policy models that might seek to provide similar economic benefits to local communities have far less global cachet, far less cultural capital, than proven policies from elsewhere, backed by material elements such as policy documents, white papers, and benchmarking schemes. What is valued about certain policy ideas, how that value is learned in the institutional context of specific governance regimes is, we argue, tied closely to mobility (Temenos and McCann 2013).

Although the policy mobilities approach highlights the discursive and representational elements of policy learning, it also appreciates that policymaking is intertwined with an array of physical materials. By this way of thinking, policy mobility can be construed as more than just a human endeavor. Policy actors' intentionality is shaped by their engagements with materials, including documents, facilities, and places, for example, as well as the systems of physical and informational infrastructure that sometimes facilitate and sometimes constrain the movement of policy ideas.

Constraints are also important to consider when examining the ways in which policy is mobilized. Particularly important are institutional landscapes that might impede an idea whose time might seem to have come. In the Australian context, by contrast to the United Kingdom, TIF is also an example of a moment when "the movement of ideas gets stuck" (Cresswell 2012, 651). It has failed to be introduced into Australia, despite work being done by analysts, consultants, policymakers, and politicians in both Australia and the United States to render it "introduction-ready." Financial, governmental, and legal conditions in the country have proved insurmountable, although discussions over the future introduction of TIF in Australia continue. In certain political contexts, spurning a policy that comes from elsewhere is politically expedient, or materially so. Ideas from elsewhere can be powerful. Yet when those interests are working against existing, already-territorialized ones, barriers might appear, and a failure to "land" could be the outcome. These immobilities and failures are important to consider not only to examine how neoliberalization does or does not continue to appear in locations but also to recognize the political motivations that might provide a crack in the armature of dominant political economic arrangements, allowing light to shine on spaces for alternative urban-economic development.

Acknowledgments

The authors acknowledge all those that have contributed to this article through their comments at various conferences and on various papers over the years. More specifically, we thank Graham Bowden at the University of Manchester for generating Figure 1. The paper is better for the comments from three anonymous reviewers and the Editor, Mei-Po Kwan. Responsibility for the arguments here are ours alone.

Notes

1. More details are available at http://www.cdfa.net/cdfa/cdfaweb.nsf/pages/about.html (last accessed 10 September 2014).
2. This is in contrast to borrowing against already realized revenue streams, which is the case for the issuing of general obligation (GO) bonds that are backed by the full faith and credit of the issuing (borrowing) government.
3. The focus of this article is not the arguments for and against the use of TIF (on which see Man 2001; Jonas and McCarthy 2009; Briffault 2010).
4. The full speech is available at http://www.libdemvoice.org/full-text-nick-cleggs-speech-to-liberal-democrat-autumn-conference-21236.html (last accessed 8 September 2014). This was an attempt by Clegg on behalf of the Liberal Democrats to distinguish their approach to financial decentralization and the empowering of local government from that of their coalition partners, the Conservatives.

References

Adey, P., D. Bissell, K. Hannam, P. Merriman, and M. Sheller, eds. 2013. *The Routledge handbook of mobilities.* London and New York: Routledge.

Baker, T. 2014. Cities from elsewhere: Chronic homelessness and globalising urban policy. PhD thesis, School of Environmental and Life Sciences, University of Newcastle, Callaghan, Australia.

Barber, L. 2013. (Re)making heritage policy in Hong Kong: A relational politics of global knowledge and local innovation. *Urban Studies* 51 (6): 1179–95.

Bok, R. 2015. Airports on the move? The policy mobilities of Singapore Changi Airport at home and abroad. *Urban Studies* 52:2724–40.

Briffault, R. 2010. The most popular tool: Tax increment financing and the political economy of local government. *The University of Chicago Law Review* 77 (1): 65–95.

Bunnell, T. 2015. Antecedent cities and inter-referencing effects: Learning from and extending beyond critiques of neoliberalisation. *Urban Studies* 52:1983–2000.

Clarke, N. 2011. Urban policy mobility, anti-politics, and histories of the transnational municipal movement. *Progress in Human Geography* 36 (1): 25–43.

Cochrane, A., and K. Ward. 2012. Researching the geographies of policy mobility: Confronting the methodological challenges. *Environment and Planning A* 44 (1): 5–12.

Cohen, D. 2015. Grounding mobile policies: Ad hoc networks and the creative city in Bandung, Indonesia. *The Singapore Journal of Tropical Geography* 36:23–37.

Cook, I. R. 2008. Mobilising urban policies: The policy transfer of US business improvement districts to England and Wales. *Urban Studies* 45 (4): 773–95.

Cook, I. R., and K. Ward. 2011. Trans-urban networks of learning, mega events and policy tourism: The case of Manchester's Commonwealth and Olympic Games projects. *Urban Studies* 48 (12): 2519–35.

———. 2012. Conferences, informational infrastructures and mobile policies: The process of getting Sweden "BID ready." *European Urban and Regional Studies* 19 (2): 137–52.

Cook, I. R., S. V. Ward, and K. Ward. 2014. A springtime journey to the Soviet Union: Postwar planning and policy mobilities through the Iron Curtain. *International Journal of Urban and Regional Research* 38 (3): 805–22.

———. 2015. Policy tourism and postwar planning: The international study tours of the Town and Country Planning Association 1947–61. *Planning, Theory and Practice* 16 (2): 184–205.

Cresswell, T. 2001. Introduction. *Mobilities* 43 (Spring): 3–25.

———. 2006. *On the move: Mobility in the modern western world*. London and New York: Routledge.

———. 2010. Towards a politics of mobility. *Environment and Planning D: Society and Space* 28 (1): 17–31.

———. 2012. Mobilities II: Still. *Progress in Human Geography* 36 (5): 645–53.

Cresswell, T., and P. Merriman, eds. 2011. *Geographies of mobilities: Practices, spaces, subjects*. Aldershot, UK: Ashgate.

Faist, T. 2013. The mobility turn: A new paradigm for the social sciences. *Ethnic and Racial Studies* 36 (11): 1637–46.

Faulconbridge, J. R. 2013. Mobile "green" design knowledge: Institutions, bricolage and the relational production of embedded sustainable building designs. *Transactions of the Institute of British Geographers* 38 (2): 339–53.

Fisher, S. 2014. Exploring nascent climate policies in Indian cities: A role for policy mobilities? *International Journal of Urban Sustainable Development* 6 (2): 154–73.

Geddie, K. 2014. Policy mobilities in the race for talent: Competitive state strategies in international student mobility. *Transactions of the Institute of British Geographers* 40 (2): 235–48.

González, S. 2011. Bilbao and Barcelona "in motion." How urban regeneration "models" travel and mutate in the global flows. *Urban Studies* 48 (10): 1397–1418.

Hannam, K., M. Sheller, and J. Urry. 2006. Mobilities, immobilities and moorings. *Mobilities* 1 (1): 1–22.

Healey, P., and R. Upton, eds. 2010. *Crossing borders: International exchange and planning practise*. London and New York: Routledge.

Jacobs, J. M. 2012. Urban geographies I: Still thinking cities relationally. *Progress in Human Geography* 36 (3): 412–22.

Jacobs, J. M., and L. Lees. 2013. Defensible space on the move: Revisiting the urban geography of Alice Coleman. *International Journal of Urban and Regional Research* 37 (5): 1559–83.

Jonas, A. E. G., and L. McCarthy. 2009. Urban management and regeneration in the United States: State intervention or redevelopment at all costs. *Local Government Studies* 35 (3): 299–314.

Kingfisher, C. 2013. *A policy travelogue: Tracing welfare reform in Aotearoa/New Zealand and Canada*. New York: Bergahn.

Klacik, J. D., and K. A. Kriz. 2001. A review of state tax increment financing laws. In *Tax increment financing and economic development: Uses, structures and impact*, ed. C. Johnson and J. Man, 15–30. Albany: State University of New York Press.

Kuus, M. 2014. *Geopolitics and expertise*. Malden, MA: Wiley-Blackwell.

MacLeod, G. 2013. New urbanism/smart growth in the Scottish Highlands: Mobile policies and post-politics in local development planning. *Urban Studies* 50 (11): 2196–2221.

Man, J. 2001. Determinants of the municipal decision to adopt tax increment financing. In *Tax increment financing and economic development: Uses, structures and impact*, ed. C. Johnson and J. Y. Man, 87–100. Albany: State University of New York Press.

McCann, E. 2008. Expertise, truth, and urban policy mobilities: Global circuits of knowledge in the development of Vancouver, Canada's "four pillar" drug strategy. *Environment and Planning A* 40 (4): 885–904.

———. 2011. Urban policy mobilities and global circuits of knowledge: Toward a research agenda. *Annals of the Association of American Geographers* 101 (1): 107–30.

McCann, E., and C. Temenos. 2015. Mobilizing drug consumption rooms: Inter-place networks and harm reduction drug policy. *Health & Place* 31 (2): 216–23.

McCann, E., and K. Ward. 2010. Relationality/territoriality: Toward a conceptualization of cities in the world. *Geoforum* 41 (2): 175–84.

———, eds. 2011. *Mobile urbanism: Cities and policymaking in the global age*. Minneapolis: University of Minnesota Press.

———. 2012a. Assembling urbanism: Following policies and "studying through" the sites and situations of policy making. *Environment and Planning A* 44 (1): 42–51.

———. 2012b. Policy assemblages, mobilities and mutations: Toward a multidisciplinary conversation. *Political Studies Review* 10 (3): 325–32.

———. 2013. A multi-disciplinary approach to policy transfer research: Geographies, assemblages, mobilities and mutations. *Policy Studies* 34 (1): 2–18.

———. 2015. Thinking through dualisms in urban policy mobilities. *International Journal of Urban and Regional Research* forthcoming.

McFarlane, C. 2011. *Learning the city: Knowledge and translocal assemblage.* Malden, MA: Wiley-Blackwell.

Mukhtarov, F. 2014. Rethinking the travel of ideas: Policy translation in the water sector. *Policy & Politics* 42 (1): 71–88.

Müller, M. 2015. (Im-)mobile policies: Why sustainability went wrong in the 2014 Olympics in Sochi. *European Urban and Regional Studies* 22 (2): 191–209.

Peck, J. 2002. Political economies of scale: Fast policy, interscalar relations, and neoliberal workfare. *Economic Geography* 78 (3): 331–60.

———. 2005. Struggling with the creative class. *International Journal of Urban and Regional Research* 29 (4): 740–70.

———. 2011. Geographies of policy: From transfer-diffusion to mobility-mutation. *Progress in Human Geography* 35 (6): 773–97.

Peck, J., and N. Theodore. 2001. Exporting workfare/importing welfare-to-work: Exploring the politics of Third Way policy transfer. *Political Geography* 20 (4): 427–60.

———. 2010. Mobilizing policy: Models, methods, and mutations. *Geoforum* 41 (2): 169–74.

———. 2015. *Fast policy.* Minneapolis: University of Minnesota Press.

Phelps, N. A., T. Bunnell, M. A. Miller, and J. Taylor. 2014. Urban inter-referencing within and beyond a decentralized Indonesia. *Cities* 39 (1): 37–49.

Prince, R. 2010. Policy transfer as policy assemblage: Making policy for the creative industries in New Zealand. *Environment and Planning A* 42 (1): 169–86.

———. 2012. Policy transfer, consultants and the geographies of governance. *Progress in Human Geography* 36 (2): 188–203.

Quark, A. A. 2013. Institutional mobility and mutation in the global capitalist system: A neo-Polanyian analysis of a transnational cotton standards war, 1870–1945. *Environment and Planning A* 45 (7): 1588–1604.

Rapoport, E. 2015. Globalising sustainable urbanism: The role of international masterplanners. *Area* 47 (2): 110–15.

Robinson, J. 2011. The travels of urban neoliberalism: Taking stock of the internationalization of urban theory. *Urban Geography* 32 (8): 1087–1109.

———. 2013. "Arriving at" the urban/urban policy: Traces of elsewhere in making city futures. In *Critical mobilities*, ed. O. Söderström, S. Randeria, D. Ruedin, G. D'Amato, and F. Panese, 1–28. London and New York: Routledge.

Roy, A. 2012. Ethnographic circulations: Space–time relations in the worlds of poverty management. *Environment and Planning A* 44 (1): 31–41.

Roy, A., and A. Ong, eds. 2011. *Worlding cities: Asian experiments and the art of being global.* Malden, MA: Wiley-Blackwell.

Sheller, M. 2014. *Aluminium dreams: The making of light modernity.* Cambridge, MA: MIT Press.

Shore, C., and S. Wright, eds. 1997. *Anthropology of policy: Critical perspectives on governance and power.* London and New York: Routledge.

Shore, C., S. Wright, and S. Però, eds. 2011. *Policy worlds: Anthropology and analysis of contemporary power.* New York: Bergahn Books.

Söderström, O. 2014. *Cities in relations: Trajectories of urban development in Hanoi and Ouagadougou.* Malden, MA: Wiley-Blackwell.

Temenos, C., and E. McCann. 2012. The local politics of policy mobility: Learning, persuasion, and the production of a municipal sustainability fix. *Environment and Planning A* 44 (6): 1389–1406.

———. 2013. Policies. In *The Routledge handbook of mobilities*, ed. P. Adey, D. Bissell, K. Hannam, P. Merriman, and M. Sheller, 575–84. London and New York: Routledge.

Urban Task Force. 1999. *Towards an urban renaissance.* London and New York: Routledge.

———. 2005. Towards a strong urban renaissance. http://www.integreatplus.com/sites/default/files/towards_a_strong_urban_renaissance.pdf (last accessed 30 November 2015).

Ward, K. 2006. "Policies in motion," urban management and state restructuring: The trans-local expansion of business improvement districts. *International Journal of Urban and Regional Research* 30 (1): 54–75.

———. 2007. Business improvement districts: Policy origins, mobile policies and urban liveability. *Geography Compass* 1 (3): 657–72.

Weber, R. 2002. Extracting value from the city: Neoliberalism and urban redevelopment. *Antipode* 34 (3): 519–40.

———. 2010. Selling urban futures: The financialization of urban redevelopment policy. *Economic Geography* 86 (3): 251–74.

Wedel, J. R., C. Shore, G. Feldman, and S. Lathrop. 2005. Toward an anthropology of public policy. *The Annals of the American Academy of Political and Social Science* 600 (1): 30–51.

Wilson, D., and C. Sternberg. 2012. Changing realities: The new racialized redevelopment rhetoric in Chicago. *Urban Geography* 33 (7): 979–99.

Wood, A. 2014. Moving policy: Global and local characters circulating bus rapid transit through South African cities. *Urban Geography* 35 (8): 1238–54.

Zhang, J. 2012. From Hong Kong's capitalist fundamentals to Singapore's authoritarian governance: The policy mobility of neo-liberalising Shenzhen, China. *Urban Studies* 49 (13): 2853–71.

Temporal Trends of Intraurban Commuting in Baton Rouge, 1990–2010

Yujie Hu and Fahui Wang

Department of Geography and Anthropology, Louisiana State University

Based on the 1990–2010 Census Transportation Planning Package data of Baton Rouge, Louisiana, this research analyzes the temporal trends of commuting patterns in both time and distance. In comparison to previous work, commuting length is calibrated more accurately by Monte Carlo–based simulation of individual journey-to-work trips to mitigate the zonal effect. First, average commute distance kept climbing between 1990 and 2010, whereas average commute time increased between 1990 and 2000 but then slightly dropped toward 2010. Second, urban land use remained a good predictor of commuting pattern over time (e.g., explaining up to 90 percent of mean commute distance and about 30 percent of mean commute time). Finally, the percentage of excess commuting increased significantly between 1990 and 2000 and stabilized afterward.

本研究根据路易斯安那州巴顿鲁治于 1990 年至 2010 年间的运输规划套装数据, 分析通勤模式于时间和距离方面的时态趋势。与过往的研究相较之下, 本研究透过根据蒙地卡罗方法对个人工作旅次所进行的模拟, 更精确地测定通勤长度, 以减轻区块效应。首先, 平均通勤距离在 1990 年和 2010 年间持续增加, 而平均通勤时间在 1990 年和 2000 年间增加, 但在接近2010年时却稍微减少。再者, 城市土地使用仍然作为通勤模式随着时间改变的良好预测 (例如解释近百分之九十的通勤距离平均数和约百分之三十的通勤时间平均数)。最后, 过剩通勤的百分比在 1990 年和 2000 年之间显着增加, 随后则呈现稳定状态。

Con base en datos del Censo para el Paquete de Planificación del Transporte de 1990-2010 de Baton Rouge, Luisiana, esta investigación analiza las tendencias temporales de los patrones de viaje pendular, tanto en tiempo como en distancia. En comparación con el trabajo precedente, la longitud del viaje pendular se calibró con mayor exactitud por medio de simulación basada en Monte Carlo de viajes individuales al trabajo para suavizar el efecto zonal. Primero, la distancia promedio del viaje pendular se mantuvo en aumento entre 1990 y 2010, mientras que el tiempo promedio de ese viaje se incrementó entre 1990 y 2000, pero luego cayó ligeramente hacia el 2010. Segundo, el uso urbano del suelo se sostuvo como un buen predictor del patrón de desplazamiento pendular sobre el tiempo (e.g., explicando hasta el 90 por ciento de la distancia promedio del viaje pendular y cerca del 30 por ciento de la media del tiempo de desplazamiento). Finalmente, el porcentaje del exceso de desplazamiento pendular aumentó significativamente entre 1990 y 2000 para estabilizarse de ahí en adelante.

ommuting is a daily human mobility behavior for employed individuals, and it has a significant influence on society. Commuting is strongly connected with some practical issues on which many public policies focus (Sultana and Weber 2014). It is a major contributor to traffic congestion, air pollution, and greenhouse gas emissions. Clearly, the most congested period in a day occurs at the morning and afternoon commuting peaks, even though commuting only represents 20 to 25 percent of all-purpose trips in the United States (Horner 2004). In line with the worsened traffic congestion, commuters are spending more time in their daily commute trips all across the United States. For example, one out of twelve U.S. workers in 2001 spent an hour or more in their one-way commute trips, compared to one out of twenty in 1995. It was also reported that the average commute speed declined about 10 percent in midsize metropolitan areas from 1990 to 2009 (Santos et al. 2011). Among all economic sectors, transportation (including commuting) is the "second-largest contributor to total U.S. emissions" (next to industry) but with the fastest growth rate (U.S. Department of State 2010, 36). Understanding the temporal change of commuting and its underlying causes is a step toward the larger goals of traffic congestion mitigation and carbon emission control.

Background and Related Literature

Commuting varies across areas and by sociodemographic groups. To understand its variability, some focus on spatial factors (Horner 2004), and others emphasize aspatial factors such as race (Kain 2004), wage (Wang 2003) or income (Horner and Schleith 2012), and gender (Kwan and Kotsev 2015). In other words, commuting could be explained by where the commuters are and who they are (Wang 2001). Due to limited space, this article focuses on the former. The latter will be examined elsewhere.

Specifically, researchers with a spatial perspective have a sustained interest in explaining intraurban variation of commuting by land use pattern. In essence, commuting is for a worker to overcome the spatial barrier from home to workplace; therefore explanation of commuting naturally begins with a focus on the spatial separation of resident (population) and employment locations. Past attempts include modeling how far a residential location is from a job concentration area such as the central business district (CBD) or subcenters (Wang 2000) or from the overall job market such as job accessibility (Wang 2003) or measuring the need of commuting beyond a local area (e.g., captured by the jobs–housing balance ratio; Sultana 2002). There is no shortage of doubters, however, on whether commuting could be predicted by urban land use pattern (e.g., Giuliano and Small 1993).

Excess commuting is another line of research closely related to the paradigm of interrelatedness between land use and commuting. It is the proportion of actual commute over minimum (optimal or required) commute when assuming that people could freely swap their homes and jobs in a city (Hamilton 1982; White 1988). Instead of focusing on the variation in commuting across areas, it highlights how much overall commuting could be reduced based on the preceding assumption. In other words, the concept captures the potential (or lack of potential) for a city to optimize commuting without altering the existing land use and, to some extent, reflects efficiency in its land use layout. Excess commuting was mostly based on a homogeneous work group assumption, however, and some studies made efforts to relax such an assumption by considering individual utility needs, household structure, gender, employment class, and other constraints (Cropper and Gordon 1991; S. Kim 1995; Horner 2002, 2004; Yang 2008). Some studies then extended the idea of measuring minimum commute and proposed other commuting efficiency metrics, such as random commute, maximum commute, and proportionally matched commute (Horner 2002; Yang 2008; Murphy and Killen 2011).

A few recent studies examined the temporal change in commuting patterns and related policy implications. Horner (2007) explained the spatial–temporal pattern of intraurban commuting (mileage and multiple commuting efficiency metrics) from the jobs–housing balance perspective in Tallahassee, Florida, from 1990 to 2000. Similarly, Chen, Zhan, and Wu (2010) investigated the change in commuting patterns from analyzing the residential and employment distributions in central Texas between 1990 and 2000. Yang (2008) examined the temporal change in excess commuting in Boston and Atlanta between 1980 and 2000 and concluded that urban structure alone could lead to change in excess commuting when controlling individuals' preferences.

Finally, measures of commute length merit some discussion. Commuting studies often use commute time, as it is directly available from survey data; however, distance could provide a more consistent measure of commuting length (Sultana and Weber 2007). Among the few studies on commute distance, most used a zonal centroid-to-centroid approach to reconstruct either Euclidean or network distance (e.g., Wang 2001; Yang 2008). Such an approach could bias the estimate, particularly in large zones and also by omitting intrazonal distance (Hewko, Smoyer-Tomic, and Hodgson 2002). For this reason, Horner and Schleith (2012) used a small analysis unit such as census block to mitigate the scale effect. Commuting data at the block level, however, are not widely available for most cities in the United States. Different unit scales and unit zone definitions cause inconsistency in analysis results, particularly common in comparison analysis over time, and thus lead to the *modifiable areal unit problem* (MAUP; Niedzielski, Horner, and Xiao 2013). Some recent studies have shown great promise in using Global Positioning System data and activity-travel surveys of individual trip makers in commuting studies (Shen, Kwan, and Chai 2013; Kwan and Kotsev 2015). Such data, however, might not be representative of all commuters and are also not universally available. More accessible and accurate measures of commute length remain very much needed.

This article examines the aforementioned issues that have been central to the study of commuting and urban structure. The contributions of our research can be highlighted in three aspects. Foremost, we measure commute length by Monte Carlo–based simulation of individual resident workers and jobs and the trips between them, a significant improvement over the zonal-level centroid-to-centroid approach, to mitigate the aforementioned aggregation error and scale effect. Second, the new and longer temporal trends for commuting are detected by analyzing the newly available 2006 to 2010 Census Transportation Planning Package (CTPP) data. Finally, when defining excess commute, the optimal commuting pattern is now formulated as an integer programming problem based on simulated individual trips instead of the conventional linear programming approach for zonal-level trips.

Data Sources and Commute Distance Calibration

The study area is East Baton Rouge Parish in Louisiana (hereafter simply referred to as Baton Rouge). With a population of 440,000 in 2010, it is the core of the Baton Rouge metropolitan area (other surrounding parishes are mostly rural). The parish is a county equivalent unit in Louisiana. The major data sources are the 1990 and 2000 CTPPs extracted from the long-form decennial census and the most recent 2006 to 2010 CTPP based on the five-year American Community Survey. All CTPPs consist of three parts: Part 1 on residential places (e.g., number of resident workers, worker breakdowns by transportation mode or by wage range), Part 2 on workplaces (e.g., number of jobs), and Part 3 on journey-to-work flow (e.g., number of commuters and mean travel time on origin–destination trips). For simplicity, the 2006 to 2010 data and corresponding analyses are hereafter referred to as 2010.

The CTPP data in Baton Rouge used different area units over the years: traffic analysis zone (TAZ) in 1990, multiple zones in 2000 (census tract, census block group, and TAZ for Parts 1 and 2, only census tract for Part 3), and census tract and TAZ in 2010. We chose census tract as the unified unit. Luckily, in

Baton Rouge, the 1990 TAZs were mostly components of census tracts for easy aggregation with very few minor adjustments. There were 85, 89, and 91 census tracts in Baton Rouge in 1990, 2000, and 2010, respectively. As shown in Figure 1, the major job concentrations of the study area in 2010 are (1) downtown (CBD); (2) the largest employer, Louisiana State University (about 3 miles south of the CBD); and (3) a large area with several major hospitals and health clinics (3–5 miles southeast of downtown). The pattern is generally consistent with that in 2000 (Antipova and Wang 2010; Antipova, Wang, and Wilmot 2011).

The CTPP provides information on commute time but not on commute distance. We use Monte Carlo simulation to obtain journey-to-work trips between individual points. Briefly, the first task is to randomly generate points of resident workers and jobs within tracts so that their total numbers at the zonal level are proportional to the observed patterns of resident workers and jobs in the CTPP. The second task is to pair the origins (workers) and destinations (jobs) to form OD trips that follow a discrete frequency distribution that is also consistent with the reported journey-to-work flows in the CTPP and then measure the network distance for each OD trip. For more technical detail, see Hu and Wang (2015b).

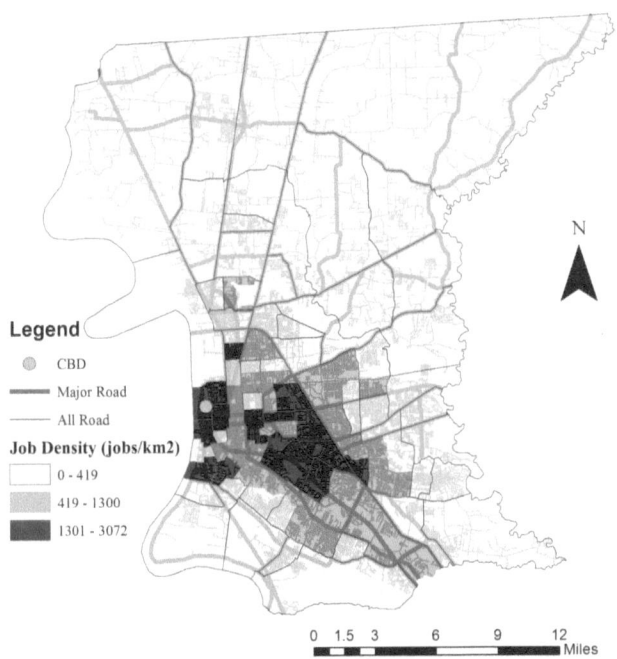

Figure 1. Employment density in Baton Rouge 2010. *Note:* CBD = central business district. (Color figure available online.)

By doing so, the zonal journey-to-work flow is disaggregated into individual trips. It permits more accurate estimation of commute distances and mitigates the aggregation error and zonal effect. For example, the average commute distance in Baton Rouge in 2000 was 5.95 miles by the zonal centroid-to-centroid approach and 6.17 miles by our simulation-based approach. The longer estimates by the simulation technique are validated through a one-tailed t test. In addition, the centroid-to-centroid approach could assign the intrazonal travel distance uniformly as 0, whereas our approach yields a more realistic measure that varies with the zone size (e.g., 7.94 miles in the largest tract in the northeast corner). This is especially important in excess commuting analysis because the "optimal" commuting pattern based on the zonal data often includes a large number of intrazonal trips when assuming a zero intrazonal commute. Our approach yields the optimal pattern with more interzonal trips when a trip is shorter for a resident worker travels to a job in a neighboring tract than within the same tract, which is more realistic (Hu and Wang 2015a).

Overall Temporal Trend of Commuting

People commute by multiple modes. As shown in Table 1, the majority commuted by automobile (including driving alone and carpooling), and the percentage was steady over time (i.e., 94–95 percent). Therefore, we estimated commute distance based on the road network without taking transit, bicycle, or pedestrian routes into consideration. Mean commute distance (time) in a tract is the average of travel distances (times) from this tract to all tracts weighted by corresponding number of commuters. Specifically, a tract's mean commute time was based on the zonal commuter flow matrix and corresponding travel time reported in the CTPP. Mean commute distance,

although not reported, was based on the Monte Carlo simulation of individual trips, however.

Also, as shown in Table 1, the mean commute distance on average increased steadily from 5.95 miles in 1990 to 6.17 miles in 2000 and further to 6.25 miles in 2010. This is consistent with some previous studies (Levinson and Kumar 1994; Cervero and Wu 1998) and reflects a long-standing trend of more workers moving farther from their jobs either to search farther for jobs for maximizing their earnings or to move their residences for better housing. The increasing rate was higher in 1990 to 2000 than in 2000 to 2010, however. One possible reason might be the economic recession that began in 2008 (Horner and Schleith 2012).

The mean commute time for overall population increased from 1990 to 2000 and then declined to 2010. Other studies also found that the average commute time stayed stable or even dropped over time (e.g., C. Kim 2008). They ascribed the declining time to colocation of jobs and housing; that is, people relocate their residence or jobs to cut back commute time as traffic becomes more congested. An increasing use of suburban roads that are usually newer and wider than roads in the central city makes it achievable to commute longer distances in a shorter duration. Taking driving alone, for example, the implied average commuting speed was 24.3 mph in 1990, dropped to 21.9 mph in 2000, and only recovered slightly to 22.8 mph. Therefore, a small increment in commute distance from 1990 to 2000 came with a much larger climb in commute time, and it was only after 2000 that the colocation theory became relevant and led to a small drop in commute time. The significant climb in commute time for 1990 to 2000 might reflect people's increasing endurance of long commutes or traffic congestion that grew much more rapidly than people could adapt (Levinson and Wu 2005), and the relocation adjustment came afterward.

Table 1. Modal splits and commute lengths in Baton Rouge, 1990–2010

Year	Modal splits				Mean commute distance (mile)	Mean commute time (min)
	Drove alone	Carpool	Public transit	Others[a]		
1990	82.35	11.76	1.29	4.60	5.95	16.73
2000	83.16	11.91	1.40	3.54	6.17	18.73
2010[b]	83.77	11.08	1.75	3.40	6.25	17.98

[a]Others includes taxi, motorcycle, bicycle, walk, and so on.
[b]Based on 2006–2010 Census Transportation Planning Package; 2010 is used here and in all other tables for simplicity.

Table 2. Regression models of mean commute distance across census tracts, 1990–2010

	1990	2000	2010	1990	2000	2010	1990	2000	2010
Intercept	2.40***	2.91***	3.01***	10.37***	14.87***	13.65***	0.88***	0.99***	1.10***
	(10.03)	(10.52)	(9.93)	(29.14)	(20.75)	(21.34)	(4.60)	(7.06)	(5.74)
D_{CBD}	0.57***	0.50***	0.48***						
	(17.38)	(13.67)	(12.22)						
JWR				−3.88***	−10.38***	−8.05***			
				(−14.07)	(−8.35)	(−7.57)			
JWR^2				0.37***	2.60***	1.75***			
				(13.93)	(5.33)	(4.36)			
JobP							0.67***	0.67***	0.64***
							(29.44)	(40.26)	(29.22)
No. of observations	85	89	91	85	89	91	85	89	91
R^2	0.78	0.68	0.63	0.72	0.78	0.76	0.91	0.95	0.91

Note: t statistics are given in parentheses. D_{CBD} = distance from central business district; JWR = jobs-to-workers ratio; JobP = job proximity index.
***Significant at the 0.001 level.

Commuting and Land Use Patterns

There has been a long tradition of attempts to explain intraurban variability of commuting by land use patterns. The analysis begins by examining the impact of the CBD with the highest employment concentration. Table 2 shows that distance from the CBD (denoted by D_{CBD}) explained the variation in mean commute distance across census tracts by 78 percent in 1990, 68 percent in 2000, and 63 percent in 2010. The declining explanatory was attributable to dispersion of jobs beyond the CBD area (e.g., percentage of workers commuting to the area within a 3-mile radius of the CBD was 42 percent in 1990, 32 percent in 2000, and 31 percent in 2010). The effect of distance

from CBD remained significant on mean commute time in all models (Table 3) but much weaker than on mean commute distance (Table 2). The lower performance of regression models on commute time was attributable to the nonuniform modal distribution across tracts. In fact, for drove-alone commuters alone, the pattern of mean commute time was largely consistent with that of mean commute distance, and the corresponding regression models yielded $R^2 = 0.60$, 0.41, and 0.50 in 1990, 2000, and 2010, respectively (details not reported here). Adding the square term D_{CBD}^2 did not improve the explanatory power of the regression models and thus was not included.

Second, we look into the *jobs–housing balance* hypothesis proposed by Cervero (1989). An

Table 3. Regression models of mean commute time across census tracts, 1990–2010

	1990	2000	2010	1990	2000	2010	1990	2000	2010
Intercept	14.72***	16.96***	15.29***	19.69***	27.86***	25.36***	13.32***	14.42***	13.63***
	(26.82)	(23.06)	(25.01)	(28.25)	(13.51)	(17.12)	(20.44)	(16.12)	(21.15)
D_{CBD}	0.32***	0.27**	0.40***						
	(4.28)	(2.80)	(5.04)						
JWR				−2.51***	−12.26***	−8.45***			
				(−4.65)	(−3.43)	(−3.43)			
JWR^2				0.21***	3.60*	1.99*			
				(4.12)	(2.57)	(2.15)			
JobP							0.25***	0.30***	0.31***
							(5.74)	(5.22)	(7.41)
No. of observations	85	89	91	85	89	91	85	89	91
R^2	0.18	0.08	0.22	0.21	0.26	0.35	0.28	0.24	0.38

Note: t-statistics are given in parentheses. D_{CBD} = distance from central business district; JWR = jobs-to-workers ratio; JobP = job proximity index.
*Significant at the 0.05 level.
**Significant at the 0.01 level.
***Significant at the 0.001 level.

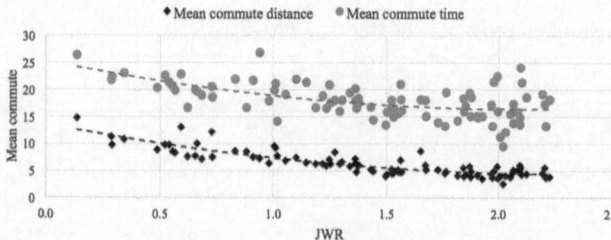

Figure 2. Mean commute distance and time versus jobs-to-workers ratio (JWR) in 2010.

imbalanced area has far more resident workers than jobs and thus more workers need to commute outside of the area for their jobs and tend to incur more commuting. Following Wang (2000), this research used the floating catchment area method to define a circular area around each census tract centroid and calculated the jobs-to-workers ratio (JWR) within each catchment area. A higher JWR implies less need of commuting beyond the catchment area and thus is expected to correlate with less commuting. We explanatory with radii ranging between 1.5 and 7.5 miles and settled with 5 miles for its best explanatory power.

As an example, Figure 2 shows the relationship between mean commute distance and time versus JWR in 2010. Both display a quadratic trend, but the trend is much clearer for distance than time. This is confirmed by a regression model with the added square term of JWR, as reported in Tables 2 and 3. In 1990, 2000, and 2010, the negative sign of JWR and the positive sign of JWR^2 indicate that mean commute distance or time at the tract level declined with JWR, but the declining slope got flatter in higher JWR areas. Both terms are statistically significant in all models. The models for mean commute distance in Table 2 performed well with R^2 that was 0.72 in 1990, peaked at 0.78 in 2000, and dropped slightly to 0.76 in 2010. The models for mean commute time in Table 3 also confirmed the quadratic trend in all years, although with lower R^2 values (0.21, 0.26, and 0.35 in 1990, 2000, and 2010, respectively). In short, the results confirm the importance of jobs–housing imbalance in affecting commuting pattern, much more significant in Baton Rouge than in large metropolitan areas reported in other studies (e.g., Wang 2000).

Either the emphasis on the role of CBD or the jobs–housing balance approach does not consider all job locations in explaining commuting patterns. The *job proximity index* (JobP) captures the spatial separation

between a worker's residence and all potential job sites (Wang 2003), formulated as

$$JobP_i = \sum_{n}^{j=1} (P_{ij}d_{ij}), \qquad (1)$$

where $P_{ij} = (J_j d_{ij}^{-\beta}) / \sum_{k=1}^{n} (J_k d_{ik}^{-\beta})$.

Similar to the notion of the Huff (1963) model, the probability of workers residing in zone i and going to work in zone j (denoted by P_{ij}) is predicted as the gravity kernel of job site j out of all job sites k ($= 1, 2, \ldots, n$). Each gravity kernel is positively related to the number of jobs there J_j (or J_k) and negatively to the distance or time between them d_{ij} (or d_{ik}) powered to the distance friction coefficient β. Then JobP at zone i is simply the aggregation of all distances (time) d_{ij} with corresponding probabilities P_{ij} over all job sites ($j = 1, 2, \ldots, n$).

Calibration of JobP requires defining the value for the distance friction coefficient β in Equation 1. In this research, the β value was computed from the log-transformed regression based on the classic gravity model such as

$$C_{ij} = aW_i J_j d_{ij}^{-\beta}, \qquad (2)$$

where C_{ij} is the number of commuters from a tract with W_i resident workers and to a tract with J_j jobs for a distance (time) of d_{ij} (Wang 2015). Based on the CTPP data, the derived β value was 0.404 in 1990, 0.547 in 2000, and 0.475 in 2010 if the journey-to-work trips were measured in distance; and the β value was 0.295 for 1990, 0.353 for 2000, and 0.385 for 2010 if measured in time.

The regression results for mean commute distance and time by JobP are again reported in Tables 2 and 3, respectively. The mean commute distance at the tract level was well explained by JobP with $R^2 = 0.91, 0.95,$ and 0.91 in 1990, 2000, and 2010, respectively. This is a significant improvement over other factors; that is, D_{CBD} and JWR. Similarly, regression models on mean commute time returned lower R^2 values ranging between 0.24 and 0.38. We also did not add the square term $JobP^2$ because it did not improve the explanatory power of the regression models. Note that we also run a series of regression models on mean commuting distance (calibrated by the zonal centroid-to-centroid approach), and results indicate weaker R^2 than the simulation-calibrated distance (due to space limits, results are not shown here). This again demonstrates the value of our simulation approach that yielded a more accurate estimate of commute distance.

Temporal Change in Excess Commuting

Although the land use pattern in Baton Rouge explained the commuting variations to some extent, there was still a proportion of variations unexplained (particularly commute time). Part of the gap could be attributable to the *excess commuting* behavior that individuals do not necessarily optimize their journey-to-work trips as suggested by the spatial arrangements of land uses; that is, jobs and houses. Even if individual workers make every effort to minimize their commuting, the outcome might still differ from what is required by minimizing total commute collectively for the whole study area.

Most studies adopt the linear programming (LP) approach by White (1988) to define the minimum commute based on zonal data such as

$$\text{Min } T_{\min} = \sum_{i=1}^{m} \sum_{j=1}^{n} (x_{ij} d_{ij}) \quad (3)$$

subject to

$$\sum_{j=1}^{n} x_{ij} = W_i \quad (4)$$

$$\sum_{i=1}^{m} x_{ij} = J_j, \quad (5)$$

where x_{ij} is the (nonnegative) optimal number of commuters from a resident worker tract i to a job tract j to be solved, and d_{ij}, W_i, and J_j are the same as defined in Equation 2.

There are a couple of concerns on this zonal model. The first is the lack of spatial accuracy when using average commute time (distance) between zones. Another involves inconsistent and unreliable strategies for defining intrazonal commute lengths (e.g., some use 0, and others use reported time or estimated distance). Both are especially problematic in large zones or termed the MAUP, as discussed in Horner and Murray (2002). As explained previously, this research uses the Monte Carlo approach to simulate the individual locations for workers and jobs to

improve the accuracy of estimation. Index the locations for simulated workers and jobs as k and l, respectively. The total number of simulated workers is the same as that of simulated jobs, denoted by n. It is formulated as an integer programming problem such as

$$\text{Min } T_{\min} = \sum_{k=1}^{n} \sum_{l=1}^{n} (x_{kl} d_{kl}) \quad (6)$$

subject to

$$\sum_{l=1}^{n} x_{kl} = 1 \quad (7)$$

$$\sum_{k=1}^{n} x_{kl} = 1 \quad (8)$$

Note that x_{kl} is a binary integer variable (1 when a journey-to-work flow is chosen by the optimization algorithm between a simulated worker location and a simulated job location and 0 otherwise). The objective function (Equation 6) is to minimize the total commute by n simulated commuters, and Equations 7 and 8 ensure that each worker can be assigned to one unique job and vice versa.

Figure 3 summarizes the results on excess commuting in Baton Rouge between 1990 and 2010. Once again, the average commute distance in the study area increased steadily from 1990 to 2000 and again to 2010, and the average commute time increased from 1990 to 2000 but dropped slightly to 2010. The portion of excess commuting measured in time was consistently higher than that in distance in corresponding years (67.66 percent vs. 54.27 percent in 1990, 75.41 percent vs. 63.99 percent in 2000, and 75.34 percent vs. 63.48 percent in 2010). This discrepancy is understandable because actual commute time included those by slower transportation modes and the optimal time was estimated solely by driving alone. As the concept of excess commuting is proposed mainly to assess the potential of commuting reduction given the land use pattern of a city, our discussion here focused on the results in terms of commute distance.

The minimum (required) commute was 3.32 miles in 1990, dropped to 2.62 in 2000, and inched up slightly to 2.71 in 2010. It suggested that land uses in Baton Rouge might have changed in a way toward more efficiency in terms of commuting need from 1990 to 2000 (e.g., improved proximity between jobs and resident workers in general) and stayed largely

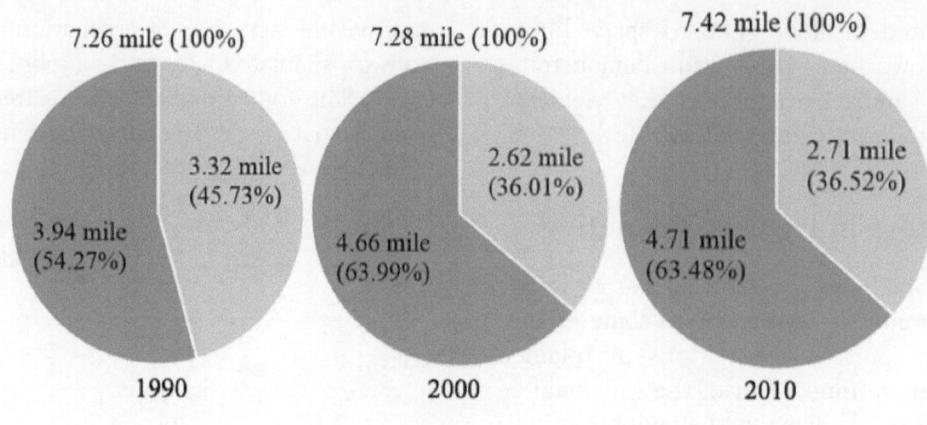

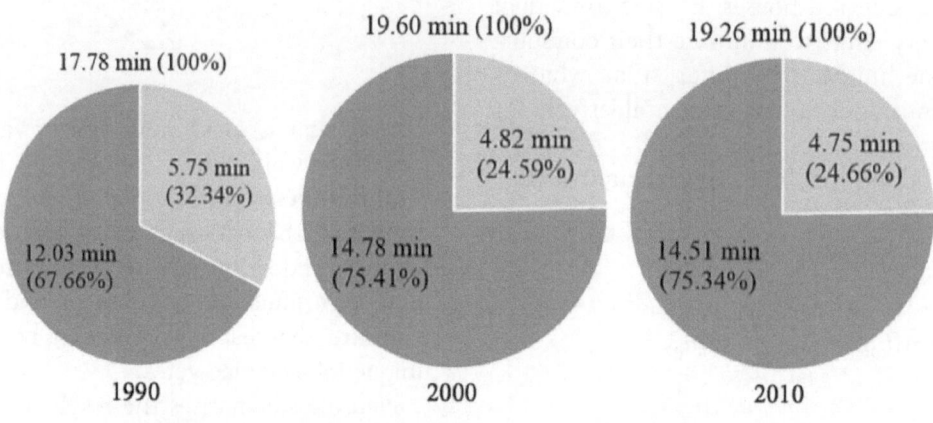

Figure 3. Excess commuting in 1990 through 2010.

stable until 2010. The resident workers did not take advantage of the change, however, and actually increased their trip lengths on average from 7.26 miles in 1990 to 7.28 miles in 2000 and then again to 7.42 miles in 2010. This led to the rise in excess commuting distance from 54.27 percent in 1990 to 63.99 percent in 2000 and stayed at 63.48 percent in 2010. Many factors could have contributed to this trend of largely increasing excess commuting, such as an increasing female labor participation rate (and thus more multi-worker households) and a small increase in carpool modal share from 1990 to 2000 (Table 1). A more in-depth discussion on the impact of other aspatial factors is beyond the scope of this article.

Conclusion

This research analyzes the temporal trends of commuting patterns in both time and distance, explains the observed commuting patterns from a spatial perspective (i.e., land use), and measures the extent of excess commuting. Findings are summarized as follows.

First, mean commute distance steadily increased over time in Baton Rouge, and mean commute time rose along with mean commute distance between 1990 and 2000 but dropped slightly afterward. The gap between the two in 2000 to 2010 was attributable to a nonuniform modal distribution across tracts. Future research could expand the study area by incorporating neighboring parishes to see whether there are more pronounced changes in the metropolitan area and use high temporal resolution data to examine the impact of major external factors (e.g., significant fluctuation in gas price) on commuting.

Second, on the interrelatedness between intraurban commuting and land use patterns, our research indicates that the spatial variability in mean commute distance in this midsize city can be well explained by distance from the CBD, jobs–housing balance ratio, and even more than 90 percent by the job proximity index. The models on mean commute time also show improvement over existing studies. The better results might be attributable to improved measure of commute distance and a moderate city size in this study. The finding lends support to the effectiveness of planning policies that are aimed at trip reduction by improving jobs–housing balance and job proximity.

Finally, our study finds that Baton Rouge in its entirety experienced an increase of excess commuting from 1990 to 2000 in both commute distance and time and stayed at about the same levels toward 2010. This indicates that the land-use configuration changed in such a way that jobs collectively moved closer to residences and thus became better balanced from 1990 to 2000, but the resident workers did not take advantage of that and incurred more excess commuting. The trend of rising excess commuting was largely halted between 2000 and 2010. The economic downturn beginning in 2008 might be one reason underlying the new trend (Horner and Schleith 2012). The low temporal resolution data (i.e., five-year pooled 2006–2010 CTPP) prevent us from validating this speculation.

On the methodological front, this research proposes a Monte Carlo simulation–based approach for improved modeling of commuting patterns. Granted, a more disaggregated data (e.g., block level) would incur minor aggregation error, but such detailed data in terms of both spatial and temporal resolution are not always available in commuting studies. Given the more widely available zonal-level commuting data (e.g., CTPP in the United States), our approach mitigates the zonal effect and permits a more accurate estimate of commute length and subsequently a more reliable measure of excess commuting. The formulation of integer programming also has good potential for wider adoption in modeling the optimal commuting pattern of individual trip makers.

References

Antipova, A., and F. Wang. 2010. Land use impacts on trip chaining propensity for workers and non-workers in Baton Rouge, Louisiana. *Annals of GIS* 16 (3): 141–54.

Antipova, A., F. Wang, and C. Wilmot. 2011. Urban land uses, socio-demographic attributes and commuting: A multilevel modeling approach. *Applied Geography* 31 (3): 1010–18.

Cervero, R. 1989. Jobs–housing balancing and regional mobility. *Journal of the American Planning Association* 55 (2): 136–50.

Cervero, R., and K. L. Wu. 1998. Sub-centering and commuting: Evidence from the San Francisco Bay area, 1980–90. *Urban Studies* 35 (7): 1059–76.

Chen, X., F. B. Zhan, and G. Wu. 2010. A spatial and temporal analysis of commute pattern changes in central Texas. *Annals of GIS* 16 (4): 255–67.

Cropper, M. L., and P. L. Gordon. 1991. Wasteful commuting: A re-examination. *Journal of Urban Economics* 29 (1): 2–13.

Giuliano, G., and K. A. Small. 1993. Is the journey to work explained by urban structure? *Urban Studies* 30 (9): 1485–1500.

Hamilton, B. W. 1982. Wasteful commuting. *The Journal of Political Economy* 90 (5): 1035–53.

Hewko, J., K. E. Smoyer-Tomic, and M. J. Hodgson. 2002. Measuring neighbourhood spatial accessibility to urban amenities: Does aggregation error matter? *Environment and Planning A* 34 (7): 1185–1206.

Horner, M. W. 2002. Extensions to the concept of excess commuting. *Environment and Planning A* 34 (3): 543–66.

———. 2004. Spatial dimensions of urban commuting: A review of major issues and their implications for future geographic research. *The Professional Geographer* 56 (2): 160–73.

———. 2007. A multi-scale analysis of urban form and commuting change in a small metropolitan area (1990–2000). *The Annals of Regional Science* 41 (2): 315–32.

Horner, M. W., and A. T. Murray. 2002. Excess commuting and the modifiable areal unit problem. *Urban Studies* 39 (1): 131–39.

Horner, M. W., and D. Schleith. 2012. Analyzing temporal changes in land-use–transportation relationships: A LEHD-based approach. *Applied Geography* 35 (1): 491–98.

Hu, Y., and F. Wang. 2015a. Decomposing excess commuting: A Monte Carlo simulation approach. *Journal of Transport Geography* 44:43–52.

———. 2015b. Monte Carlo method and application in urban traffic simulation. In *Quantitative methods and socio-economic applications in GIS*, ed. F. Wang, 259–77. Boca Raton, FL: CRC.

Huff, D. L. 1963. A probabilistic analysis of shopping center trade areas. *Land Economics* 39:81–90.

Kain, J. F. 2004. A pioneer's perspective on the spatial mismatch literature. *Urban Studies* 41 (1): 7–32.

Kim, C. 2008. Commute time stability: A test of a co-location hypothesis. *Transportation Research Part A: Policy and Practice* 42 (3): 524–44.

Kim, S. 1995. Excess commuting for two-worker households in the Los Angeles metropolitan area. *Journal of Urban Economics* 38 (2): 166–82.

Kwan, M. P., and A. Kotsev. 2015. Gender differences in commute time and accessibility in Sofia, Bulgaria: A study using 3D geovisualisation. *The Geographical Journal* 181 (1): 83–96.

Levinson, D. M., and A. Kumar. 1994. The rational locator: Why travel times have remained stable. *Journal of the American Planning Association* 60 (3): 319–32.

Levinson, D., and Y. Wu. 2005. The rational locator reexamined: Are travel times still stable? *Transportation* 32 (2): 187–202.

Murphy, E., and J. E. Killen. 2011. Commuting economy: An alternative approach for assessing regional commuting efficiency. *Urban Studies* 48 (6): 1255–72.

Niedzielski, M. A., M. W. Horner, and N. Xiao. 2013. Analyzing scale independence in jobs—Housing and commute efficiency metrics. *Transportation Research Part A: Policy and Practice* 58:129–43.

Santos, A., N. McGuckin, H. Y. Nakamoto, D. Gray, and S. Liss. 2011. *Summary of travel trends: 2009 National Household Travel Survey.* Washington, DC: U.S. Department of Transportation, Federal Highway Administration. http://nhts.ornl.gov/2009/pub/stt.pdf (last accessed 16 November 2015).

Shen, Y., M. P. Kwan, and Y. Chai. 2013. Investigating commuting flexibility with GPS data and 3D geovisualization: A case study of Beijing, China. *Journal of Transport Geography* 32:1–11.

Sultana, S. 2002. Job/housing imbalance and commuting time in the Atlanta metropolitan area: Exploration of causes of longer commuting time. *Urban Geography* 23 (8): 728–49.

Sultana, S., and J. Weber. 2007. Journey-to-work patterns in the age of sprawl: Evidence from two midsize Southern metropolitan areas. *The Professional Geographer* 59 (2): 193–208.

———. 2014. The nature of urban growth and the commuting transition: Endless sprawl or a growth wave? *Urban Studies* 51 (3): 544–76.

U.S. Department of State. 2010. *U.S. climate action report 2010.* Washington, DC: Global Publishing Services.

Wang, F. 2000. Modeling commuting patterns in Chicago in a GIS environment: A job accessibility perspective. *The Professional Geographer* 52 (1): 120–33.

———. 2001. Explaining intraurban variations of commuting by job proximity and workers' characteristics. *Environment and Planning B* 28 (2): 169–82.

———. 2003. Job proximity and accessibility for workers of various wage groups. *Urban Geography* 24 (3): 253–71.

———. 2015. *Quantitative methods and socioeconomic applications in GIS.* 2nd ed. Boca Raton, FL: CRC.

White, M. J. 1988. Urban commuting journeys are not "wasteful." *The Journal of Political Economy* 96:1097–1110.

Yang, J. 2008. Policy implications of excess commuting: Examining the impacts of changes in US metropolitan spatial structure. *Urban Studies* 45 (2): 391–405.

A Location-Centric Network Approach to Analyzing Epidemic Dynamics

Shiran Zhong and Ling Bian

Department of Geography, University at Buffalo, The State University of New York

Recent health threats, such as the SARS, H1N1, and ebola pandemics, have stimulated great interest in network models to study the transmission of communicable diseases through human interaction and mobility. Most current network models have focused on an individual-centric perspective where individuals are represented as nodes and the interactions among them as edges. Few of these models are concerned with the discovery of the spatial patterns and dynamics of epidemics. We propose a location-centric, transmission network approach, in which nodes denote locations and edges denote possible disease transmissions between locations. We then identify the dynamics of transmission flows, the dynamics of critical locations, and the spatial–temporal extent of transmission pathways to assess the impact of these spatial dynamics on the evolution of an epidemic. Results show that transmission flows shift from elementary schools to middle schools and finally universities and professional schools at different phases of an epidemic. Critical locations, identified using network analysis, are responsible for the upsurge in transmission flows during the peaks of the epidemic. The length of transmission pathways shows a power law distribution and their spatial extent is rather small. Insights gained from this study will help devise spatially sensitive strategies to control communicable diseases.

晚近的健康威胁, 诸如 SARS、H1N1 和埃博拉流行病等, 刺激了网络模型研中研究传染性疾病透过人际互动与移动造成传染的巨大兴趣。目前既有的网络模型, 多半聚焦以个人为核心的观点, 其中个人被视为节点, 而人际互动则被视为边缘。这些模型, 鲜少考量空间模式的发掘和疫情动态。我们则提出一个以地点为中心的传染网络方法, 其中节点指涉地点, 而边缘则指涉疾病于地点之间的可能传染途径。我们接着指认传染流动的动态, 关键地点的动态, 以及传染途径的时空范围, 以评估这些空间动态对于传染病演化的影响。研究结果显示, 传染病在不同阶段中, 其传染流动从小学转移至中学, 最终并扩及大学与专业学校。运用网络分析所指认的关键地点, 则是流行病高峰期间传染流动暴增的原因。传染途径的长度显示出幂率分布, 而其空间范围是相当小的。本研究中获得的洞见, 将能够协助策划对空间敏感的策略, 以控制传染性疾病。

Las recientes amenazas a la salud, como las representadas por las SARS, H1N1 y la pandemia del ébola, han estimulado gran interés por los modelos de redes para estudiar la trasmisión de enfermedades contagiosas por medio de la interacción y movilidad humanas. La mayoría de los modelos actuales de redes están enfocados a una perspectiva centrada en el individuo, donde los individuos se representan como nodos y las interacciones entre ellos como bordes. Pocos de estos modelos tienen interés en el descubrimiento de los patrones espaciales y la dinámica de las epidemias. Lo que nosotros proponemos es un enfoque de red de trasmisión centrado en localización, en el que los nodos denotan lugares y los bordes denotan posibles trasmisiones de la enfermedad entre los lugares. Luego identificamos la dinámica de los flujos de trasmisión, la dinámica de lugares críticos y el alcance espacial-temporal de las rutas de trasmisión para evaluar el impacto de estas dinámicas espaciales sobre la evolución de una epidemia. Los resultados muestran que los flujos de trasmisión cambian desde las escuelas elementales a las escuelas de educación media y finalmente a las universidades y escuelas profesionales en las diferentes fases de una epidemia. Las localizaciones críticas, identificadas por medio de análisis de redes, son responsables del incremento significativo de los flujos de trasmisión durante los picos de la epidemia. La longitud de las rutas de trasmisión muestra una distribución de ley de potencia y su alcance espacial es bastante pequeño. Las percepciones ganadas con este estudio ayudarán a idear estrategias espacialmente sensibles para controlar enfermedades contagiosas.

Human interaction and mobility behaviors are essential to the transmission of communicable diseases (Gushulak and MacPherson 2000). Interactions at homes, workplaces, and schools could cause local transmission, whereas individuals' mobility between these places could propagate the transmission to an area-wide epidemic (Eubank et al. 2004; Cauchemez et al. 2008). Recent health threats, such as the SARS, H1N1, and ebola pandemics, have stimulated great interest in network models to study communicable disease transmissions (Salathé et al. 2010; Bian et al. 2012; Gomes et al. 2014; Liu et al. 2015).

A network consists of nodes and edges. Most recent network models have focused on an individual-centric perspective where individuals are denoted as nodes, and the interactions among them are denoted as edges. Individuals' behaviors (e.g., interaction, mobility) and attributes (e.g., demographics, infection status) are used to simulate an individualized transmission process. Collectively, these individual transmissions contribute to a dispersion pattern at the population level. Although the attributes might include spatial information, such as the location of transmission, it is treated as a spatial stamp of the transmission process because these models focus primarily on health outcomes, such as the number of infected cases during an epidemic (Cooley et al. 2008; Salathé et al. 2010).

Few of these models are concerned with the discovery of the spatial patterns and dynamics of epidemics; for example, how do disease transmissions "flow" in space? Are there critical locations of transmission? To what extent do transmissions expand spatially and temporally? Ultimately, how do these spatial dynamics influence an epidemic? It is well understood that this information is crucial to devising spatially informed policies to control and prevent communicable diseases (Bian 2013). Current individual-centric networks are inadequate to address the aforementioned questions. An alternative, location-centric perspective is required to easily represent and identify spatial dynamics of the disease transmission process.

The objective of this study is twofold. First, we propose a location-centric, transmission network approach, in which nodes denote locations and edges denote possible disease transmissions between locations. The spatial and temporal cooccurrence and mobility information embedded in the discrete infection cases are used to construct the networks. Second, we (1) examine the dynamics of transmission flows drawn from the reconstructed location-centric networks, (2) identify critical locations by applying network centrality analysis, and (3) extract transmission pathways based on the spatial and temporal continuity in the transmission flows. The influence of these spatial dynamics on the epidemic is analyzed. Results are visualized using 2D and 3D approaches.

The study area is a metropolitan area in midwest China that was affected by both a seasonal influenza epidemic and the H1N1 pandemic at the same time in 2009 (Smith et al. 2009; Centers for Disease Control and Prevention [CDC] 2011). Both diseases shared many similar symptoms, such as cough, fever, and headache, except that the symptoms of H1N1 were unusually severe (Fiore et al. 2010). The seasonal influenza infection cases reported during the epidemic are used to support the intended study because of their representativeness and data availability. Insights gained from this study will help design data models in network forms, so that spatial patterns can be easily discerned. From the perspective of disease control, the identification of transmission flows, critical locations, and transmission pathways will help policymakers develop location-oriented prevention and control strategies.

Study Area and Data

The study area encompasses 60 km^2 with a population of about 890,000. The data set used for this study consists of 4,315 anonymous cases; each was clinically diagnosed as seasonal influenza. The epidemic extended over seventy-three days between 1 September and 12 November. Each case is associated with a residential address; a workplace address; the type of workplace; the symptom onset date; and the gender, age, and occupation of the infected person. Although each address has a specific location, the data available for this study only identify the residential address within a named residential community and identify the workplace address in an area equivalent to the U.S. census block group. Based on the occupation information, most cases involve students. Their workplace type typically includes elementary schools, middle schools (Grades 7–12), professional schools, or universities. This data set was obtained from the China Information System for Diseases Control and Prevention.

Methods

To investigate spatial characteristics of influenza transmission, a total of seventy-three daily location-centric transmission networks are constructed using the cooccurrence and mobility information observed in the discrete influenza cases. Critical nodes and transmission pathways are then derived and visualized. The impacts of spatial dynamics on the epidemic are discussed. Each is detailed next.

Construct Location-Centric Transmission Networks

Spatial and temporal cooccurrence and mobility are key components in the construction of location-centric transmission networks. Spatial and temporal cooccurrence refers to the spatial and temporal range shared between discrete events. Social ties have been inferred based on a small number of high-frequency cooccurrences (Crandall et al. 2010). Because interactions between individuals at homes and workplaces have a long duration and high frequency, cooccurrences at these locations are deemed to foster localized disease transmission (Cauchemez et al. 2011; Bian et al. 2012). Presently, the cooccurrence concept has been limited to the construction of static, single-link social ties. Our study extends the concept to the construction of dynamic, extended transmission networks. Mobility, in the context of this study, refers to individuals' daily travel between homes and workplaces. Mobility facilitates the disease transmission between locations, thus propagating localized transmissions to an area-wide epidemic.

By projecting the location of the 4,315 cases into space, a total of 1,026 distinct locations are identified (including residential and workplace locations). These locations serve as nodes in the seventy-three daily transmission networks and the disease transmission flows between locations serve as edges. The transmission flows are reconstructed by a spatial cooccurrence constraint and a temporal cooccurrence constraint, as explained next.

First, an influenza case is denoted using the following formula:

$$C_n = (T_n, R_n, W_n),$$

where C_n is the nth case, T_n denotes the symptom onset date of Case n, and R_n and W_n denote the residential address and workplace address of Case n, respectively.

Based on the denotation of each case, a transmission flow from C_i to C_j occurs if

$$(R_i \cup W_i) \cap (R_j \cup W_j) \neq NULL \qquad (1)$$

and

$$0 < T_j - T_i \leq 7. \qquad (2)$$

The $(R_i \cup W_i)$ and $(R_j \cup W_j)$ in the first constraint represent individuals' mobility paths between their residences and workplaces for individuals i and j, respectively. The intersection of the two mobility paths indicates that individuals i and j must share at least one location. For example, they might live or work together. Along with the spatial cooccurrence defined in the first constraint, the second constraint focuses on the temporal cooccurrence. This constraint requires that individual j shows symptoms within seven days after individual i shows symptoms. This indicates that Case j falls within the infectious period of Case i, which is typically seven days. The infectious period of seven days is adopted from the literature (Heyman 2004; Fiore et al. 2010; CDC 2011), including a review of seventy-one influenza reports (Carrat et al. 2008) and our preliminary sensitivity analysis that experimented with the infectious period from two to ten days. The infectious period of seven days best supports the actual epidemic patterns.

The reconstructed transmission edges derived from the preceding two constraints are projected into space. They are directed straight lines between two locations rather than the transportation routes actually used by individuals. Of the two location nodes associated with each transmission edge, one is a source location and the other is a target location. The target location is where C_i and C_j colocate, either at a residence or a workplace, and C_i's other location on the mobility path is the source location. Each transmission edge starts on the symptom onset date of C_i at the source location and ends on the symptom onset date of C_j at the target location.

Multiple transmission edges can share the same mobility path when multiple individuals travel along the same path and bring infections between the two locations. This situation is hereafter called multiedge flows. Internal transmissions at a single location are considered transmission edges without spatial length and direction. The number of cases is represented as an attribute of location nodes. The assemblage of location nodes and the reconstructed transmission edges constitute the location-centric transmission networks, one network for each of the seventy-three days of the epidemic. The 1,026 distinct locations remain the same in the seventy-three daily networks, but the edges between them change over time, as the transmission flows could change locations across time.

Identify Critical Locations

Two network centrality indexes, degree centrality and bridging centrality, are used to identify critical nodes in the transmission networks. Network centrality indexes measure how central individual nodes are in a network (Newman 2010). The degree centrality is the count of a node's connecting edges, which is a measure of a node's ability to generate outflows to or receive inflows from neighboring nodes. A higher value means more active flows (Scott and Carrington 2011). Degree centrality seems the best indicator for disease transmission probability in the context of interaction- and mobility-induced transmissions, whereas the role of betweenness and closeness centralities remains inconclusive (Christley et al. 2005; Bian et al. 2012).

The bridging centrality is the product of the bridging coefficient and the betweenness centrality. The bridging coefficient indicates whether a node is positioned between high-degree neighboring nodes, and the betweenness centrality measures the extent to which a node is located on the shortest paths of any pairs of other nodes (Borgatti, Everett, and Johnson 2013). A high bridging centrality value implies a node's ability to control flows between high-degree clusters in a network (Hwang et al. 2008).

Degree centrality and bridging centrality are calculated for all nodes of all seventy-three daily networks. For each index, 74,898 values are derived (73 days × 1,026 locations = 74,898). Summary statistics of each index, including the sum and standard deviation, are calculated for each location throughout the seventy-three day period to identify critical nodes in the transmission networks. Locations with continuously high degree centrality values (i.e., great sums and relatively small standard deviations) are considered the critical locations in terms of the magnitude of their transmission flows. Similarly, locations with continuously high bridging centrality values are considered the critical locations in terms of their ability to control transmission flows. Both sets of critical locations are visualized using a 3D approach to show their spatial dynamics (Kwan 2000, 2004).

Extract Transmission Pathways

A transmission pathway is composed of a series of spatially and temporally continuous transmission edges, including the internal transmission edges. Two spatial transmission edges are considered continuous if both of the following constraints are met: The target location of one transmission edge is identical to the source location of the other transmission edge and,

further, one edge starts before or on the day the other edge ends. In the situation of multiedge flow, each edge represents a segment of a distinct pathway.

The spatial extent and temporal duration of the transmission pathways are analyzed to examine whether the pathway duration has an impact on their spatial extent. The spatial extent of a transmission pathway is measured by the area of the rectangular convex hull that encloses the path. The temporal duration of a path is the total number of days the path lasts, and is referred to as the pathway length. The transmission pathways are displayed in 3D to disclose the spatial and temporal patterns of transmission flows during the epidemic, following the visualization method of Kwan (2000, 2004).

Results and Discussion

Transmission Flows

The result includes seventy-three daily location-centric transmission networks, one network for each day during the entire epidemic. Figure 1 shows the daily networks of five days: 1 September, 22 September, 5 October, 22 October, and 31 October (hereinafter 9/1, 9/22, 10/5, 10/22, and 10/31). The five dates correspond to the dynamics of the epidemic: two minor peaks in the early phase, one major trough, and then two major peaks at the later phase.

According to local news sources, the first reported case of the epidemic in the region occurred at the center of the study area in mid-August (a case not included in the data set). By the time the epidemic had evolved into the first peak on 9/1, the transmission had expanded from its initial location outward to cover a large area. Most target locations are elementary schools because when schools start on 9/1, students already infected with influenza during the summer bring the disease to schools, thus creating the first peak of the epidemic on the school starting date. The elementary school at Location 20 contributes the most to the peak in terms of both the number of cases and flows (Figure 1A). Among all flows, few are outflows, indicating extensive internal transmissions within the school (Table 1).

After a series of mild fluctuations, the epidemic reached its second peak on 9/22. The transmission flows are more active and more focused than the first peak (Figure 1B and Table 1), as shown by a greater number of inflows to a target location, a greater

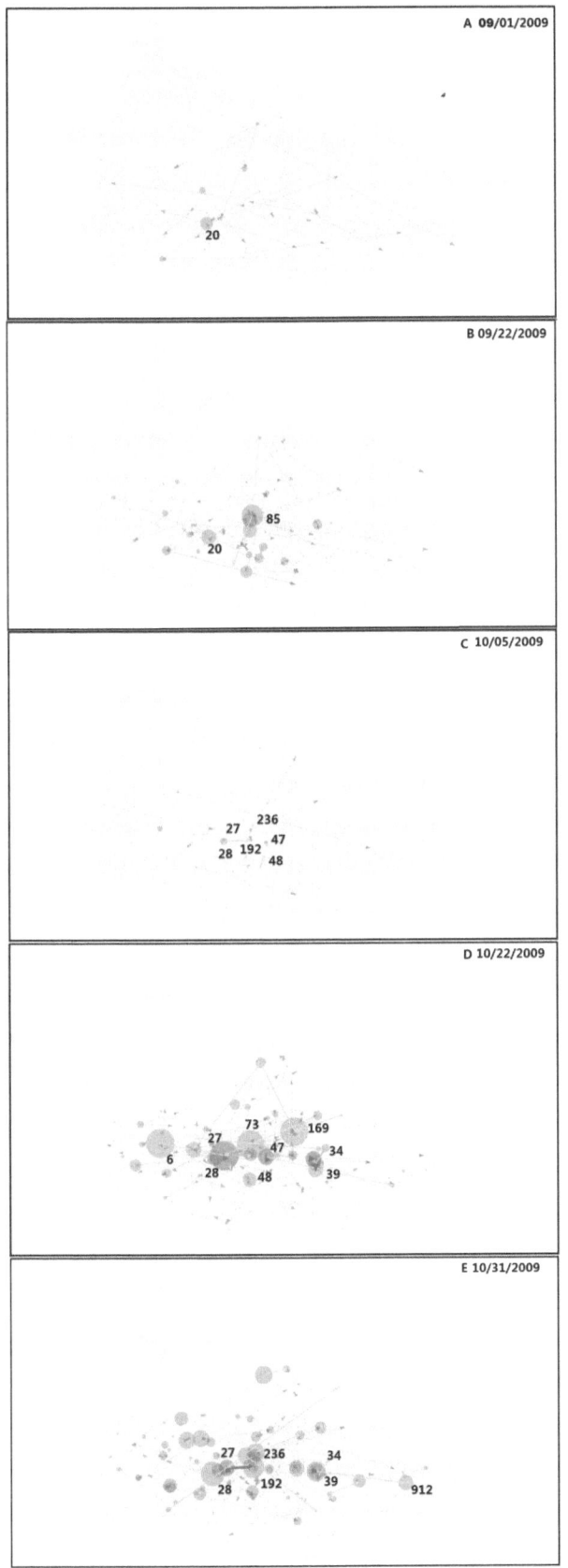

number of multiedge flows, and a greater number of large-case target locations. They remain as elementary schools, such as the one at Location 85 that has the most cases and flows. The second peak could be the result of further spread of the first peak. Extensive internal transmissions at schools might have brought infections to students' residential locations, as shown by increased outflows, especially multiedge ones.

After affecting mostly elementary schools, the epidemic arrived at a trough on 10/5, reaching an epidemic-wide low number of transmission cases and flows (Figure 1C and Table 1). Flows remain active between the campus–dorm location pairs of two universities and a professional school at Locations 27 and 28, 192 and 236, and 47 and 48. The trough might be the result of depletion of susceptible students in the elementary schools after the two initial peaks.

By late October, the magnitude of transmission rose again and the epidemic reached its highest peak on 10/22 (Figure 1D, Table 1). There is a widespread upsurge of transmission flows between a large number of locations over a large area. Most target locations are middle schools—such as those at Locations 73, 6, and 169—and universities and professional schools—such as Locations 27, 47, and 39. High schools have larger pools of susceptible students, as their student population is two or more times as great as elementary schools in the study area, and universities and professional schools are typically five to ten times as great as middle schools (or ten to twenty times as great as elementary schools).

The most active transmission flows are between campus–dorm pairs of universities and professional schools (e.g., Locations 27–28, 47–48, 39–34). Unlike elementary schools and middle schools where students commute between schools and home, students in university and professional schools reside in dorms (a mandate in the study area). The dorms are either on or near a campus, as shown on the maps. This mobility behavior significantly increases the likelihood of mass

Figure 1. Transmission networks on (A) 9/1, (B) 9/22, (C) 10/5, (D) 10/22, and (E) 10/31. Blue circles denote location nodes. Their size indicates the number of cases at the location. Arrows denote transmission edges. The yellow, orange, and red arrows indicate 1, 2 through 10, and ≥11 transmission edges, respectively. Note a river runs through the northern part of the study area from west to east. (Color figure available online.)

Table 1. Number of transmission flows and cases on the five days

Date	Total transmission edges	External transmission edges (outflow)	Internal transmissions	Internal transmission rate
1 September	105	26	79	0.75
22 September	208	52	156	0.75
5 October	24	11	13	0.54
22 October	493	105	388	0.79
31 October	305	75	230	0.75

interaction among a large number of students and might explain the extensive transmission once universities and professional schools are affected.

Shortly after this major peak began to decline, another major peak rose around 10/31 (Figure 1E and Table 1) that is associated with another considerable increase in transmission flows. The transmission flows, although fewer in number, are active at more focused locations with a greater magnitude, effectively maintaining the same magnitude of cases observed at the 10/22 peak (Table 1). Transmissions at middle schools have reduced considerably. Flows between the campuses and dorms of universities and professional schools dominate the transmission, such as those at Locations 27 and 28, 192 and 236, and 39 and 34. Particularly, the two universities (at Locations 27 and 192) share the same dorms, thus generating a considerably large number of cases and flows. This significantly contributes to the second highest peak of the epidemic before the epidemic declined to its end around mid-November.

Critical Locations

The top ten critical locations in terms of the degree centrality that measures the magnitude of transmission flow are listed in Table 2 and shown in Figure 2. These locations have the highest sum of degree centrality and a relatively low standard deviation. Locations identified earlier as having high flows (see the previous section) are among those listed. For example, the elementary school at Location 20 has the highest degree at the 9/1 peak, whereas the elementary school at Location 85 is the highest at the 9/22 peak. Middle schools at Location 73, along with 99, 6, and 41, are among the highest at the 10/22 peak. The university at Location 27 ranks the highest at the last peak on 10/31. If intervention strategies, such as travel restrictions or vaccination, are applied to these schools, the transmission flows might be controlled effectively and reduce their impact on other locations.

The critical locations with high degree centrality values extend through the center of the study area (Figure 2). This observation concurs with previous findings that nodes that are central in a network (e.g., have a high degree centrality) tend to be at the geographic center as well (Barthélemy 2011). This is because spatial constraints, such as mobility cost, often favor geographically central locations for important nodes that have a high spatial accessibility in addition to its high network accessibility.

The top ten critical locations in terms of the bridging centrality are listed in Table 3 and shown in Figure 2. These locations possess the highest sum of bridging centrality, and a relatively low standard deviation. These

Table 2. Top ten critical locations ranked by degree centrality and their sum and standard deviation of degree centrality

Rank	Location ID	Sum	Std	9/1	9/22	10/5	10/22	10/31
1	99	838	18.91656				Active	
2	20	746	9.874556	Active				
3	73	635	10.47575				Active	
4	27	583	5.157231				Active	Active
5	85	529	7.395158		Active			
6	41	525	7.811419				Active	Active
7	26	470	5.030425				Active	Active
8	6	429	9.858647				Active	Active
9	142	427	5.633903				Active	
10	35	401	5.270788	Active				

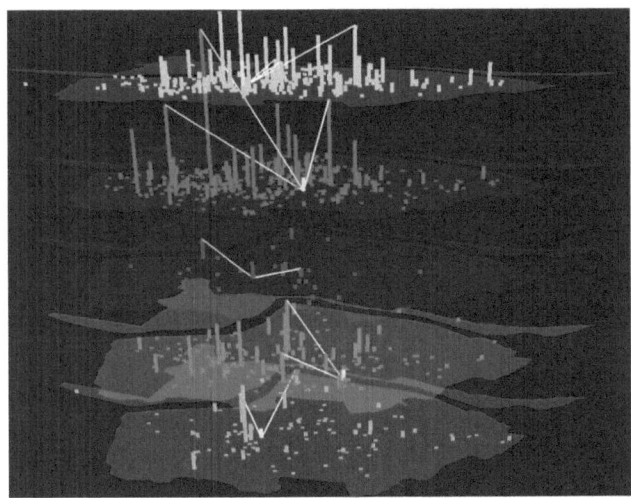

Figure 2. Spatial distribution of degree centrality in colored vertical bars for five days from bottom to top: 9/1, 9/22, 10/5, 10/22, and 10/31. Also shown is the location of the highest bridging centrality (yellow vertical bar) for the five days and locations bridged by the high-bridging location (white links). (Color figure available online.)

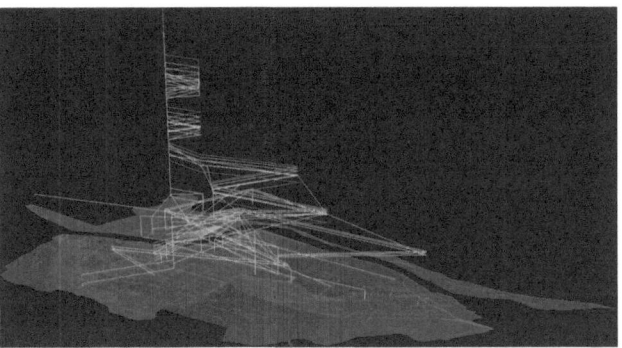

Figure 3. Top 5 percent longest transmission pathways. The vertical dimension shows the temporal range of seventy-three days. The color-encoded seven longest pathways are slightly shifted in the vertical direction to distinguish among them. Other pathways are displayed in light gray. (Color figure available online.)

divide-and-conquer effect to dissect spatially continuous disease transmissions into localized transmissions.

Transmission Pathways

In total, 17,985 transmission pathways are generated. For viewing clarity, the top 5 percent (955 in total) longest pathways are visualized in 3D (Figure 3). The value in the third dimension records the date when the transmission flow reaches a location. The horizontal segments are transmission flows between locations. The vertical segments indicate internal transmissions lingering at a single location for multiple days.

The seven longest pathways (forty to forty-two days) converge to one location, the university at Location 27, and form the longest internal transmission of the epidemic. This is due to the large student population on campus and the mass flow between the campus and dorms.

locations, mostly residential communities, bridge between high-flow middle schools, fostering cross-transmissions. For example, the top ranked residential community at Location 191 (Table 3) bridges between the top ranked high-flow middle schools at Locations 99, 73, and 41 (Table 2). All ten critical locations are active at the 10/22 peak, each connecting up to seven middle schools with each other. These might explain why middle schools account for a considerable proportion of the target locations during the major peak on 10/22. For a clearer view, only the location with the highest bridging centrality is shown for each day (Figure 2). The high-bridging locations play an influential role in sustaining the momentum of an epidemic. Spatially oriented intervention strategies at these locations might have the

Table 3. Top ten critical locations by bridging centrality and the sum and standard deviation of bridging centrality

Rank	Location ID	Sum	Std	9/1	9/22	10/5	10/22	10/31
1	191	9,417.908	17.3513				Active	
2	88	6,378.168	17.36493				Active	
3	222	6,364.241	17.71181				Active	Active
4	28	4,566.322	12.20597				Active	
5	759	4,081.006	15.98819				Active	
6	255	4,025.136	11.84818				Active	
7	26	4,006.323	9.782303		Active		Active	
8	97	3,710.142	12.72711				Active	Active
9	163	3,321.843	9.417358				Active	
10	27	3,150.414	9.508776				Active	Active

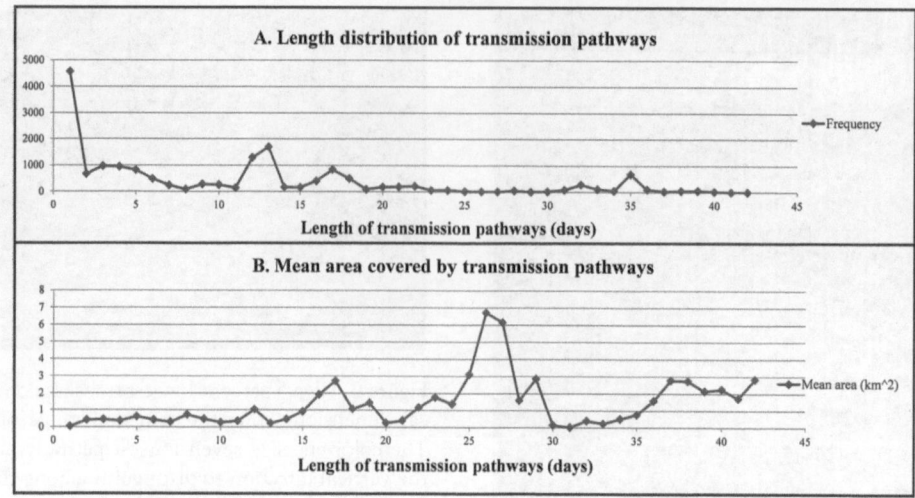

Figure 4. (A) Frequency distribution of transmission pathway length, $p(Y) \sim X^{-2.69}$. (B) Average area covered by pathways.

Approximately 25 percent of transmission pathways last only one day. The next length range of two to ten days accounts for another 25 percent. The third length range of eleven to twenty days represents more than 30 percent of all pathways. The longest 5 percent of pathways are from thirty-five to forty-two days. The length of the pathways follows a trend closely resembling a power law distribution ($\alpha = 2.69$, p value from the Kolmogorov–Smirnov test = 0.99; Figure 4A). This implies that although most flows are short-lived, a few have a long duration. This observation is similar to the reported human travel trajectory patterns (Gonzalez, Hidalgo, and Barabasi 2008; Barthélemy 2011) and reflects the individual mobility pattern behind the transmission pathways.

The average area covered by the rectangular convex hull of transmission pathways (Figure 4B) does not increase monotonically with the pathway length. Most pathways, short or long, are within an area of 3 km^2. It implies that the long pathways tend to move between locations (e.g., campus and dorms) repeatedly over time, thus having a long duration within a small area. The greatest degree of spatial spread is caused by several individuals who have long mobility paths. This coincides with the previous finding that the spread of the disease in space is usually caused by a few long-distance travelers (Eubank et al. 2004).

Critical locations with high-degree flows, by both internal and external transmission, could lengthen an epidemic, as in the situation of Location 27 shown in Figure 3. This in turn gives rise to the possibility of long-distance travelers' infections expanding the spatial extent of the epidemic. Thus, closing high-flow

locations during an epidemic can not only control the internal transmissions but also reduce the spatial and temporal extent of the epidemic.

Conclusions

The location-centric approach proposed in this study is critical to improve our understanding of transmission dynamics of communicable diseases in space, as well as to stimulate new designs for epidemiological modeling. The location-centric network approach is progress toward this goal, but it is also in a great need of devising spatially and temporally sensitive network indexes that could quantitatively characterize the spatial dynamics of epidemics.

Acknowledgment

The use of the case data has been approved by the institutional review board at the authors' institution.

References

Barthélemy, M. 2011. Spatial networks. *Physics Reports* 499 (1): 1–101.

Bian, L. 2013. Spatial approaches to modeling dispersion of communicable diseases—A review. *Transactions in GIS* 17 (1): 1–17.

Bian, L., Y. Huang, L. Mao, E. Lim, G. Lee, Y. Yang, M. Cohen, and D. Wilson. 2012. Modeling individual vulnerability to communicable diseases: A framework and design. *Annals of the Association of American Geographers* 102 (5): 1016–25.

Borgatti, S. P., M. G. Everett, and J. C. Johnson. 2013. *Analyzing social networks.* Thousand Oaks, CA: Sage.

Carrat, F., E. Vergu, N. M. Ferguson, M. Lemaitre, S. Cauchemez, S. Leach, and A. J. Valleron. 2008. Time lines of infection and disease in human influenza: A review of volunteer challenge studies. *American Journal of Epidemiology* 167 (7): 775–85.

Cauchemez, S., A. Bhattarai, T. L. Marchbanks, R. P. Fagan, S. Ostroff, N. M. Ferguson, D. Swerdlow, and the Pennsylvania H1N1 Working Group. 2011. Role of social networks in shaping disease transmission during a community outbreak of 2009 H1N1 pandemic influenza. *Proceedings of the National Academy of Sciences* 108 (7): 2825–30.

Cauchemez, S., A. J. Valleron, P. Y. Boelle, A. Flahault, and N. M. Ferguson. 2008. Estimating the impact of school closure on influenza transmission from Sentinel data. *Nature* 452 (7188): 750–54.

Centers for Disease Control and Prevention (CDC). 2011. The 2009 H1N1 pandemic: Summary highlights, April 2009–April 2010. http://www.cdc.gov/h1n1flu/cdcresponse.htm (last accessed 2 August 2014).

Christley, R. M., G. L. Pinchbeck, R. G. Bowers, D. Clancy, N. P. French, R. Bennett, and J. Turner. 2005. Infection in social networks: Using network analysis to identify high-risk individuals. *American Journal of Epidemiology* 162 (10): 1024–31.

Cooley, P., L. Ganapathi, G. Ghneim, S. Holmberg, W. Wheaton, and C. R. Hollingsworth. 2008. Using influenza-like illness data to reconstruct an influenza outbreak. *Mathematical and Computer Modelling* 48 (5): 929–39.

Crandall, D. J., L. Backstrom, D. Cosley, S. Suri, D. Huttenlocher, and J. Kleinberg. 2010. Inferring social ties from geographic coincidences. *Proceedings of the National Academy of Sciences* 107 (52): 22436–41.

Eubank, S., H. Guclu, V. A. Kumar, M. V. Marathe, A. Srinivasan, Z. Toroczkai, and N. Wang. 2004. Modelling disease outbreaks in realistic urban social networks. *Nature* 429 (6988): 180–84.

Fiore, A. E., T. M. Uyeki, K. Broder, L. Finelli, G. L. Euler, J. A. Singleton, J. K. Iskander, and N. J. Cox. 2010. Prevention and control of influenza with vaccines. *Morbidity and Mortality Weekly Report* 59:1–62.

Gomes, M. F., A. P. Piontti, L. Rossi, D. Chao, I. Longini, M. E. Halloran, and A. Vespignani. 2014. Assessing the international spreading risk associated with the 2014 West African ebola outbreak. *PLOS Currents Outbreaks* 1.

Gonzalez, M. C., C. A. Hidalgo, and A. L. Barabasi. 2008. Understanding individual human mobility patterns. *Nature* 453 (7196): 779–82.

Gushulak, B. D., and D. W. MacPherson. 2000. Population mobility and infectious diseases: The diminishing impact of classical infectious diseases and new approaches for the 21st century. *Clinical Infectious Diseases* 31 (3): 776–80.

Heyman, D. L. 2004. *Control of communicable diseases manual.* Washington, DC: American Public Health Association.

Hwang, W., T. Kim, M. Ramanathan, and A. Zhang. 2008. Bridging centrality: Graph mining from element level to group level. In *Proceedings of the 14th ACM SIGKDD international conference on knowledge discovery and data mining,* 336–44. New York: ACM.

Kwan, M. P. 2000. Interactive geovisualization of activity-travel patterns using three-dimensional geographical information systems: A methodological exploration with a large data set. *Transportation Research Part C: Emerging Technologies* 8 (1): 185–203.

———. 2004. GIS methods in time–geographic research: Geocomputation and geovisualization of human activity patterns. *Geografiska Annaler: Series B, Human Geography* 86 (4): 267–80.

Liu, Y., X. Liu, S. Gao, L. Gong, C. Kang, Y. Zhi, G. Chi, and L. Shi. 2015. Social sensing: A new approach to understanding our socio-economic environments. *Annals of the Association of American Geographers* 105 (3): 512–30.

Newman, M. 2010. *Networks: An introduction.* New York: Oxford University Press.

Salathé, M., M. Kazandjieva, J. W. Lee, P. Levis, M. W. Feldman, and J. H. Jones. 2010. A high-resolution human contact network for infectious disease transmission. *Proceedings of the National Academy of Sciences* 107 (51): 22020–25.

Scott, J., and P. J. Carrington, eds. 2011. *The Sage handbook of social network analysis.* Thousand Oaks, CA: Sage.

Smith, G. J., J. Bahl, D. Vijaykrishna, J. Zhang, L. L. Poon, H. Chen, and Y. Guan. 2009. Dating the emergence of pandemic influenza viruses. *Proceedings of the National Academy of Sciences* 106 (28): 11709–12.

Another Tale of Two Cities: Understanding Human Activity Space Using Actively Tracked Cellphone Location Data

Yang Xu,* Shih-Lung Shaw,* Ziliang Zhao,* Ling Yin,[†] Feng Lu,[‡] Jie Chen,[‡] Zhixiang Fang,[§] and Qingquan Li[¶]

*Department of Geography, University of Tennessee, Knoxville
[†]Shenzhen Institutes of Advanced Technology, Chinese Academy of Sciences, Shenzhen
[‡]State Key Laboratory of Resources and Environmental Information System, Institute of Geographic Sciences and Natural Resources Research, Chinese Academy of Sciences, Beijing
[§]State Key Laboratory of Information Engineering in Surveying, Mapping and Remote Sensing, Wuhan University
[¶]Shenzhen Key Laboratory of Spatial Smart Sensing and Services, Shenzhen University

Activity space is an important concept in geography. Recent advancements of location-aware technologies have generated many useful spatiotemporal data sets for studying human activity space for large populations. In this article, we use two actively tracked cellphone location data sets that cover a weekday to characterize people's use of space in Shanghai and Shenzhen, China. We introduce three mobility indicators (daily activity range, number of activity anchor points, and frequency of movements) to represent the major determinants of individual activity space. By applying association rules in data mining, we analyze how these indicators of an individual's activity space can be combined with each other to gain insights of mobility patterns in these two cities. We further examine spatiotemporal variations of aggregate mobility patterns in these two cities. Our results reveal some distinctive characteristics of human activity space in these two cities: (1) A high percentage of people in Shenzhen have a relatively short daily activity range, whereas people in Shanghai exhibit a variety of daily activity ranges; (2) people with more than one activity anchor point tend to travel further but less frequently in Shanghai than in Shenzhen; (3) Shenzhen shows a significant north–south contrast of activity space that reflects its urban structure; and (4) travel distance in both cities is shorter around noon than in regular work hours, and a large percentage of movements around noon are associated with individual home locations. This study indicates the benefits of analyzing actively tracked cellphone location data for gaining insights of human activity space in different cities.

活动空间是地理学中的重要概念。晚近能感知位置的科技革新, 已为研究大型人口的人类活动空间生产了诸多有用的时空数据集。我们于本文中, 运用两组动态追踪的手机位置数据集, 这两组数据集涵盖了一整个工作日, 用以描绘人们在中国上海与深圳的空间使用。我们引介三项能动性指标 (每日活动范围, 活动定着点数量, 以及移动的频率) 来呈现个人活动空间的决定因素。我们透过将关联规则运用至数据挖掘, 分析这些个人活动空间的指标, 如何能够相互结合, 以获得这两座城市中的能动性模式之洞见。我们进一步检视这两座城市中的累计能动性模式的时空变异。我们的研究结果, 显示出这两座城市中的人类活动的若干特徵: (1) 在深圳, 有高比率的人口具有相对而言较短的每日活动范围, 而在上海, 人们则展现出多样的每日活动范围; (2) 在上海, 具有两个以上活动定着点的人们, 较具有两个以上活动定着点的深圳人更倾向拥有较为远距、却较不频繁的移动; (3) 深圳展现出显着的南–北活动空间对比, 并反映出其城市结构; (4) 两座城市中的旅行距离, 在中午期间较正常上班时间为短, 而午间移动有大幅的比例与个人家户位置有关。本研究指出, 分析动态追踪的手机位置数据, 有助于获得不同城市中的人类活动空间之洞见。

El espacio de actividad es un concepto importante en geografía. Los recientes avances en tecnologías inteligentes de localización han generado muchos conjuntos de datos espaciotemporales útiles para estudiar el espacio de actividad humana para poblaciones grandes. En este artículo usamos dos conjuntos de datos de localización del teléfono celular activamente rastreado que cubren un día de la semana para caracterizar el uso del espacio por la gente en Shanghai y Shenzhen, China. Introdujimos tres indicadores de movilidad (ámbito cotidiano de actividad, número de puntos de anclaje de la actividad, y frecuencia de los movimientos) para representar los principales determinantes del espacio de actividad individual. Aplicando reglas de asociación en la minería de datos, analizamos la forma como estos indicadores del espacio de actividad de un individuo pueden combinarse entre sí para ganar entendimiento sobre los patrones de movilidad en estas dos ciudades.

Adicionalmente examinamos variaciones espaciotemporales de patrones agregados de movilidad en las dos ciudades. Nuestros hallazgos revelan algunas características distintivas del espacio de la actividad humana en las dos urbes: (1) Un alto porcentaje de la gente de Shenzhen tiene un ámbito de actividad cotidiana relativamente corto, mientras la gente de Shanghai exhibe una variedad de ámbitos de actividad cotidiana; (2) la gente que tiene más de un punto de anclaje de la actividad tiende a viajar más lejos pero menos frecuentemente en Shanghai que en Shenzhen; (3) Shenzhen muestra un contraste significativo de espacio de actividad en sentido norte-sur que refleja su estructura urbana, y (4) la distancia de viaje en ambas ciudades es más corta alrededor del mediodía que en las horas regulares de trabajo, y un alto porcentaje de los movimientos alrededor del mediodía están asociados con las localizaciones individuales de los hogares. Este estudio señala los beneficios de analizar los datos de la localización del teléfono celular activamente rastreado para ganar entendimiento del espacio de la actividad humana en diferentes ciudades.

Human activities and movements generate the pulses of our cities. Studying human activity space could yield important insights into many socioeconomic phenomena and facilitate our understanding of human behavior and its relationships with the built environment. Activity space is an important concept in geography that describes the spatial extent, frequent locations, and movements of people's daily activities (Golledge and Stimson 1997; Schönfelder and Axhausen 2003). In the past several decades, studies of human activity space were mainly based on travel surveys and Global Positioning System data (Hanson 1980; Dijst 1999; Kwan 1999, 2000; Axhausen et al. 2002; Shoval and Isaacson 2007; Shaw, Yu, and Bombom 2008; Zheng et al. 2008; Chen et al. 2011; Shen, Kwan, and Chai 2013). Recent advancements of location-aware technologies have made it possible to collect large individual tracking data sets for studying the whereabouts of people over space and time. These newly emerging data sources, like social media and cellphone location data, provide us with opportunities to investigate human activity space for large populations. Although various methods have been suggested to measure people's use of space (Candia et al. 2008; Isaacman et al. 2010; Song et al. 2010; Cheng et al. 2011; Cho, Myers, and Leskovec 2011; Becker et al. 2013; Silm and Ahas 2014), several research challenges remain to be better addressed. For example, many previous studies examined important determinants of human activity space independently. It remains unclear how different determinants of an individual activity space are related to each other. Although some studies have used methods such as clustering to identify mobility patterns based on multiple characteristics of individual activity space, it can be difficult to interpret the major characteristics of each population group. In this study, we develop some intuitive individual mobility indicators (IMIs) to represent individual activity space from three critical perspectives (i.e., spatial extent, frequent locations, and movements). We then introduce several approaches to uncover the interrelationships of these mobility indicators and compare activity space patterns among different cities or population groups.

We use two large, actively tracked cellphone location data sets collected in two major Chinese cities, Shanghai and Shenzhen, on a workday to investigate and compare human activity space patterns between these two cities as an example to illustrate the usefulness of our proposed IMIs. Different from call detail records (CDRs) that are passively collected when people engage in communication activities such as phone calls and text messages (Song et al. 2010; Becker et al. 2013; Xu et al. 2015), actively tracked cellphone location data provide locations of each cellphone at a regular time interval by detecting where a cellphone is located. Because many people make infrequent use of their cellphones, in a day and cellphone usage tends to have a natural biased spatiotemporal pattern (e.g., more cellphone communications after work than before work in a day), actively tracked cellphone location data generally offer better spatiotemporal coverage of individual activity space than CDR data. The main objective of this article is to develop a method that can measure the major characteristics of individual activity space based on actively tracked cellphone location data such that we can effectively compare aggregate activity space patterns among different cities. To achieve the objective, we develop three IMIs—daily activity range, number of activity anchor points, and frequency of movement—to answer critical questions of individual activity space (i.e., how far, how many, and how frequent). We then apply association rules in data mining (Han, Kamber, and Pei 2011) to examine how the three indicators are related to each other among the activity spaces of different individuals. We further investigate spatial and temporal variations of major characteristics of aggregate human activity patterns between Shanghai and Shenzhen.

Literature Review

Activity space and its related concepts (Lynch 1960; Brown and Moore 1970; Horton and Reynolds 1971; Lenntorp 1977; Golledge and Stimson 1997) have been widely used in geography to examine people's use of space. Various approaches including, but not limited to, standard deviational ellipse (Yuill 1971), confidence ellipse (Schönfelder and Axhausen 2003), and daily potential path area (Kwan 1998) have been proposed to measure individual space usage from perspectives of spatial extent, frequent locations, and movements. Over the past several decades, many studies have applied these approaches to study human activity space and its relationships with sociodemographic characteristics (Hanson and Hanson 1981; Newsome, Walcott, and Smith 1998; Dijst 1999; Kwan 1999; Axhausen et al. 2002; Buliung and Kanaroglou 2006). Most of these studies involved activity-travel surveys that can be expensive to collect and often limited in sample size. As we move into the big data era, many new data sources have emerged. For example, there have been several studies that used actively tracked mobile phone location data to solve problems related to mobility prediction (Gao, Tang, and Liu 2012), recognition of place categories (Zhu et al. 2012), and estimation of demographic attributes (Brdar, Culibrk, and Crnojevic 2012). Such data sets provide new opportunities for understanding people's use of space in their daily lives. Large data volumes present new challenges to the study of human activity space, however. In recent years, research has been conducted to study human activity space using cellphone location data. Measures such as radius of gyration (Gonzalez, Hidalgo, and Barabási 2008; Song, Blumm, and Barabási 2010), activity anchor points (Phithakkitnukoon et al. 2010; Cho, Myers, and Leskovec 2011), and daily activity range (Becker et al. 2013) have been used to reflect major characteristics of individual activity space. Most of the studies analyzed these characteristics separately, which could lead to a partial view of individual activity space.

Although clustering methods have been applied to address some research issues, such as identifying individuals with similar location sequence (Li et al. 2008), commuting flexibility (Shen, Kwan, and Chai 2013),

and spatiotemporal activity patterns (Chen et al. 2011), it sometimes can be difficult to interpret the major characteristics of each population group derived from the clustering algorithms. Moreover, these clustering methods (e.g., hierarchical clustering) are computationally intensive and often perform inefficiently over very large data sets. This study attempts to develop some easy-to-compute and yet effective approaches to gain insights into activity space patterns. We build three mobility indicators to represent the most important determinants of individual activity space. By combining activity space theory and association rules in data mining, this study aims at providing a multidimensional view of individual activity space and facilitating a comparison of human activity spaces across different cities.

Study Area and Data Sets

Shanghai and Shenzhen are two major cities in China with their gross domestic products ranked the first and fourth, respectively, among all Chinese cities (National Bureau of Statistics of China 2012). Shanghai is a century-old metropolis, with a population of 24 million as of 2013. It has eighteen administrative districts and covers an area of 6,340 km^2 (Figure 1A and 1B). Shenzhen, which is located in southern China adjacent to Hong Kong, has six administrative districts covering 1,952 km^2 and a population of 15 million as of 2012 (Figure 1C). Shenzhen was a small fishing village when it was chosen as China's first Special Economic Zone (SEZ) in 1979. Fast economic growth and urbanization have transformed Shenzhen into a major migrant city. As of 2011, the migrant population accounted for more than 70 percent of the total population in Shenzhen (Gazette of the People's Government of Shenzhen Municipality 2011). According to recent travel surveys (Lu and Gu 2011; Urban Planning Land & Resources Commission of Shenzhen Municipality 2013), nonmotorized trips accounted for a large percentage of total trips in Shanghai (walking: 26.2 percent; bicycle or moped: 28.7 percent) and in Shenzhen (walking: 50.0 percent; bicycle or moped: 6.2 percent). Comparing people's daily activity space in these two cities can help us better understand their urban dynamics that could be useful for urban design, transportation planning, business studies, and other applications.

This article uses two actively tracked cellphone data sets[1] collected on a weekday in Shenzhen (23 March 2012) and Shanghai (3 September 2012), respectively. The Shenzhen data set covers 5.8 million cellphones, with their locations reported approximately once every

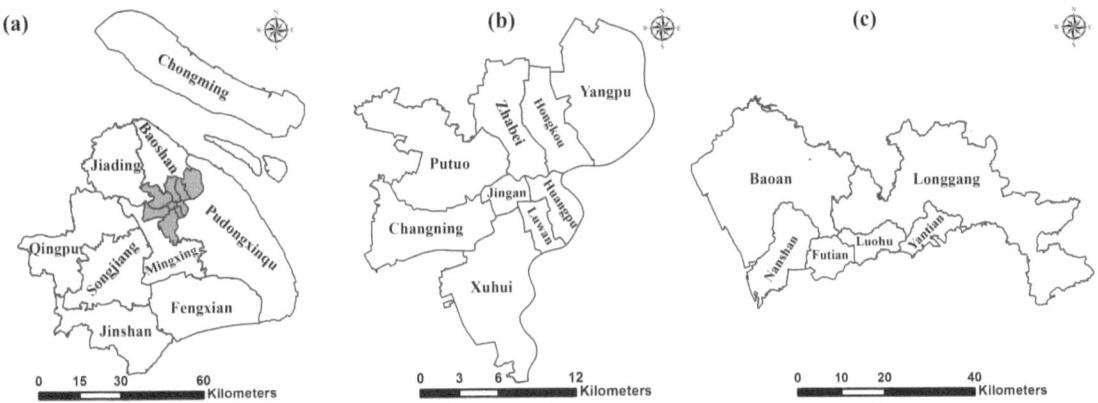

Figure 1. Study areas: (A) administrative districts of Shanghai; (B) inset map of the central part of Shanghai; (C) administrative districts of Shenzhen.

hour as (x, y) coordinates of the cellphone tower to which a cellphone is assigned. This data set includes cellphone location records between 00:00 and 23:00 during the study day; each cellphone therefore has twenty-three observations. The Shanghai data set consists of 0.69 million cellphones. To be comparable, we removed records of the 23:00–24:00 time window in Shanghai's data set. Table 1 shows an example of the data format of the two cellphone data sets. The average nearest distance among the cellphone towers in Shanghai is 0.21 km, as compared to 0.19 km in Shenzhen.

Method

This section first introduces three IMIs, followed by estimation of each individual's home location that will be used as a reference point when we analyze individual activity space. We then describe how association rules are used to summarize and compare people's activity spaces in Shanghai and Shenzhen.

Individual Mobility Indicators

As shown in Table 1, an individual's cellphone trajectory T can be represented as

$$T = \{P_1(x_1, y_1, t_1), \ P_2(x_2, y_2, t_2), \ldots, P_i(x_i, y_i, t_i)\},$$

(1)

where P_i denotes the ith ($i = 1, 2, \ldots, 23$) cellphone location record; x_i and y_i denote the longitude and latitude of a cellphone tower; and t_i represents a one-hour time window in which each location was recorded. We develop three IMIs, which are the number of activity anchor points, daily activity range, and frequency of movement, to capture the major characteristics of an individual activity space represented by T.

Measures such as standard deviational ellipse (Yuill 1971) and radius of gyration (Gonzalez, Hidalgo, and Barabási 2008) have been used in previous studies to represent the spatial dispersion of an individual's daily activities. In our study, we introduce *daily activity range*, which is defined as the maximum distance

Table 1. Example of an individual's cellphone records in both data sets

User ID	Record ID	Time window in which location was reported (t)	Longitude of cellphone tower (x)	Latitude of cellphone tower (y)
932*****	1	00:00–01:00	113.*****	22.*****
932*****	2	01:00–02:00	113.*****	22.*****
932*****	3	02:00–03:00	113.*****	22.*****
...	...	...	113.*****	22.*****
932*****	23	22:00–23:00	113.*****	22.*****

between all pairs of cellphone towers in T, to describe the spatial extent of an individual's activity space.[2]

Activity anchor points have been frequently used in the literature (e.g., Dijst 1999; Schönfelder and Axhausen 2003; Ahas et al. 2010) to denote a person's major activity locations such as home, workplace, favorite restaurants, and so on. The meaning of an activity anchor point could vary due to the context of each study, however. In this article, we define an *activity anchor point* as a set of cellphone towers that are geographically concentrated and where an individual spent a certain amount of time. One challenge of using cellphone location data to determine an individual's activity anchor points and movements among the anchor points is that an individual's cellphone location could switch among adjacent cellphone towers due to either cellphone load balancing (Csáji et al. 2013) or cellphone signal strength variation (Isaacman et al. 2012). Hence, to derive activity anchor points for T, we first extract all cellphone towers traversed by T, and calculate the frequency (i.e., number of time windows) each cellphone tower was visited. We then select the most visited cellphone tower, and group all the cellphone towers that are located within 0.5 km of the selected tower into a cluster. We then select the next most visited cellphone tower and perform the same grouping process. The process is repeated until all of the cellphone towers in T are processed. Finally, we calculate the number of cellphone location records (i.e., observations) assigned to each cluster. In this study, any cluster with two or more cellphone location records is identified as an activity anchor point. Figure 2 gives an example of an individual's trajectory in a three-dimensional space–time system proposed by Hägerstrand (1970). This individual's cellphone tower locations are grouped into four clusters, with three of them (clusters A, B, and C) being identified as activity anchor points.

Note that we choose a constant distance threshold of 0.5 km to derive individual activity anchor points for the two cellphone data sets, and the reasons are as follows. First, although we are aware that cellphone tower densities could vary within a city, choosing a constant threshold enables us to consistently evaluate the space usage of individuals in a city. Second, as Shanghai and Shenzhen share a similar average nearest distance among cellphone towers (0.21 km and 0.19 km, respectively), choosing 0.5 km can not only address the problem of signal switches among nearby cellphone towers but also facilitates the comparison of human activity space between these two cities.

Movement is another important characteristic of human activity space. When deriving *frequency of movement* in T, we only consider the movements that occurred between clusters (green lines in Figure 2) because it is difficult to determine whether the movements within clusters (i.e., red lines in Figure 2) represent an individual's actual movements or are simply caused by load balancing or signal switches. By choosing a threshold of 0.5 km, we minimize the impact of load balancing and signal switches while maintaining all major movements in an individual's trajectory. Here, the frequency of movement is defined as the number of intercluster movements in T. This indicator measures how actively an individual travels among different activity locations in a day. For example, the individual in Figure 2 has a frequency of movement of 5 (i.e., the number of green segments in Figure 2).

Estimation of Individual Home Location

Considering people's daily routines in most big cities in China (Long, Zhang, and Cui 2012), we define home location of an individual as the activity anchor point with a minimum of four hours of stay at the location before 7:00 a.m. Based on this rule, we are able to estimate the home location for 97 percent of the sampled population in both Shanghai and Shenzhen. For each city, we compare our estimated home locations by administrative districts against the most recent census data (Gazette of the Sixth National Population Census for Shanghai Municipality 2010; Gazette of the People's Government of Shenzhen Municipality 2011). We find that our estimates are in agreement with the census data according to the population distribution

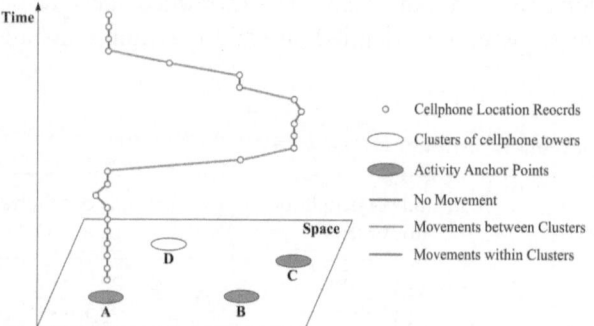

Figure 2. An individual's cellphone trajectory T and key concepts in individual mobility indicators represented by a space–time system proposed by Hägerstrand (1970). (Color figure available online.)

by administrative districts (with Pearson coefficients of 0.95 and 0.99 for Shanghai and Shenzhen, respectively).

Building Association Rules

Association rules have been widely used in business research to uncover items that are frequently purchased together. They also have been used to describe associations between quantitative items or attributes (Han, Kamber, and Pei 2011). To uncover the major characteristics of individual activity space in the two cities, one key challenge is to analyze how the three IMIs are associated with each other to characterize each individual's activity space. Note that the IMIs can be represented as

$$IMIs \rightarrow (N, R, F), \qquad (2)$$

where N denotes the number of activity anchor points, R describes an individual's daily activity range, and F denotes the frequency of movement. The value of each indicator can be partitioned into several intervals:

$$N \rightarrow (N_1, N_2, \ldots, N_a) \qquad (3)$$

$$R \rightarrow (R_1, R_2, \ldots, R_b) \qquad (4)$$

$$F \rightarrow (F_1, F_2, \ldots, F_c), \qquad (5)$$

where a, b, and c represent the number of intervals or classes defined for each corresponding indicator. For each individual X, the IMIs can be represented by their specific characteristics based on the defined intervals:

$$X \rightarrow (N_i, R_j, F_k) \text{ (where } 0 \leq i \leq a, 0 \leq j \leq b, \ 0 \leq k \leq c). \qquad (6)$$

We then introduce association rules to summarize the characteristics of human activity space for each city. The association rules are formulated as

$$(X, \text{``}N_i\text{''}) \Rightarrow (X, \text{``}R_j,\text{''} \text{ and ``}F_k\text{''}). \qquad (7)$$

These rules describe how different intervals of the three IMIs are associated with each other in each individual's activity space. For each city, the support and the confidence of the association rules are calculated as follows:

$$\text{support}(X, \text{``}N_i\text{''}) \Rightarrow (X, \text{``}R_j,\text{''} \text{ and ``}F_k\text{''})$$
$$= \frac{\text{number of individuals with } (X, \text{``}N_i\text{''})}{\text{total population in the cellphone data set}} \qquad (8)$$

$$\text{confidence}(X, \text{``}N_i\text{''}) \Rightarrow (X, \text{``}R_j,\text{''} \text{ and ``}F_k\text{''})$$
$$= \frac{\text{number of individuals with } (X, \text{``}N_i\text{''} \text{ and ``}R_j\text{''} \text{ and ``}F_k\text{''})}{\text{number of individuals with } (X, \text{``}N_i\text{''})} \qquad (9)$$

The *support* of a rule denotes the amount of individuals meeting the left-hand-side (LHS) condition divided by the total population of the data set. The *confidence* of a rule denotes the amount of individuals meeting both sides of the rule divided by the number of individuals meeting the LHS condition. Both support and confidence indexes describe important characteristics of human activity spaces extracted from a particular data set. Note that we use N_i as the LHS of the association rules because the number of activity anchor points for an individual X is a discrete variable, which can be directly derived from individual cellphone trajectories.

Analysis Results

General Statistics

We first derive the general statistics of IMIs for the two cities. As shown in Figure 3A, the majority of people in Shenzhen had only one or two activity anchor points in the study day (38.8 percent and 38.5 percent of the population, respectively), whereas people in Shanghai were more diversified regarding the number of activity anchor points (N). For daily activity range (R), a large percentage of people in Shenzhen traveled within a very short distance during the day, as illustrated in Figure 3B. The cumulative distribution shows that nearly 50 percent of the people in Shenzhen traveled within 1.0 km and about 82 percent traveled within 5.0 km, as compared to 26 percent of people in Shanghai who traveled within 1.0 km and 60 percent who traveled within 5.0 km. The medians of R in Shenzhen and Shanghai are 1.1 km and 3.1 km, respectively.[3] For frequency of movement (F), people on average made 3.76 movements in Shenzhen, as compared to 4.34 in Shanghai. The results indicate that (1) people in Shanghai had more major activity locations (i.e., N) in a day than people in Shenzhen; (2) the spatial extent

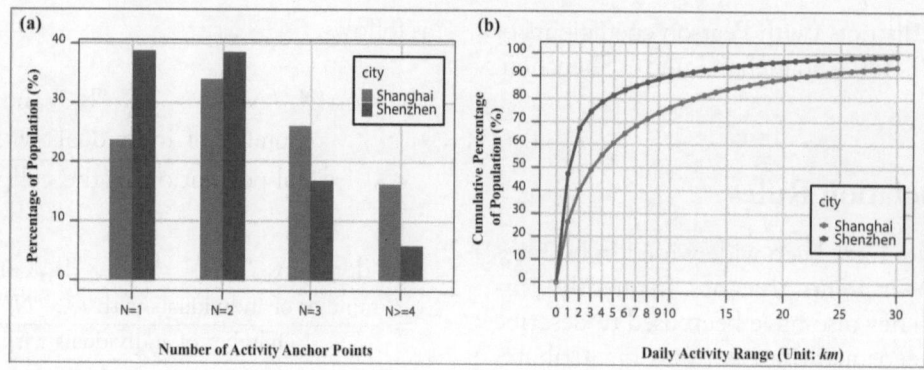

Figure 3. Distribution patterns of (A) number of activity anchor points and (B) daily activity range in Shanghai and in Shenzhen. (Color figure available online.)

of people's activities in Shanghai was generally larger than that of people in Shenzhen; and (3) people in Shanghai were more "active" in terms of the movements among their daily activity locations. It is still unclear, however, how the three determinants (N, R, and F) are related to each other in an individual's activity space. For example, do people with the same number of activity anchor points in Shenzhen and Shanghai have similar daily activity range or movement frequency? In the next section, we discuss the interrelationships of (N, R, F) based on the association rules to further understand the differences and similarities of individual activity space in the two cities.

Association Rules of IMIs

To generate the association rules, we first partition the three IMIs into intervals. As shown in Table 2, we partition N, R, and F into four, five, and five intervals, respectively. Each interval (N_i, R_j, or F_k) represents a particular value or range of values for the corresponding indicator. By mining the associations among the three indicators using the defined intervals, we are able to uncover the major characteristics of individual activity space for particular population groups within each city.

The support and confidence of the association rules are calculated to compare individual activity spaces in the two cities. As illustrated in Figure 4A and 4B, although there are more people with one activity anchor point in Shenzhen (support = 38.8 percent) than in Shanghai (support = 23.6 percent), people in these two subsets ($N = 1$) had very similar activity space characteristics. The two subsets are dominated by individuals with very short daily activity range (R_1) and low movement frequency (F_1 and F_2). Only a very small percentage of people traveled very far (R_4 and R_5) and frequently (F_4 and F_5). The result indicates that quite a few people in both cities stayed around one particular location during the day. The "immobility" of these individuals reflects an interesting perspective of human activity spaces in the two cities and calls for further investigation of its driving force and related societal implications.

When $N = 2$ (Figure 4A and 4B), the percentages of people with different travel ranges distribute relatively evenly within each interval of R in Shanghai, whereas Shenzhen shows a decay with increasing travel range. The subset of Shenzhen is dominated by people with short daily activity ranges (e.g., 63.2 percent of people with $R \leq 2$ km), whereas in the

Table 2. Intervals (classes) defined for the association rules

Number of activity anchor points (N)		Daily activity range (R)		Frequency of movement (F)	
Intervals	Values	Intervals	Values	Intervals	Values
N_1	$N = 1$	R_1	$R \leq 1$ km	F_1	$F = 0$
N_2	$N = 2$	R_2	1 km $< R \leq 2$ km	F_2	$1 \leq F \leq 3$
N_3	$N = 3$	R_3	2 km $< R \leq 5$ km	F_3	$4 \leq F \leq 7$
N_4	$N \geq 4$	R_4	5 km $< R \leq 10$ km	F_4	$8 \leq F \leq 11$
		R_5	$R > 10$ km	F_5	$F \geq 12$

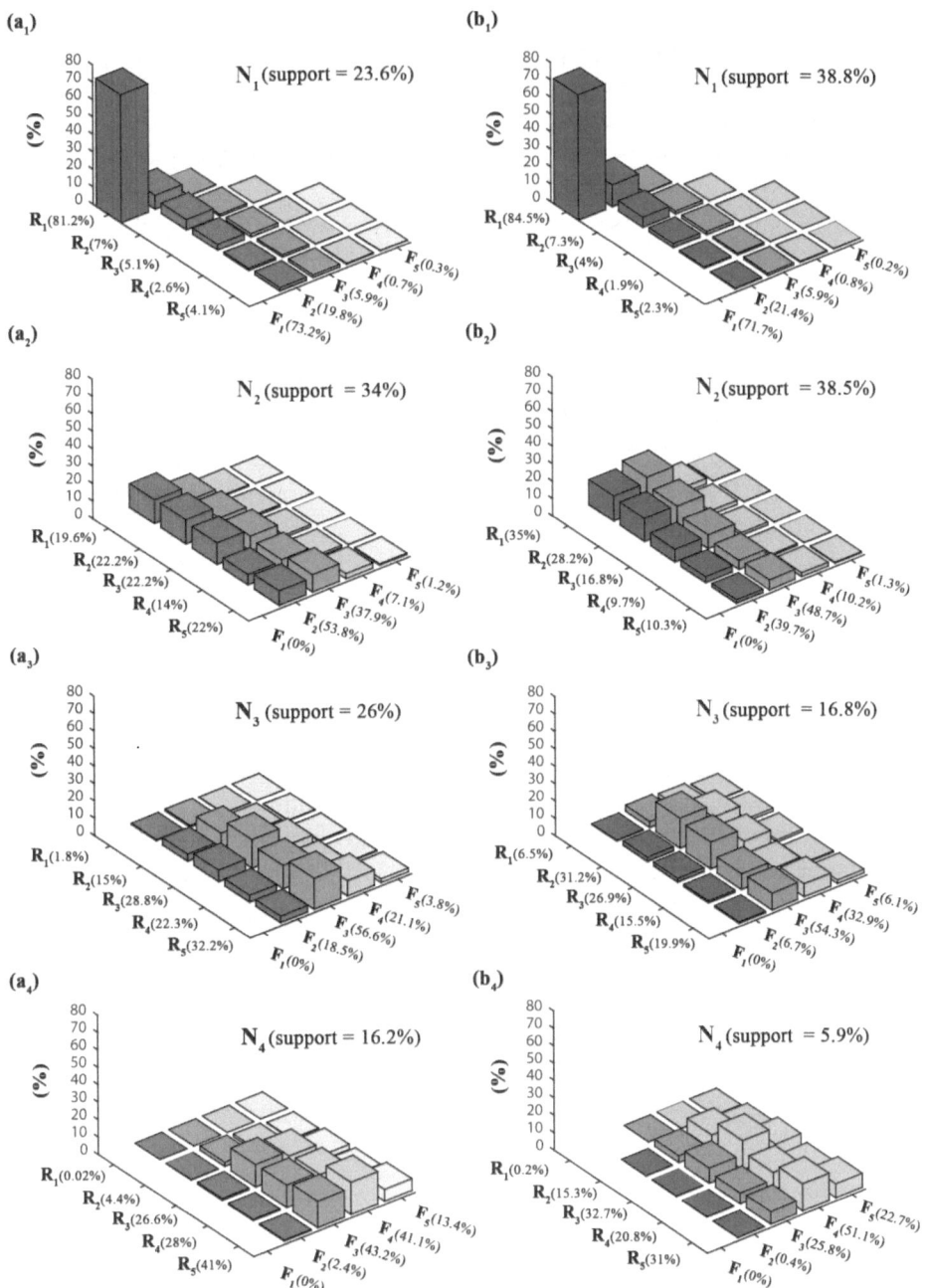

Figure 4. Association rules of individual activity space: (A₁–A₄) confidence of rules with left-hand-side (LHS) organized by N_i in Shanghai; (B₁–B₄) confidence of rules with LHS organized by N_i in Shenzhen. For each individual graphic, the percentages next to interval labels denote the sum of confidence for the corresponding rows or columns. (Color figure available online.)

subset of Shanghai, many people traveled very far in a day (e.g., 36 percent with $R > 5$ km). The observed difference could be potentially explained by the home–work relationships of people in the two subsets considering that home and workplace are two primary activity locations for most people. How frequently people moved serves as an important indicator of urban dynamics. According to our

observation, although the majority of people in both subsets fall within F_2 and F_3, people in Shenzhen traveled more frequently (39.7 percent and 48.7 percent within F_2 and F_3) than people in Shanghai (53.8 percent and 37.9 percent within F_2 and F_3). Note that we analyze the temporal variations of people's movement patterns in the two cities later in this section to further examine when (and where) people were more active in their daily activity spaces.

There is a notable change of the distribution of association rules as N increases from 2 to 3 for people in Shanghai. The majority of people with $N = 3$ (Figure 4A) in Shanghai had a very large daily activity range (54.5 percent with $R > 5$ km), which is quite different from the relatively even distribution of R when $N = 2$ (Figure 4A). The result indicates that the third activity anchor point might have a significant effect on people's travel range in Shanghai. Shenzhen, however, still has a large proportion of people with a short daily activity range (64.6 percent with $R \leq 5$ km, as shown in Figure 4B), which is similar to what we observe when $N = 2$ (Figure 4B).

According to the comparisons among the three subgroups in the two cities, we can see that in Shenzhen, a small activity space was usually enough to fulfill various purposes of people's daily activities such as work, dining, recreation, and so forth. In Shanghai, activity locations were more widely distributed in an individual's activity space. People were more likely to travel far from their primary activity locations (i.e., home and workplace) for certain travel and activity purposes. For $N \geq 4$ (Figure 4A and 4B), we see an increase in both travel range and movement frequency in both cities as compared to the previous three subgroups. People in Shanghai still traveled further but less frequently, as compared to the same population group in Shenzhen. Note that we have tested other partition schemes to generate different intervals for IMIs, and the corresponding association rules reveal similar patterns of people's activity spaces in the two cities.

Spatial Variations of Human Activity Space

Analyzing the geographic patterns of people's activity space within the context of the built environment could produce an improved understanding of their daily activity patterns. For example, it would be meaningful to explore the geographic distributions of people with a small daily activity range (R), which is an important feature of individual activity space in both cities, especially in Shenzhen. Figure 5 illustrates the geographic distributions of people with a daily activity range $R \leq 2$ km. Specifically, we divide the study areas into 2-km grids and aggregate individuals based on their estimated home locations. Each grid cell represents the number of individuals with $R \leq 2$ km, normalized by the total number of individuals in that grid cell.

As shown in Figure 5A, many grid cells in the core areas of Shanghai have a higher percentage of people with $R \leq 2$ km (i.e., green cells in Huangpu, Luwan, and Jingan districts; readers can refer to the inset map in Figure 1B) as compared to the grid cells in suburbs (i.e., orange and red cells) such as Jinshan, Songjiang, Qingpu, Jiading, and Pudongxinqu districts. Note that we also observe grid cells with higher percentages (i.e., green cells) in certain suburbs such as Minxing and Fengxian districts. It is interesting to find that the observed patterns are in general agreement with the analysis results by Sun, Pan, and Ning (2008), who studied job–housing balance in Shanghai. They indicated that core areas such as Huangpu, Luwan, and Jingan have more job opportunities as compared to the number of residents, so more people would have a

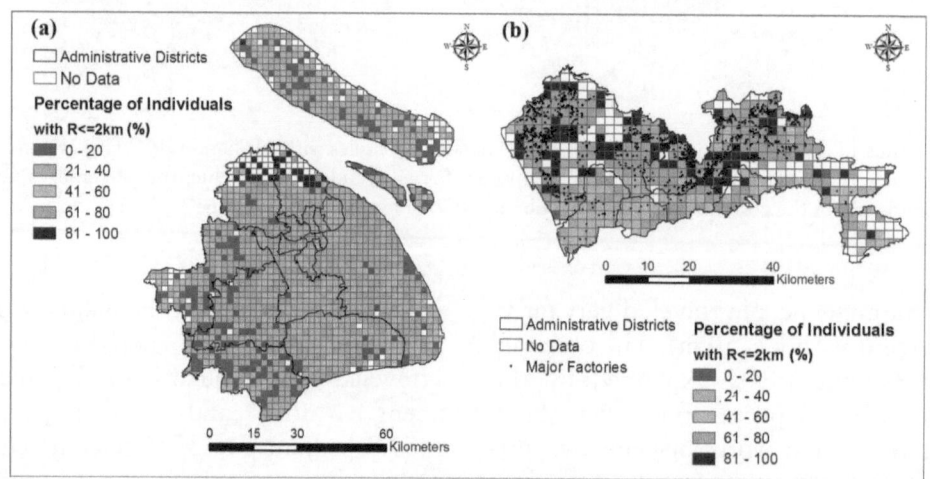

Figure 5. Geographic distributions of individuals with daily activity range ≤ 2 km in the two cities: (A) Geographic patterns in Shanghai; (B) geographic patterns in Shenzhen. (Color figure available online.)

relatively shorter commuting distance. Some suburbs around the core areas are more housing-oriented, so more people would have a longer commuting distance. We avoid making any further statement because the daily activity range examined in this study does not reflect people's actual commuting distances.

Figure 5B shows that there is a general north–south divide in Shenzhen and the proportion of people with $R \leq 2$ km in most grid cells of the two northern districts (Baoan and Longgang) is larger than 60 percent or even 80 percent, which indicates that most people who live around these cells have a small daily activity range during the study day. In the southern part of Shenzhen, the percentages are generally lower. To explore potential causes of the identified patterns, we also display the locations of major factories in Shenzhen. It appears that grid cells with a high percentage are generally colocated with major factories in Shenzhen. Many factories in Shenzhen provide workers with dormitories adjacent to their workplace. In addition, many immigrants tend to rent apartments near their workplace to save commuting time and cost. The findings suggest that the geographic patterns of people's activity space in Shenzhen and in Shanghai are quite different and that the identified patterns are likely to be related to the underlying socioeconomic characteristics.

Temporal Variations of Aggregate Movement Patterns

We further analyze the temporal variations of aggregate movement patterns to understand when people were more active in their daily activity spaces. Figure 6 shows the percentages of people who moved

through a day in the two cities, organized by the number of individual activity anchor points (N). People with $N = 1$ in the two cities did not move much in a day. The percentages are relatively stable over time (less than 10 percent). As expected, movement patterns of the other three subgroups in these two cities exhibit two peaks during the morning and afternoon rush hours. There is a local peak around time intervals 12 and 13 for people in Shenzhen, however, which indicates that people in Shenzhen moved more frequently around noon than other work hours. More importantly, the difference of aggregate movement patterns around noon between the two cities explains our previous finding that people in Shenzhen generally move more frequently than people in Shanghai when controlling the number of activity anchor points (N).

We further explore the temporal variations of average movement distances in the two cities. As shown in Figure 7, people's movement distances in both cities are generally lower around noon than the work hours. By further analyzing movements around noon (i.e., time intervals 12 and 13), we find that 43 percent of individuals travel from or to their home locations around noon in Shanghai, as compared to 66 percent in Shenzhen. The shorter movement distance around noon reveals an interesting aspect of people's lifestyle in both cities. Although people in Shanghai have a longer travel distance in general, short-range movements still dominate in both cities. As described previously, the share of nonmotorized trips accounts for more than 50 percent of all trips in both Shanghai and Shenzhen. Our analysis results suggest that travel mode, such as walking and bicycling, should receive more attention in urban and transportation planning that has been mentioned in

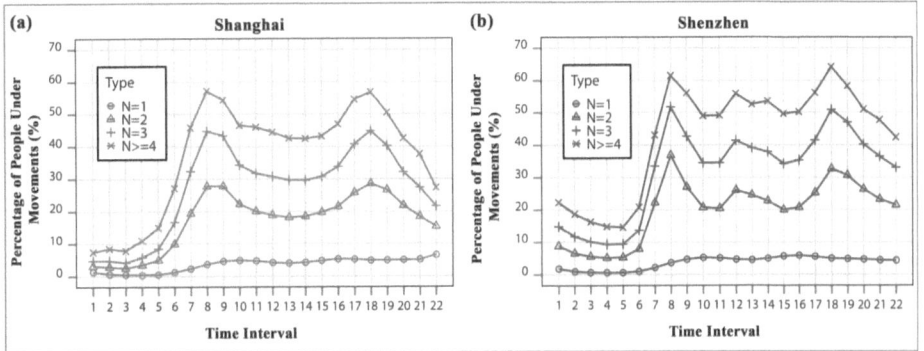

Figure 6. Temporal variations of aggregate movement patterns in the two cities, organized by the number of individual activity anchor points: (A) Temporal patterns in Shanghai; (B) temporal patterns in Shenzhen. (Each time interval is associated with two consecutive time windows as shown in Table 1. For example, time interval 1 in this figure shows movement patterns from 00:00–01:00 to 01:00–02:00). (Color figure available online.)

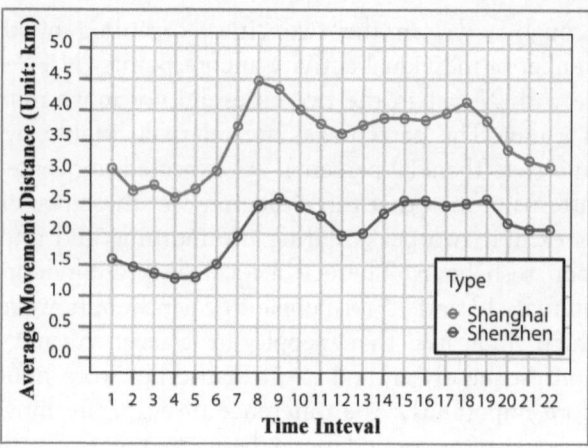

Figure 7. Temporal variation of average movement distance in Shanghai and Shenzhen. (Color figure available online.)

some government reports in recent years (e.g., Urban Planning Land & Resources Commission of Shenzhen Municipality 2013). Appropriate transportation services should be deployed to accommodate medium- and short-range trips in cities such as Shenzhen and Shanghai.

Discussion and Conclusion

The emergence of big individual tracking data sets brings new opportunities and challenges to the understanding of human activity space in urban environments. In this study, we develop several intuitive IMIs using cellphone location data to describe the major determinants of individual activity space. Different from previous studies (Kang et al. 2010; Isaacman et al. 2012; Becker et al. 2013) that investigate determinants of human activity space independently, we analyze how these mobility indicators (e.g., daily activity range, number of activity anchor points, and frequency of movement) can be combined with each other to analyze an individual's activity space. The association rules of IMIs are able to uncover the complexities of individual activity space for a given population or a geographic region. The support and confidence of the derived rules serve as a signature of people's daily activity patterns and enable us to compare human activity spaces systematically across different geographic regions. By using active tracking cellphone location data sets collected in Shanghai and Shenzhen, we summarize and compare the major characteristics of activity space patterns in these two cities. The association rules and spatiotemporal analysis of aggregate human activity patterns allow us to better understand the socioeconomic

characteristics of these cities and yield some insights into transportation planning and urban design.

Our analysis results reveal several interesting aspects of human activity space in the two cities, and the implications are worth discussing. First, quite a few people in both cities stay around one particular location for the whole day. Such unique activity patterns might reflect some societal issues such as "urban villages" (Wei and Yan 2005) in cities that consist of low-income communities of migrant population. Additional efforts are needed to further examine the "immobility" of these people and the potential driving forces related to land use planning (Pan, Shen, and Zhang 2009) and social segregation (Schönfelder and Axhausen 2003; Silm and Ahas 2014). Second, for the majority of people in Shenzhen, a small activity space was usually enough to fulfill the needs of people's daily activities, which is consistent with the government's goal of building a compact city with sustainable urban form. In Shanghai, however, activity locations are more widely distributed in an individual's activity space, and people are more likely to travel far from their home and workplaces for certain activity purposes. Shenzhen and Shanghai, one being a city with a large migrant population and the other being a century-old metropolis with many local residents, have very different sociodemographic characteristics and urban forms, which play an important role in shaping people's daily activity patterns. Third, the geographic disparity of people's travel range in Shenzhen is significant. The difference between the north and south could be partially explained by the socioeconomic divide in Shenzhen. In Shanghai, the geographic disparity is less obvious, and our analysis suggests that the identified patterns could be potentially explained by people's commuting patterns and the job–housing relationships in the city.

Currently, the research findings only reflect people's activity space in the two cities for a day. In the future, we plan to further investigate the temporal variations of individual activity space (e.g., seasonality, and difference between workdays and weekends) by using actively tracked cellphone data sets that cover longer time periods. It would also be meaningful to compare the analysis results derived from active and passive cellphone location data (e.g., CDRs), for example, to examine whether they reveal similar or different patterns of people's activity space. This will help us better understand the strengths and weaknesses of each data type and the intrinsic characteristics of human activity space. Nevertheless, the research findings in this article enhance our understanding of the geographies of human mobility in a space–time context. We believe that the proposed methods are useful to other types of large individual tracking data sets for data-intensive analyses of human activity space.

Notes

1. The mobile phone location data sets used in this study were acquired from research collaborators in China. The research was approved by the Institutional Review Board (IRB) at the University of Tennessee, Knoxville.
2. Radius of gyration (Gonzalez, Hidalgo, and Barabási 2008) is another frequently used measure that describes the spatial dispersion of an individual's activity space. In this study, we calculated both daily activity range and radius of gyration, and we found that they are highly correlated with each other (Pearson coefficient = 0.96). We thus use daily activity range in this study to represent the spatial extent of an individual's activity space due to its intuitive meaning.
3. The median of R in Shenzhen and Shanghai (1.1 km and 3.1 km, respectively) are much lower than that of the New York and Los Angeles regions (6.08 km and 8.0 km, respectively) computed by Isaacman et al. (2010) using cellphone location data. It is not surprising to see that the two U.S. cities have a larger daily activity range than Shanghai and Shenzhen because U.S. cities are more automobile oriented.

Funding

This research was jointly supported by the National Natural Science Foundation of China (41231171, 41301440, 41271408, 41571431, 41371377), Fundamental Research Funding of Shenzhen (JCYJ20130401170306842), Shenzhen Scientific Research and Development Funding Program (ZDSY20121019111146499), and Shenzhen Dedicated Funding of Strategic Emerging Industry Development Program (JCYJ20121019111128765).

References

Ahas, R., S. Silm, O. Järv, E. Saluveer, and M. Tiru. 2010. Using mobile positioning data to model locations meaningful to users of mobile phones. *Journal of Urban Technology* 17 (1): 3–27.

Axhausen, K. W., A. Zimmermann, S. Schönfelder, G. Rindsfüser, and T. Haupt. 2002. Observing the rhythms of daily life: A six-week travel diary. *Transportation* 29 (2): 95–124.

Becker, R., R. Cáceres, K. Hanson, S. Isaacman, J. M. Loh, M. Martonosi, J. Rowland, S. Urbanek, A. Varshavsky, and C. Volinsky. 2013. Human mobility characterization from cellular network data. *Communications of the ACM* 56 (1): 74–82.

Brdar, S., D. Culibrk, and V. Crnojevic. 2012. Demographic attributes prediction on the real-world mobile data. Paper presented at the Nokia Mobile Data Challenge 2012 Workshop, Newcastle, UK.

Brown, L. A., and E. G. Moore. 1970. The intra-urban migration process: A perspective. *Geografiska Annaler Series B: Human Geography* 52 (1): 1–13.

Buliung, R. N., and P. S. Kanaroglou. 2006. A GIS toolkit for exploring geographies of household activity/travel behavior. *Journal of Transport Geography* 14 (1): 35–51.

Candia, J., M. C. González, P. Wang, T. Schoenharl, G. Madey, and A.-L. Barabási. 2008. Uncovering individual and collective human dynamics from mobile phone records. *Journal of Physics A: Mathematical and Theoretical* 41 (22): 224015.

Chen, J., S.-L. Shaw, H. Yu, F. Lu, Y. Chai, and Q. Jia. 2011. Exploratory data analysis of activity diary data: A space–time GIS approach. *Journal of Transport Geography* 19 (3): 394–404.

Cheng, Z., J. Caverlee, K. Lee, and D. Z. Sui. 2011. Exploring millions of footprints in location sharing services. Paper presented at the 5th International Conference on Weblogs and Social Media, Barcelona, Spain.

Cho, E., S. A. Myers, and J. Leskovec. 2011. Friendship and mobility: User movement in location-based social networks. In *Proceedings of the 17th ACM SIGKDD International Conference on Knowledge Discovery and Data Mining*, ed. C. Apte, 1082–90. New York: ACM. http://dl.acm.org/citation.cfm?id=2020408&picked=prox (last accessed 13 January 2016).

Csáji, B. C., A. Browet, V. A. Traag, J.-C. Delvenne, E. Huens, P. Van Dooren, Z. Smoreda, and V. D. Blondel. 2013. Exploring the mobility of mobile phone users. *Physica A: Statistical Mechanics and its Applications* 392 (6): 1459–73.

Dijst, M. 1999. Two-earner families and their action spaces: A case study of two Dutch communities. *GeoJournal* 48 (3): 195–206.

Gao, H., J. Tang, and H. Liu. 2012. Mobile location prediction in spatio-temporal context. Paper presented at the Nokia Mobile Data Challenge 2012 Workshop, Newcastle, UK.

Gazette of the People's Government of Shenzhen Municipality. 2011. Issue No. 17, Serial No. 741. http://www.sz.gov.cn/zfgb/2012_1/gb785/201204/t20120423_1844697.htm (last accessed 28 October 2014).

Gazette of the Sixth National Population Census for Shanghai Municipality. 2010. http://www.stats-sh.gov.cn/sjfb/201105/218819.html (last accessed 28 October 2014).

Golledge, R. G., and R. J. Stimson. 1997. *Spatial behavior: A geographic perspective*. New York: Guilford.

Gonzalez, M. C., C. A. Hidalgo, and A.-L. Barabási. 2008. Understanding individual human mobility patterns. *Nature* 453 (7196): 779–82.

Hägerstrand, T. 1970. What about people in regional science? *Papers in Regional Science* 24 (1): 7–24.

Han, J., M. Kamber, and J. Pei. 2011. *Data mining: Concepts and techniques*. 3rd ed. Amsterdam: Morgan Kaufmann.

Hanson, S. 1980. The importance of the multi-purpose journey to work in urban travel behavior. *Transportation* 9 (3): 229–48.

Hanson, S., and P. Hanson. 1981. The travel-activity patterns of urban residents: Dimensions and relationships to sociodemographic characteristics. *Economic Geography* 57 (4): 332–47.

Horton, F. E., and D. R. Reynolds. 1971. Effects of urban spatial structure on individual behavior. *Economic Geography* 47 (1): 36–48.

Isaacman, S., R. Becker, R. Cáceres, S. Kobourov, J. Rowland, and A. Varshavsky. 2010. A tale of two cities. In *Proceedings of the Eleventh Workshop on Mobile Computing Systems & Applications*, ed. A. Dalton, 50–51. New York: http://dl.acm.org/citation.cfm?id=1734583&picked=prox (last accessed 13 January 2016).

Isaacman, S., R. Becker, R. Cáceres, M. Martonosi, J. Rowland, A. Varshavsky, and W. Willinger. 2012. Human mobility modeling at metropolitan scales. Paper presented at the 10th International Conference on mobile Systems, Applications, and Services, Low Wood Bay, Lake District, UK. http://www.sigmobile.org/mobisys/2012/index.php (last accessed 13 January 2016).

Kang, C., S. Gao, X. Lin, Y. Xiao, Y. Yuan, Y. Liu, and X. Ma. 2010. Analyzing and geo-visualizing individual human mobility patterns using mobile call records. In *Proceedings of the 18th International Conference on Geoinformatics*, ed. Y. Liu and A. Chen, 1–7. Beijing: IEEE. http://ieeexplore.ieee.org/xpl/mostRecentIssue.jsp?reload=true&punumber=5559273 (last accessed 13 January 2016).

Kwan, M.-P. 1998. Space–time and integral measures of individual accessibility: A comparative analysis using a point-based framework. *Geographical Analysis* 30 (3): 191–216.

———. 1999. Gender, the home–work link, and space–time patterns of non-employment activities. *Economic Geography* 75 (4): 370–94.

———. 2000. Interactive geovisualization of activity-travel patterns using three-dimensional geographical information systems: A methodological exploration with a large data set. *Transportation Research Part C: Emerging Technologies* 8 (1): 185–203.

Lenntorp, B. 1977. Paths in space–time environments: A time-geographic study of movement possibilities of individuals. *Environment and Planning* A 9 (8): 961–72.

Li, Q., Y. Zheng, X. Xie, Y. Chen, W. Liu, and W.-Y. Ma. 2008. Mining user similarity based on location history. Paper presented at the 16th ACM SIGSPATIAL International Conference on Advances in Geographic Information Systems, Irvine, CA, USA.

Long, Y., Y. Zhang, and C. Cui. 2012. 利用公交刷卡数据分析北京职住关系和通勤出行 [Identifying commuting pattern of Beijing using bus smart card data]. *Journal of Geographical Sciences* 67 (10): 1339–52.

Lu, X., and X. Gu. 2011. The fifth travel survey of residents in Shanghai and characteristics analysis. *Urban Transport of China* 9 (5): 1–7.

Lynch, K. 1960. *The image of the city*. Cambridge, MA: MIT Press.

National Bureau of Statistics of China. 2012. 主要城市年度GDP [Annual GDP for major cities in China]. http://data.stats.gov.cn/workspace/index?m=csnd (last accessed 28 October 2014).

Newsome, T. H., W. A. Walcott, and P. D. Smith. 1998. Urban activity spaces: Illustrations and application of a conceptual model for integrating the time and space dimensions. *Transportation* 25 (4): 357–77.

Pan, H., Q. Shen, and M. Zhang. 2009. Influence of urban form on travel behaviour in four neighbourhoods of Shanghai. *Urban Studies* 46 (2): 275–94.

Phithakkitnukoon, S., T. Horanont, G. Di Lorenzo, R. Shibasaki, and C. Ratti. 2010. Activity-aware map: Identifying human daily activity pattern using mobile phone data. In *Human behavior understanding*, ed. A. A. Salah, T. Gevers, N. Sebe, and A. Vinciarelli, 14–25. New York: Springer.

Schönfelder, S., and K. W. Axhausen. 2003. Activity spaces: Measures of social exclusion? *Transport Policy* 10 (4): 273–86.

Shaw, S. L., H. Yu, and L. S. Bombom. 2008. A space–time GIS approach to exploring large individual–based spatiotemporal datasets. *Transactions in GIS* 12 (4): 425–41.

Shen, Y., M.-P. Kwan, and Y. Chai. 2013. Investigating commuting flexibility with GPS data and 3D geovisualization: A case study of Beijing, China. *Journal of Transport Geography* 32:1–11.

Shoval, N., and M. Isaacson. 2007. Sequence alignment as a method for human activity analysis in space and time. *Annals of the Association of American Geographers* 97 (2): 282–97.

Silm, S., and R. Ahas. 2014. Ethnic differences in activity spaces: A study of out-of-home nonemployment activities with mobile phone data. *Annals of the Association of American Geographers* 104 (3): 542–59.

Song, C., Z. Qu, N. Blumm, and A.-L Barabási. 2010. Limits of predictability in human mobility. *Science* 327 (5968): 1018–21.

Sun, B., X. Pan, and Y. Ning. 2008. 上海市就业与居住间衡对交通出行的影响分析 [Analysis on influence of job-housing balance on commute travel in Shanghai]. *Urban Planning Forum* 1:77–82.

Urban Planning Land & Resources Commission of Shenzhen Municipality. 2013. 深市步行和自行车交通规及设计导 [Guidelines of transport planning and design for pedestrian and bicycle systems in Shenzhen]. http://www.szpl.gov.cn/xxgk/ztzl/zxcgh/jtghcgg.pdf (last accessed 31 August 2015).

Wei, L., and X. Yan. 2005. Transformation of "urban village" and feasible mode. *City Planning Review* 7:9–13.

Xu, Y., S.-L. Shaw, Z. Zhao, L. Yin, Z. Fang, and Q. Li. 2015. Understanding aggregate human mobility patterns using passive mobile phone location data: A home-based approach. *Transportation* 42 (4): 625–46.

Yuill, R. S. 1971. The standard deviational ellipse; an updated tool for spatial description. *Geografiska Annaler Series B: Human Geography* 53 (1): 28–39.

Zheng, Y., L. Liu, L. Wang, and X. Xie. 2008. Learning transportation mode from raw GPS data for geographic applications on the web. Paper presented at the 17th International Conference on World Wide Web, Beijing.

Zhu, Y., E. Zhong, Z. Lu, and Q. Yang. 2012. Feature engineering for place category classification. Paper presented at the Nokia Mobile Data Challenge 2012 Workshop, Newcastle, UK.

Information for Authors

The *Annals of the American Association of Geographers* (hereafter *Annals*) publishes original, timely, and innovative articles that advance knowledge in all facets of the discipline. Four section editors are responsible for articles in each of four major areas: Environmental Sciences (Mark Fonstad, University of Oregon); Methods, Models, and Geographic Information Sciences (Mei-Po Kwan, University of Illinois at Urbana–Champaign); Nature and Society (James McCarthy, Clark University); and People, Place, and Region (Nik Heynen, University of Georgia). Papers submitted for publication should address significant research problems and issues, and be attuned to the sensibilities of a diverse scholarly audience.

Submission. All manuscripts should be submitted electronically through AAG Manuscript Central (http://mc.manuscriptcentral.com/aag). Submissions must include a cover letter containing a statement that the manuscript has not been submitted for publication elsewhere, and will not be submitted elsewhere until a decision has been rendered by the editor. Authors are required to select the appropriate section to which the paper should be assigned. Any questions should be directed to the Managing Editor at annals@aag.org.

Review. After evaluation by the section editor, manuscripts are normally sent to at least two outside reviewers. The review process normally takes four to six months and every effort is made to respond as quickly as possible.

Manuscripts. Prepare your manuscript using Microsoft Word. Do not put any identifying information in your manuscript or your file names to ensure a blind review. This includes names of the authors, their affiliations and bio sketches, or acknowledgments (this information can be added to your final files if your paper is accepted for publication). All author contact information will be saved separately when uploading your submission to Manuscript Central. Please include the title, abstract, and keywords in the body of your manuscript even though you will be asked to save them separately as part of the upload process. Reviewers will only have access to these if they are part of your manuscript file.

Do *not* embed tables or figures in your manuscript file. They should be placed at the end of the document or uploaded separately. Figure files can be uploaded as .gif, .jpg, .eps, and .tif; however, only .eps or .tif files can be used for publication purposes.

Manuscripts should be no longer than 11,000 words total, including abstract, references, notes, tables, and figure captions. Text must be in a 12-point font with 1 inch margins. All parts of the manuscript (abstract, text, notes, references, tables, and figure captions) must be double-spaced and paginated. Format manuscript as follows, starting each section with a new page: (1) title page, (2) abstract, (3) text, (4) notes, (5) references, (6) appendix, (7) tables, (8) figure captions, (9) figures.

Title Page. The title serves as the author's invitation to a diverse audience; it should be chosen with care. Do not include any identifying information on the title page, including the names of the authors, their affiliations or contact information, bio sketches, or acknowledgments.

Abstract/Key Words. Include an abstract of 250 words or less that summarizes the purpose, methods, and major findings of the paper. All authors should provide three to five key words or phrases by which an article can be indexed in periodical references. These should appear in italics at the end of the abstract.

Acknowledgments. Do not include any acknowledgments (including grants/financial support) in your submission. This information can be added to your final files if your paper is accepted for publication.

Units of Measure. The *Annals* uses the International System of Units (metric); other units should be noted in parentheses.

Equations. Equations should appear in the text in an appropriate type of style (Greek letters, bold type, etc.). Authors should carefully distinguish between capital and lowercase letters, Roman and Greek characters, and letters and numerals. Number equations sequentially, in parentheses on the right edge of the text. All constituent terms should be defined when they initially appear.

Notes. Specific arguments or single points may be amplified by concise notes numbered sequentially in the text. The list of notes appears immediately after the text.

References. References should be cited parenthetically in the text in this order: author's last name, year of publication, and page number. All sources in the text of a paper must be listed in the references section and vice versa. List all references alphabetically by the author's last name and chronologically, and if possible, please include full names for all authors. Provide the full, unabbreviated title of books and periodicals. Personal communications can be cited either in the endnotes or in the references section. All newspaper articles and articles from weekly magazines should be fully cited in the references section rather than worked into the text of the paper. For examples of the correct style for various forms of publication, see recent articles in the *Annals*, the Annals Style Sheet (available at http://www.aag.org/cs/publications/journals/annals), and the 15th edition of the *Chicago Manual of Style*, published by the University of Chicago Press.

Tables. All tables must be referenced in the text. Each table must be typed double-spaced on a separate page in the same12-point font as the text, and numbered sequentially with Arabic numerals. Each table must have a descriptive title as well as informational column and row headings. Decimals appearing in tables should include leading zeros: 0.1273 rather than .1273. For examples of preferred table style, see recent articles in the *Annals*.

Graphics. Maps, graphs, and photos should convey ideas efficiently and tastefully. Graphics must be legible, concise, and referenced in the text. Illustrations must be designed to fit the page and column format of the *Annals*. A full page graphic must not exceed 7.0 inches by 9.5 inches (3.3 by 9.5 inches for a single column figure) *including* space for the figure caption which will be typeset by the printer.

For printing purposes, all graphics must be computer generated and will be accepted in the following formats only: EPS (Encapsulated Postscript) for most illustrations or TIFF (Tagged-Image File Format) for raster images. Type sizes below 6 point should be avoided. For further instructions, please see the Annals Graphics Guidelines (available at http://www.aag.org/cs/publications/journals/annals).

Authors may print four color figures at no charge. Authors wishing to include additional color figures will be expected to pay a fee of $500 (an invoice will be sent upon article publication).

Copyright. Authors must sign a copyright transfer agreement before their paper can be published. In accordance with copyright laws, the manuscript must not duplicate substantial portions of previously published material. If a manuscript incorporates previously published material (in the form of text, tables, or figures), the author is responsible for obtaining written permission from the copyright holder and for any costs associated with the use of the copyrighted materials. Further advice and information is available on the Taylor and Francis Author Services website (http://http://journalauthors.tandf.co.uk/copyright/usingThirdPartyMaterial.asp).

Final Submission. Final Submission Guidelines will be provided by the Managing Editor upon acceptance of the paper by the Editor. The final version of each article (including all tables, figures, etc.) should be submitted electronically through AAG Manuscript Central.

Copyediting will be undertaken once the manuscript has been accepted and entered into production. Authors will have the opportunity to review their typeset galley proofs before the article is published. Proofs are sent to the designated corresponding author. They must be carefully checked and returned within 48 hours of receipt. Any questions should be directed to the Managing Editor at annals@aag.org.

Reprints. Each corresponding author will receive one complete copy of the issue in which the article appears and 25 complimentary reprints of their article upon registration with Rightslink, Taylor & Francis's authorized reprint provider. Authors will need to create a unique account and register with Rightslink for this free service. The link is provided at the time of page proof review. Complimentary reprints are not available post publication. A discount on additional reprints is available to authors who order before print publication.

Graphics Guidelines

All graphics (maps, graphs, and photographs) should be submitted digitally in either EPS (Encapsulated Postscript) or TIFF (Tagged-Image File Format). Graphics can be exported and saved in these formats with applications such as Adobe Illustrator, Macromedia FreeHand, and CorelDraw. When exporting to EPS or TIFF, all fonts *must* be embedded. Graphic formats such as GIF, JPEG, PDF, and images produced in PowerPoint or imported into a Word document are *not* acceptable for high-quality printing. All raster images (TIFF) must be saved with a resolution of 600 dpi.

Layout

A full-page graphic must not exceed 7.0 inches (wide) by 9.5 inches (long), or 3.3 inches (wide) by 9.5 inches (long) for a single column, *including* space for the figure caption which will be typeset by the printer. Intermediate sizes are acceptable. Please plan for possible reductions of type and symbols.

Figures that contain multiple component parts must be assembled into a single figure by the author. Normal order will utilize a full page width with A – B placed above C – D, etc. The author should insert the letters A), B), etc. slightly above and aligned with the left edge of the component.

Note: Basic map design principles can be found in the following textbooks:

> Dent, Borden D., Jeffrey S. Torguson, and Thomas W. Hodler. 2009. *Cartography—Thematic Map Design*, Sixth Edition. WCB/McGraw-Hill. ISBN 0-07-294382-5
>
> Robinson, A. H., J. L. Morrison, P. C. Muehrcke, A. J. Kimerling, and S. C. Guptill. 1995. *Elements of Cartography*, Sixth Edition. John Wiley and Sons. ISBN 0-471-55579-7
>
> Slocum, Terry A., Robert B. McMaster, Fritz C. Kessler, and Hugh H. Howard. 2009. *Thematic Cartography and Visualization*, Third Edition. Prentice Hall. ISBN 0-32-29834-1

Lettering

1. All type should be from common type families (examples: Helvetica, Arial, Times New Roman, etc.). If an unique type font is critical to the theme of the map, such as helping to engender the feelings of an historical era, then a copy of the postscript font should be included with the final figure files, and its presence noted.
2. Type must be sufficiently large to be easily read at final print size; no type smaller than 6 point should be used.
3. Appropriate cartographic lettering placement should be observed.
4. Type placed on screen tints or area patterns must be clearly readable. White type can be used over dark screen tints, and area patterns can be lightened to prevent interference with type.
5. Generally, map titles will be incorporated into the figure caption and will appear outside and below the figure. Figure captions should **NOT** be a part of the graphic. All captions will be typeset, and should be submitted on a separate page.

Figure numbers should be a part of the figure's file name (see below).

6. The International System of Units (metric) should be used; other units may be noted in parentheses.

Screen Tints and Area Patterns

1. Screen tints and area pattern fills should reflect the importance of graphic elements through visual hierarchy (i.e., more important elements appear more prominently than less important elements).
2. To insure distinction between screen tints (particularly on choropleth maps), the following percentages are suggested:

> 2 tints: 30%, 50%
> 3 tints: 10%, 25%, 50%
> 4 tints: 10%, 20%, 50%, 80%
> 5 tints: 10%, 20%, 40%, 60%, 100%
> 6 tints: 10%, 15%, 25%, 40%, 60%, 100%

Typically, graphics should not have more than 6 tints.

3. All graphic files should be submitted at a resolution of at least 600 dpi.
4. Authors may print up to four color figures at no charge. Authors wishing to include additional color figures will be expected to pay a fee of $500 (an invoice will be sent upon article publication).
5. Color images need to be submitted in CMYK format.

Final submission of graphics files

1. The figures should be labeled using the following naming convention: corresponding author's name, the figure numbers, and extension for the file type. For example: figures for an article submitted by Dr. Jan Smith will be:

> SmithFigure1.eps
> SmithFigure2.eps
> SmithFigure3.tif

The figure number must correspond to the figure sequence referenced within the manuscript.

Note: All graphics are referred to as figures. For example, do not label something as Map1, it should be labeled as Figure1.

2. In order for you to evaluate and test the quality of your figures prior to submission, print a copy of each figure using a laser printer at a print resolution of at least 600 dpi. After printing the figure, measure the width of the figure and verify that it is no wider than the journal's page width and inspect the text and image for quality and readability. Increase text size where appropriate as it is important for the understanding of your graphics.
3. List of figure captions and sources typed double-spaced.
4. Final graphics files should be submitted electronically to AAG Manuscript Central (along with the final manuscript file, tables, etc.) after the paper has been accepted.

Please contact the Cartography Editor, Dr. Thomas Hodler (cartedit@gmail.com) with questions about file formats, production alternatives, and design problems.

Index

INDEX